Smart clothes and wearable technology

The Textile Institute and Woodhead Publishing

The Textile Institute is a unique organisation in textiles, clothing and footwear. Incorporated in England by a Royal Charter granted in 1925, the Institute has individual and corporate members in over 90 countries. The aim of the Institute is to facilitate learning, recognise achievement, reward excellence and disseminate information within the global textiles, clothing and footwear industries.

Historically, The Textile Institute has published books of interest to its members and the textile industry. To maintain this policy, the Institute has entered into partnership with Woodhead Publishing Limited to ensure that Institute members and the textile industry continue to have access to high calibre titles on textile science and technology.

Most Woodhead titles on textiles are now published in collaboration with The Textile Institute. Through this arrangement, the Institute provides an Editorial Board which advises Woodhead on appropriate titles for future publication and suggests possible editors and authors for these books. Each book published under this arrangement carries the Institute's logo.

Woodhead books published in collaboration with The Textile Institute are offered to Textile Institute members at a substantial discount. These books, together with those published by The Textile Institute that are still in print, are offered on the Woodhead web site at: www.woodheadpublishing.com. Textile Institute books still in print are also available directly from the Institute's website at: www.textileinstitutebooks.com.

A list of Woodhead books on textile science and technology, most of which have been published in collaboration with The Textile Institute, can be found at the end of the contents pages.

Woodhead Publishing in Textiles: Number 83

Smart clothes and wearable technology

Edited by
J. McCann and D. Bryson

The Textile Institute

CRC Press
Boca Raton Boston New York Washington, DC

WOODHEAD PUBLISHING LIMITED
Oxford   Cambridge   New Delhi

Published by Woodhead Publishing Limited in association with The Textile Institute
Woodhead Publishing Limited, Abington Hall, Granta Park, Great Abington
Cambridge CB21 6AH, UK
www.woodheadpublishing.com

Woodhead Publishing India Private Limited, G-2, Vardaan House, 7/28
Ansari Road, Daryaganj, New Delhi – 110002, India

Published in North America by CRC Press LLC, 6000 Broken Sound Parkway, NW,
Suite 300, Boca Raton, FL 33487, USA

First published 2009, Woodhead Publishing Limited and CRC Press LLC
© Woodhead Publishing Limited, 2009
The authors have asserted their moral rights.

British Library Cataloguing in Publication Data
A catalogue record for this book is available from the British Library.

Library of Congress Cataloging in Publication Data
A catalog record for this book is available from the Library of Congress.

Woodhead Publishing ISBN 978-1-84569-357-2 (book)
Woodhead Publishing ISBN 978-1-84569-566-8 (e-book)
CRC Press ISBN 978-1-4398-0113-0
CRC Press order number N10013

The publishers' policy is to use permanent paper from mills that operate a
sustainable forestry policy, and which has been manufactured from pulp which is
processed using acid-free and elemental chlorine-free practices. Furthermore,
the publishers ensure that the text paper and cover board used have met
acceptable environmental accreditation standards.

Typeset by SNP Best-set Typesetter Ltd., Hong Kong
Printed by TJ International Limited, Padstow, Cornwall, UK

Contents

Contributor contact details

(* = main contact)

Editor, Chapters 3, 4 and 18

Jane McCann
Smart Clothes and Wearable
  Technology Research Group
School of Art, Media and Design
University of Wales Newport
Caerleon Campus, Lodge Road
Caerleon, Newport
South Wales, NP18 3QT
UK

E-mail: jane.mccann@newport.
ac.uk

Editor, Chapters 5 and 17

David Bryson
Teaching Fellow
School of Arts, Design and
  Technology
University of Derby
Markeaton Street
Derby, DE22 3AW
UK

E-mail: D.Bryson@derby.ac.uk

Chapter 1

Mikko Malmivaara
Research Assistant
Tampere University of Technology
Institute of Electronics,
  Kankaanpää Unit
Jämintie 14
FI-38700 Kankaanpää
Finland

E-mail: mikko.malmivaara@tut.fi
mikko.malmivaara@reima.fi

Chapter 2

Richard D Hurford
University of Wales Newport
School of Art, Media and Design
Caerleon Campus, Lodge Road
Caerleon, Newport
South Wales, NP18 3QT
UK

E-mail: richard.hurford1@newport.
ac.uk

Chapter 6

Fatema Saifee
Research Fellow
University of Wales Newport
Lodge Road, Newport
South Wales, NP18 3QT
UK

E-mail: fsaifee@gmail.com

Chapter 7

Laura Thomas MA RCA
Woven Textiles: Art, design and
    consultancy
2 Douglas Buildings
Royal Stuart Lane
Cardiff Bay, CF10 5EL
Wales, UK

E-mail: info@laurathomas.co.uk

Chapter 8

Dr Faith Kane
Department of Fashion and
    Textiles
Faculty of Art and Design
De Montfort University
The Gateway
Leicester, LE1 9BH
UK

E-mail: fkane@dmu.ac.uk

Chapter 9

Adam J Martin
University of Wales Newport
School of Art, Media and Design
Caerleon Campus, Lodge Road
Caerleon, Newport
South Wales, NP18 3QT
UK

E-mail: adam.martin@newport.
ac.uk

Chapter 10

Peter Lam
Wireless Edge Communications
    Limited
Bristol, UK

E-mail: peter.lam@wirelessedge.net

Chapter 11

Gao Min
School of Engineering
Cardiff University
Queen's Buildings
    Parade
Cardiff, CF24 3AA
Wales, UK

E-mail: min@Cardiff.ac.uk
min@thermoelectrics.com

Chapter 12

Jane McCann*
Smart Clothes and Wearable
    Technology Research Group
School of Art, Media and Design
University of Wales Newport
Caerleon Campus, Lodge Road
Caerleon, Newport
South Wales, NP18 3QT
UK

E-mail: jane.mccann@newport.
ac.uk

Sirpa Morsky
HAMK Poly
Finland

E-mail: sirpa.morsky@hamk.fi

Xiang Dong
China Women's University, Beijing
PR China

E-mail: wgl_xd59@126.com

Chapter 13

Ida C Agnusdei
Garment Technologist
26 Cavendish Avenue
Allestree
Derby, DE22 2AR
UK

E-mail: ida.agnusdei@ntlworld.com

Visiting Research Fellow
    (2005–2006)
Smart Clothes and Wearable
    Technology
University of Wales
Wales, UK

Chapter 14

A Taylor
University of the West of England
Frenchay Campus, Coldharbour
    Lane
Bristol, BS16 1QY
UK

E-mail: ali.taylor55@btinternet.com

Chapter 15

Dr Cathy Treadaway
University of Wales Institute
    Cardiff
Howard Gardens
Cardiff, CF24 0SP
Wales, UK

E-mail: ctreadaway@uwic.ac.uk

Chapter 16

Mike Timmins
1 Albemarle Row
Hotwells
Bristol, BS8 4LY
UK

E-mail: mwtimmins@googlemail.
com

Chapter 19

Sally Underwood
University of Wales Newport
School of Art, Media, and Design
Caerleon Campus, Lodge Road
Caerleon, Newport
South Wales, N18 3QT
UK

E-mail: sallyu72@gmail.com

Chapter 20

Johannes Birringer*
Centre for Contemporary and
    Digital Performance
School of Arts, Brunel University
West London, UB8 3PH
UK

E-mail: Johannes.Birringer@brunel.
ac.uk

Michèle Danjoux
School of Art and Design
Nottingham Trent University
Fashion, Bonington Building
Dryden Street
Nottingham, NG1 4BU
UK

E-mail: Michele.Danjoux@ntu.
ac.uk

Chapter 21

Will Stahl
PhD researcher
Information Graphics in Health
    Technology Assessment
Pen TAG, Noy Scott House
Peninsula Medical School
RD&E Hospital (Wonford)
Heavitree, Exeter
EX2 5DW, UK

E-mail: wstahl-timmins@pms.ac.uk

Woodhead Publishing in Textiles

xvi

Preface

The process of creating smart clothing and wearable technology has to consider so many factors that it has to be collaborative between end-users, textile specialists, electronics, fashion and clothing designers and manufacturers all the way from the concept for new garment or wearable device through to point of sale. This book then is designed to support the development of a shared language between designers and each of the specialisations. It takes a non-specialist approach to technical areas that are usually covered, especially well in Woodhead Publishing's publications, by whole books rather than single chapters. It is only through the development of this shared language and understanding that we can begin to collaborate and overcome the complexities of design and product development of smart clothing and wearable technology.

To provide practical guidance and support, the chapters identify and elaborate a generic critical path from fibre production through to commercial prototyping described in an accessible language for designers. Innovative technologies may be integrated in a sequence of stages to enhance functionality throughout the critical path from fibre development to product launch. The book follows this sequence in its four parts from 'The design of smart clothing and wearable technology' (Part I), 'Materials and technologies for smart clothing' (Part II), 'Production technologies for smart clothing' (Part III) through to 'Smart clothing products' (Part IV).

Clothing that has truly 'wearable' attributes should both work and look good. To bring emerging technologies to market, and promote their use, the aesthetics and comfort of the clothing must be acceptable and the technology interface simple and intuitive for an inclusive audience. The technical, aesthetic and cultural demands of the wearer should inform the selection of specific assemblies of textiles within the design process.

Part I looks first at the emergence of wearable computing (Chapter 1), types of smart clothing and wearable technology (Chapter 2) then addresses key issues designers must consider from end-user based design of innovative smart clothing (Chapter 3), the garment design process from fibre selection to product launch (Chapter 4), to considerations for designing for the body (Chapter 5).

A new shared language is needed to enable communication between those, from a disparate mix of backgrounds, who will constitute the future design development team that addresses the merging of textile and clothing technologies with wearable computing. Part II looks at critical issues from fibre, yarn and textile development through to sensors and communication, providing a blend of what are often thought of as soft and hard technologies.

Part II starts by looking at the influence of knitwear on smart wearables (Chapter 6), woven structures and their impact on function and performance (Chapter 7), non-wovens in smart clothes and wearable technologies (Chapter 8), then moves on to the more electronic aspects of sensors and computing systems in smart clothing (Chapter 9), the application of communication technologies in smart clothing (Chapter 10) and the power supplies for smart textiles (Chapter 11).

The next key step is how smart clothing and wearable technologies can be produced. Part I looked at the end-user needs and Part II at the textiles and technologies. Part III then takes all of the elements from the earlier chapters and looks at how they can be integrated into a garment that is wearable and aesthetically pleasing.

The first two chapters look at cutting and placing of materials in garment construction (Chapter 12) and developments in fabric joining for smart clothing (Chapter 13), bringing together new ways of thinking about producing clothing that is smart in terms of design and production. The next two chapters look at how decoration may also be functional in smart clothing and wearable technology with digital embroidery techniques for smart clothing (Chapter 14) and developments in digital print technology for smart textiles (Chapter 15). The last chapter in this section looks at the environmental and waste issues concerning the production of smart clothes and wearable technology (Chapter 16) as we turn items that have been traditionally easy to recycle, textiles, into hybrid textile/computer products that are difficult to recycle.

Part IV completes the picture by looking at actual products that are available in the marketplace or have been developed as concepts for research and development. However, it also dares us to imagine constructively what is needed to take smart clothes and wearable technology that stage further to what we can dream may be possible in the near future.

However, too many technologies are introduced to the market where the need is assumed to be present but which only sells in small quantities. There are few products which truly become ubiquitous, and even then, time may lead to their ultimate demise. It is important that we look at what end-users really need, even if it is not possible with current technology.

The authors in Part IV consider these themes: 'What is around now?', 'What is currently in development?' and 'What do we want in the future?' and address different aspects of the smart clothing and wearable technology market from smart clothes and wearable technology for the health and well-being market (Chapter 17), smart clothing for the ageing population (Chapter 18), smart clothing and wearable technology for people with arthritis (Chapter 19), wearable technology for performing arts (Chapter 19) and branding and presentation of smart clothing products to consumers (Chapter 21).

Jane McCann, Director of Smart Clothes and Wearable Technology
Research Group, University of Wales Newport
David Bryson, Teaching Fellow, University of Derby

Part I
The design of smart clothing and wearable technology

1

The emergence of wearable computing

M. MALMIVAARA,
Tampere University of Technology, Finland

Abstract: In the mid-1990s, the wearables community was convinced that body-worn computing devices would be a sure hit within a decade. Instead, many of the concepts initially designed wearable, such as positioning and imaging found their way into mobile phones. The following is a brief look into the history of wearable computing and it also discusses the reasons why mobile devices beat wearables at their own game.

Key words: wearable computers, wearable electronics, intelligent clothing, concepts, prototypes, mass production.

1.1   The first devices

1.1.1   What spawned wearable computing?

A large number of the innovations we use in everyday life, such as Teflon, Gore-Tex and the World Wide Web have their background in military technology. During the latter part of the cold war, military expenditure on both sides of the Atlantic grew and some developments ended up as commercial products. Computer technology certainly benefited from the arms race and even wearable technology was in the military research programs.

Although military influence in wearable technology is undeniable, it was never even an important part in making wearables successful. What was important was that, on the one hand, in the 1980s computers had developed to a point where equipment and parts were more easily available, electronics engineering and computer science were taught in many places over the world and there was a growing enthusiasm for what could be done with electronics and computing. On the other hand, the rapidly growing rate of electronics' penetration in our work, homes and everyday life brought up interesting new areas of crossover electronics research with, for example, physiology and medicine, cognitive psychology and culture and sub-cultures as an abundance of new viewpoints emerged. However, what really started it all was the World Wide Web. From the 1980s into early 1990s, wearable computing was a hobby and a playground for a small number of people in a few universities and institutes. By the mid-nineties, most North American, European and Japanese schools and universities had an Internet connection and a website (as did many in the rest of the developed countries), and research and studies could be shared all over the world. New ideas could be discussed among peers on a daily basis and web communities brought people closer to each other.

The up-and-coming area of research on wearable computing benefited greatly in many ways. First, finding others interested in something which was nascent, without having to travel across the world to an event which might or might not be relevant to a new area of research. Second, the target group being interested in computing increased the possibility of finding peers online. Third, new research was posted on the Internet and thus helped the community to stay well informed on the state of core research. Fourth, arranging the necessary physical meetings became easier because the hardcore enthusiasts could be summoned online (see also Section 1.2.3). Finally, a wearable personal computer logically has a wireless Internet connection and the most dedicated cyborgs[1] could be online and contacted almost every waking hour of their lives. This led the hardcore community to modify their websites for friendly browsing on a wearable user interface.[2] Staner of MIT has worn a computer continuously since 1993.[3]

1.1.2 Definitions

In 1998, the Tampere University of Technology and the University of Lapland and Reima Ltd, set out to explore wearable technology as a shared project. In the project, the different kinds of prototypes, concepts and the few commercial products, needed to be organized into groups for inspection. The work team, consisting of Undergraduate and Masters students of clothing design and industrial design, established the following definition.

Wearable computers

A wearable computer is a computing device assembled in a way which allows it to be worn or carried on the body while still having the user interface ready for use at all times. By constructing it to be body-worn, a wearable computer makes computing possible in situations where even a laptop would be too cumbersome to open up, boot up and interface; a wearable computer can be used all the time, wherever the user goes.

A wearable computer can be very different from a desktop computer, depending on its intended user and tasks. The user interface can allow for both input and output in many ways, depending on physical and ergonomic needs. Input devices can include full qwerty-keypads, special keying devices, and joysticks along with standard function-specific push-buttons. If output is needed, a number of graphic interfaces exist with LCD-displays and head mounted displays (HMD). The user can also be given feedback with sound or vibration.

The most distinctive feature of a wearable computer is its ability to be reprogrammed or reconfigured for another task. This may include adding or changing hardware. A wearable computer can run many programs at the same time, and tasks can be assigned or terminated during operation.

Wearable electronics

Wearable electronics are simpler than full-scale wearable computers. While a wearable computer has both input and output and is capable of adjusting to multiple tasks, wearable electronics are constructed with set tasks to fulfil one or more needs of a specific target group.

Wearable electronics differ from mobile devices by their appearance and by being fundamentally designed to be worn on the body. A true piece of wearable electronics is also required to be worn to function, i.e. conceptually linked to the wearer's body. Some wearable devices require the user interface to be present and available all the time, meaning they are more obtrusive than devices with no input (such as the wrist unit and the chest belt of a heart-rate monitor).

Intelligent clothing

A jacket with a sewn pocket for a mobile phone does not make an intelligent garment. Clothing is intelligent when it adds something traditionally unclothing-like to the garment, without taking away or compromising any traditional characteristics such as washability or wearability. Ideally, an intelligent garment offers a non-traditional garment function, such as health monitoring, in addition to its traditional function as protecting the body. It could, for example, collect data and either transfer it wirelessly and automatically to an external computing unit or process the data itself, and respond to the computed conclusions without any user interfacing.

1.1.3 The Thorp–Shannon roulette predictor

Within the wearables community, Thorp and Shannon's (1998) roulette wheel predictor is considered to be the first wearable computer. In 1961, the two mathematicians tested a cigarette pack-sized wearable computer with twelve transistors, a speaker behind the ear for output and a toe-switch for input.[4] The system required two players, one 'timer' and one 'bettor'. The timer stood at the wheel and timed the spinning of the ball on the rotor and predicted where it would land. The timer would then tap a switch under his big toe to radio the bettor a signal tone indicating the predicted octant on the rotor. The bettor sat at the far end of the roulette table to look inconspicuous. The whole ordeal required a lot of tactics and skill to avoid suspicion and the thin wiring would often break, causing malfunction.

Ed Thorp was a graduate physics student at University of California, Los Angeles (UCLA) who had spent his childhood tinkering with gadgets. Claude Shannon was a mathematics teacher at the Massachusetts Institute of Technology (MIT) and a true mathematics scientist who loved solving problems by means of logic and calculation. The two teamed up to try to solve a classic problem involving physics and mathematics, and tested their system out of boyish curiosity in June of 1961 in Las Vegas. The roulette wheel predicting system yielded a gain of +44%, but with fragile wiring and a complicated operating procedure they never got around to risking substantial bets.

As it became clear to Thorp they were not going to exploit the invention, in 1966 and after he decided to publish their work and findings.[5,6,7] This comprehensive scientific documentation probably has earned the creation the title of 'the first wearable computer'. Although the roulette wheel predictor had not much computing power, it demonstrates perfectly the emergence of wearable computing. The first examples were indeed hand-built devices constructed from scratch components, and they were conceived from curiosity of how things work and a joy of tinkering.

1.1.4 Mann's Wearcam

In the early 1980s, photographer Steve Mann built a wearable system with photography equipment. It was a system of flashlights, batteries and cameras constructed in a backpack and a helmet (see Fig. 1.1). Mann went on to study computer science at MIT and continued his interest in wearable applications. He was one of the key figures in the MIT Wearables Laboratory, which in the early 90s stood for much of the initial theoretical and practical research into wearable computing. Steve Mann was one of the founding members of the International Symposium for Wearable Computers (ISWC, see Section 1.2.3).

Dr Mann is mostly known for his work in augmented and altered/mediated reality, i.e. capturing the visible environment by a video camera and editing and projecting the image onto his custom eyeglasses.[8] In augmented reality, additional information, such as previously posted digital notes or sensory data, is added to the display in the glasses. In altered/mediated reality, the computer controls the image projected in the glasses and edits out or adds images as programmed. Offending or unwanted material, such

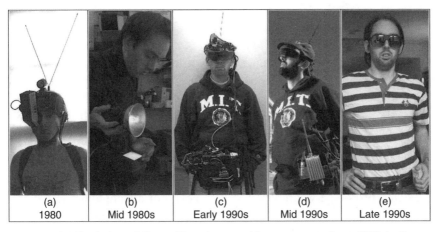

(a)	(b)	(c)	(d)	(e)
1980	Mid 1980s	Early 1990s	Mid 1990s	Late 1990s

1.1 Evolution of Steve Mann's wearable computers from 1980 to the late 1990s.

as advertising, can be replaced with natural scenery or other images. Having worn a wearable computer with a camera for years, Steve Mann has also come into contact with society's and authorities' view on imaging in public places. With the rules of imaging in public spaces being somewhat obscure, it is sometimes hard to explain one's wearable equipment and one's vision being dependent on a body-worn camera. Dr Mann has studied the opposites of personal imaging versus public security cameras found everywhere from streets to grocery stores to stations and airports.[9] He is currently a faculty member at the Department of Electrical Engineering, The University of Toronto.

1.2   The first wearable computers

1.2.1   The cyborgs

Like any established subculture, wearable computing also has its legends, the pioneers who started it all. In the world of wearables, the story very much began in North America and so the early key figures were mostly American.

The cradle of wearable computing was where research resources could be expended in new technology and where the slightly unusual pursuits of science were not weeded out. Among the most thriving habitats were Georgia Institute of Technology in Atlanta, GA (Gatech), Carnegie Mellon University in Pittsburgh, PA (CMU) and Massachusetts Institute of Technology. These universities ran research programs in exciting areas, such as computer science, computer technology, virtual reality, human sensory studies, user interfaces and new media. Some of the tech geeks at the time were irritated by the inadequacies of Personal Data Assistants (PDA) and laptop computers. The user interfaces were one problem: the screen of a PDA was too small and a laptop was too clumsy. A PDA's handwriting recognition was too slow and had its faults, and a laptop took ages to open up, boot up and find a place to rest it on. Among the first and the most famous cyborgs were a group of students at the MIT Media Lab.[10] Many of the MIT cyborgs, such as Mann, Starner, Rhodes and Orth were active members of the wearables community. Their research and publications are the foundation of wearable computing today.

The pioneers from the early MIT, Gatech and CMU, and later other schools, institutes and universities all over the world, constructed their own research prototype wearables. The complexity of a computing device and the research subjects as well as the makers' preferences all made the systems different from each other. Although someone else's rig was used as a platform, the final outcome was often very different.

The early wearable computers were maybe most inspired by the unexplored novelty of the genre. The fact that there were seemingly endless possibilities in customizing a personal wearable for showing off to your friends, but also that there seemed to be a lucrative business in equipping everybody from businessmen to single parents to school kids to maintenance technicians with a wearable computer was surely exciting. Certainly science fiction too, especially cyberpunk, affected the fresh subculture of wearables. Authors such as Isaac Asimov[11] and William Gibson[12] envisioned worlds and futures where people were constantly connected by wearable or even neural implants.

1.2.2 The first commercial wearable computers

When wearable computing had gained enough interested followers, it did not take long before someone saw commercial opportunity. In 1990, Computer Products & Services, Incorporated (CPSI) from Fairfax, Virginia, USA was renamed Xybernaut with the intention to manufacture and sell products for mobile and wearable computing. In 1996, Xybernaut[13] launched the 'Mobile Assistant', which could be delivered with custom programs and user interface for mechanics and technicians in both the military and commercial sectors, as well as for healthcare personnel. The Mobile Assistant I was followed by successive versions with improvements. The current platform, the MA V, was launched in 2003. Although many complained that the Mobile Assistants were bulky and had battery issues, the wearables research community saw Xybernaut as proof of the final commercial breakthrough. However, bad management and shady financial schemes led to the company declaring bankruptcy in 2005. In 2007, Xybernaut reemerged and operates in Chantilly, Virginia.

1.2.3 The International Symposium for Wearable Computers (ISWC)

In July, 1996 the Defense Advanced Research Projects Agency (DARPA)[14] sponsored the 'Wearables in 2005' workshop, where attendees schemed their predictions on future human personal computing. This workshop is regarded as the first organized wearables event. In August that same year, Boeing hosted a wearables conference in Seattle.[15] Some 200 people attended and a space was provided for vendors to present their research, project prototypes and products. The conference was a success and confirmed a significant interest in the subject from both the industry and academia. The enthusiasm called for a full-scale academic symposium, where relevant work could be presented to peers and where state-of-the-art research would meet industry interest for joint projects.

A year after the Boeing Seattle conference, on October 13th, 1997, CMU, MIT and Gatech co-hosted the International Symposium on Wearable Computers in Cambridge, Massachusetts. Almost 400 people registered for the event. Research in software, hardware, sensors, applications and systems was presented and the international audience included representatives from academia, military and various industries.

Because the first ISWC was such a success, a second ISWC was held the following year in Pittsburgh, hosted by the Carnegie-Mellon University, and thereafter annually in different parts of the world. The organizers of ISWC have traditionally been technology institutes and universities with computer science or electronics departments. The earlier ISWCs were predominantly software and programming oriented and most of the papers and research presented were very 'tech', but so were the first wearable computing devices! Devices and programs were also naturally there before any more in-depth systems analysis came about. The new and exciting wares exhibited and the openly future-savvy yet relaxed atmosphere at the ISWC attracted researchers from social studies, psychology and other fields of science.

Because the scope of wearable technology started to get out of hand, the events had to be kept focused amidst the increasing number of topics. Many people attending the ISWC were, in addition to wearable computers, also interested in virtual reality, context awareness and augmented reality. After the first few events, conferences in these subjects were held separately, in conjunction with ISWC, so that research less significant to wearable technology could have a chance to get communicated.

The ISWC[16] has grown to become the single most important annual conference in the wearable technology scene. Events in the subjects of pervasive computing, ubiquitous computing, smart clothes and smart textiles are held all over the world, but still the ISWC has held its position as the one not to miss.

1.3   Wearable electronics

1.3.1 Wearable electronic devices

Although the events around wearables mostly concentrated around full Personal Computer (PC) rigs and the more technical aspects, many were thinking of lighter forms of wearable technology. Smaller devices could potentially reach bigger and different target groups. On July 1st in 1979, Japanese Sony® introduced the Walkman®, a portable music cassette player with headphones.[17] Although the concept was first met with skepticism from sales as much as market, portable music became a hit and has since grown into a big business. The Polar® heart rate monitor – a wrist watch unit and a sensor belt around the chest – has been around since 1982[18] and nobody really interpreted it as a 'wearable'. Still, it has some quite wearable characteristics: it is body-worn, monitors the user and is even relatively unobtrusive in use.

The wrist watch was keenly used as a platform for a number of systems, and Dick Tracy or Michael Knight[19] telephone watches seemed to be the ultimate goal. The boundaries of miniaturization, of course, can still be stretched, but usability became the limiting factor. Even if it is technically possible to squeeze a PC into wrist watch, the user interface in a watch-sized device is very, very hard to make operable with the limited space for output and input.

There were other academic pursuits too and somehow the killer application was expected to be a little wearable device as opposed to a wearable computer or an intelligent garment. The wearables scene attracted people from the mobile device industry and some already hailed the mobile devices' era being over and wearables taking over. It was not to be so simple, especially with the mobile devices being backed up by a culture of familiarity and also the inherent difficulties of mass-producing body-worn devices (see Section 1.5). Setbacks followed with the launch of iconic mobile devices, such as the Apple iPod® (October 2001), Motorola Razr® (July 2004), Nokia N95 (March 2007) and Apple iPhone® (USA, June 29th 2007). These devices gathered so much attention that the hype strengthened the overall position of mobile devices on the consumers' 'desirable objects' list.

1.3.2 Design

When wearable computers evolved beyond PC hardware built in a back-pack, and the number of people involved in wearables increased in the late nineties, the issue of textiles became relevant. Initial collaboration was started when a wearable had to get an ergonomics update to enable long-term use and a textile or clothing design student was asked to help out. Sewing pouches for hardware meant that they had to be 'designed' and for many years, textile students and clothing design students were the 'design ingredient' or the 'softer value' trespassing in the tech geek domain.

In 1998, Francine Gemperle *et al.*, from Carnegie Mellon University in Pittsburgh, presented a paper on human ergonomics and wearable technology at the Pittsburgh, PA ISWC.[20] The 1998 CMU study on placing objects on the human body with regards to mass, size and mechanical properties of an item and guidelines for task-oriented splicing of devices into smaller modules is still relevant today, even if hardware and power supply have evolved greatly from those days. Their input is significant in more ways than one; not only did they provide the tools with which wearables could be built more comfortably, they had also explained scientifically why design is critical to the development process.

Right after the millennium change, design already had a steady foothold in wearables. The sculpted, better-looking wearables concepts attracted investors who now saw commercial potential in body-worn devices. Also academic collaboration shot to another level with design schools and technology institutes looking for opportunities to work together. Again, fancy-looking and sometimes very futuristic prototypes started to appear in the media.

1.3.3 Target groups

In the early to mid 1990s, information technology (IT) was overheating. Companies in computer and mobile technologies in particular were well funded, sometimes far over needs. The wearable technology scene was still fairly new and was maybe a little overshadowed by the hotter, more talked-about mobile gaming or Internet enterprises to catch the money train. The wearables community did naturally try to make it to the consumer market, but to succeed it would need a killer application and a target market.

The IT boom had successfully separated and targeted a group of consumers labeled 'the young, urban nomad', among other terms. This consisted of post-teenage to young adulthood aged city-dwellers, who had money to spend and were hungry for trends. The nomad-term suggests mobility and mobile telephony, in particular. In the mid nineties, mobile phone penetration increased dramatically as network coverage grew and the devices became more portable and affordable. Most people aged 15 to 25 knew how to operate computers and mobile phones (taking pleasure in knowing that most of their parents could not) and liked being the focus of attention of the trendy IT industry.

Wearable technology was also targeting the young urban nomads in research prototypes and design studies.[21,22] Although the concepts were fine tuned, commercial products did not appear. It was not necessarily a problem in marketing (see also Section 1.5). When the dust settled after the downfall of the information technology frenzy in the late 1990s and early first decade of the second millennium, academia and industry, as well as the media, started slowly recovering from the embarrassment. To keep high-tech and wearables interesting and credible, new target groups had to be named. Wearable technology sought partners in areas which had more credibility and who could afford wearables. An important step ahead was the notion of no longer relying on the trendy and cool individuals for marketing value. The following are examples of often contemplated target groups for products of wearable technology, with key advantages and problems.

Sports

Sports are never out of fashion and athletes and their sponsors spend a lot of money to get better results. Athletes, in particular, would benefit from body monitoring systems and data such as heart rate, temperature, fluid balance, acceleration and positioning. Most sports, however, benefit from losing mass in the athlete's gear and so adding electronics is a huge problem to start with. Still, wearable heart rate monitoring made it as the killer exercising application, with clear benefits for both training athletes and regular consumers following their physical condition.[23] Polar Electro was the first available wireless heart rate monitoring system, launched in 1982.[24]

Medical

Body monitoring is most accurate when done with full skin contact. This is a clear benefit for body-worn devices, especially in cases where long-term monitoring is required. Health services are a big business around the world. Public health care is naturally more tempting, with often instant large-quantity orders, but is strictly regulated in most countries. Doctors are reluctant to adopt either new devices or procedures for the amount of administrative work involved in getting them approved by officials. Changing anything in the procedures of hospitals and public health centers would only add to the stress of the often over-worked staff. On the other hand, if personal, body-worn instruments could be read and analyzed remotely, both doctors and their patients could save time and money with less routine visits. The private sector is maybe more viable, but any new technology used there needs to be approved by officials, just as in the public sector.

Official occupations

Law enforcement officers, fire fighters, maintenance workers, logistics workers and communication operators are already wearing computing technology. The benefits of having constant wireless access to networks and data bases or other people whilst attending to various tasks can definitely make working more efficient. Still, very much like in health care, getting complex electronic systems with all the maintenance, warranty, liability and procedure issues approved is often a bureaucratic exercise few want to tackle. This is probably just a question of time and, as soon as wearable systems can be mass produced with more affordable components, usability improves to lower the adoption threshold and the bureaucracy battles are conquered, wearables will make it to the tool belt.

Bradley Rhodes did suggest, in 1997, that augmented memory is the killer wearable application,[25] but wearable computing, as marketable devices and consumer trends, was still too far away from commercialization.

1.4   Intelligent clothing

1.4.1   The Philips–Levi's industrial clothing design ICD+ jacket

As an example, the Philips–Levi's collaboration's outcome ICD+ jacket was not really intelligent clothing, although interpreted as such by the media. With any new concept, people and media tend to take short cuts, but a jacket with special pockets sewn for electronics devices and their cables is not really 'intelligent'. The jacket served as a mere carrying platform for the devices and the garment and the function of mobile telephony and music were only artificially linked. The hot key words 'smart' and 'intelligent' were mostly added by media to emphasize the novelty of the concept. Both the jacket and the device could well be used separately, even with somewhat better usability! Actually, the ICD+ jacket was more a piece of wearable electronics than intelligent clothing.

The Philips–Levi's collaboration was an interesting 'world first' project,[26] where two giants put their product development teams together to form something new. Although the outcome was not a commercial success, they still managed to produce a series of jackets combining telephony, portable music and clothing.

The Philips–Levi's ICD+ jacket and its launch revealed a statistical illusion that took many company heads, analysts and investors by surprise. Although the jacket received a staggering amount of media coverage from magazines and newspapers to radio, television and Internet blogs, all the hype was never matched in sales. The audiences' interest was pure curiosity which never turned to a desire to buy the product. This statistical illusion can, in retrospect, be applied to most of the so called 'IT bubble' enterprises. A project was presented to the board and investors cited phrases and keywords of techno-jargon, an approved project was tested with target group interviews with 'high interest values' and when the product comes out the sales lingered.

Sadly, the IT bubble bursting killed off many university projects and industry jobs, but on the other hand it made competition fiercer among those dedicated institutes that still pursued research in wearable technology. Funding and resources being scarce in the mid first decade of the second millennium makes those still left work harder and actually think and create, as it is no longer possible to just take the instantly funded 'cool route' with wearable technology.

1.4.2 The Cyberia survival suit

Along with the Philips–Levi's ICD+ Jacket, the Reima Cyberia Survival Suit was one of the very first published concepts of smart clothing.[27] Unlike the ICD+ Jacket, the Cyberia was only an academic study and not intended for mass production. The academic goals of the project were to design and develop the concept and produce a prototype of a smart outfit and, in the process, study the union of clothes and electronics. The Cyberia illustrates perfectly what the first published wearable technology concepts were like, how they came about, what their purpose was and the practical difficulties and compromises associated with designing a working prototype.

The Cyberia was an interdisciplinary joint project between three Finnish partners: clothing manufacturer Reima Ltd, the University of Lapland, and the Tampere University of Technology. Reima brought their expertise in outdoor clothing, the University of Lapland stood for industrial and clothing design and the Tampere University of Technology handled the electronics and body monitoring. The three parties together operated the entire project, but later on DuPont, Nokia, Polar Electro and Suunto also participated in the project, with contributions in their respective fields of expertise. The project took 18 months and was launched at the Hannover World Expo in Germany in May 2000.[28]

The target user

As Reima's expertise is in winter outerwear and as in 1998 they still had their motor sports' division, the first choice for a subject was a snowmobiling outfit. The decision was based not only on Reima being the world leader in snowmobile garments, but even more on the fact that snowmobiling garments are padded and sturdy to begin with and so offer the designers more room to stash the electronics. A cold and hostile winter context was also appreciated. Everybody understood that the prototype would not be light or small, because it would be built from off-the-shelf components as time and funding would not allow for custom parts.

When starting to assemble the concept of a smart outfit, the team concluded that most of the target audience at the Hannover Expo probably had no idea what a snowmobile is in the first place, as snowmobiling is banned in Central Europe. This fact alone meant that the concept had to be changed.

Reima's other target groups included children and skiers, but they would both be difficult for the concept's designers. Kids' garments were too small to fit anything and the group decided it was too risky to launch a new idea of electronically enhanced garments on children. After all, Central Europeans especially were very conscious of electro smog[29] and similar hazards that close-range devices might induce. Skiers were carefully considered and they seemed like a good target group, but all proposed concepts for the slopes proved not to give enough 'real' added value to the skier. Finally, the answer presented itself from the suggested features the suit would have. Because the suit was to monitor body data and the concept should be as interesting as possible, the user should be someone who may encounter danger in what he or she does. The data would be measured; a computer would process it and take appropriate action. The Cyberia concept target user was defined as a professional person, who works alone in arctic conditions. Possible subjects are forest guards, frontier guards, reindeer herders, technicians and scientists working outside. The user works at long distances away from civilization, moving on a snowmobile, an all-terrain vehicle (ATV), skis or snowshoes. We can assume that anyone in the target group is, by profession, experienced and knowledgeable in spending long times outdoors, knows basic CPR, and can handle most situations, such as spending a night in the open by building a quick shelter. But no-one is immune to accidents and situations such as getting wet or losing consciousness in sub-zero weather, and losing a lot of blood or sustaining multiple fractures. Even a non-lethal accident can diminish the ability to act and think clearly and can lead to bad judgment and increased need of outside help.

So this was the problem to solve: to create a suit that would be comfortable enough and allow the users to work as they normally do, yet would constantly be monitoring their body and would draw conclusions from gathered data and take appropriate action if set parameters were exceeded. On top of that, the suit needed to look right, i.e. no trendy color schemes or cuts that might look silly on a technician alone in the wilderness. The user would also pack tools and equipment in pockets and wear a backpack at least every now and then.

The team interviewed a variety of people from suitable target groups. Among those was staff from a snowmobile safari company, reindeer herders and frontier guards.

Electronic functions

The suit has both electronic features and non-electronic features. The electronic features include the suit's body monitoring, communication and positioning devices. An emergency heating-panel system is also constructed into the underwear.

The suit has four kinds of sensors built into it: heart-rate sensors woven into the underwear shirt, temperature sensors inside and outside, a humidity-sensor for detecting immersion in water and motion-sensors for determining posture, movement and impacts.

A Global Positioning System (GPS) unit has one antenna on the left shoulder and whenever three or more GPS satellites are visible, the suit can be tracked down to one square meter anywhere on Earth.

For communication, the suit is equipped with a GSM (Global System for Mobile phones) mobile phone. It was decided, after much debate, that the phone module would be used only for Short Message System (SMS) text messages and not voice calls, so that the battery power would be used only for emergencies and the user would be prevented from chatting away the battery life of the phone. A computer unit runs the functions and two battery packs power the whole system.

The Yo-yo user interface

All electronic features are controlled by a palm-sized Yo-yo user interface (Fig. 1.2). The user interface (UI) is basically a handheld, aluminum-encased Liquid Crystal Display (LCD) with two push buttons; one on top and one in the side. The UI serves as an instrument panel for the suit, for input and output of all data. A power and data cable runs from the bottom of the unit to a spring-loaded cable housed in the jacket, which coils up the cable and retracts the unit to its pocket in the front of the jacket.

The Yo-yo was designed to be used with one hand, even with thick mittens on, so that operating the suit's electronic functions can easily be done even while driving a snowmobile with the other hand, or in other situations when both hands cannot be used at the same time. The user can scroll the menu on the display by pulling the Yo-yo back and forth and select by pressing a button. In an emergency, a call for help can be sent by only squeezing the unit for 10 seconds, enabling a simple self-aid action even when seriously injured.

Non-electronic features

When working on the concept of the suit, the team found situations where simple accessories (Fig. 1.3) could be very useful in an accident. While these were not 'intelligent' as in 'electronically enhanced', they certainly were a

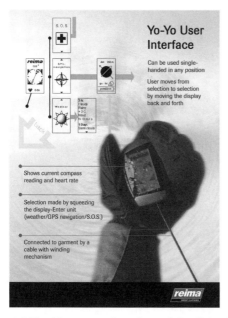

1.2 The Yo-yo user interface of the Cyberia survival suit.

smart addition to the concept. For example, falling into water through the ice produces a risk of loss of life from hypothermia. A thin, sleeping bag-like, polyurethane-coated pouch was folded into the back pocket of the jacket. Crawling into the hypothermia sack after a plunge into water in sub-zero temperature can increase the life-expectancy of a victim by up to 4 hours according to the Finnish Institute of Occupational Hazards. The sack is non-breathable and prevents moisture from evaporating and so forms a micro-climate that cools down much slower than a normally exposed wet outfit.

Another aide is the removable cargo pocket in the trouser leg. The pocket is lined with an aluminum-clad Aramid fiber, making it fire-proof and water-tight, good for melting snow into palatable water. As keeping hydrated is essential, especially in the cold, a source for fluid is necessary.

Naturally, the suit packs a pair of ice spikes in the sleeves for getting back up on thick ice, a fire kit with storm matches (a blade for cutting twigs for firewood and a piece of magnesium for setting damp or even wet wood on fire) and a first-aid kit in the pant leg (band-aids, a pressure bind and two pills of synthetic morphine for pain relief).

The electronics vest

At first, the crash pads in the elbows, knees and shoulders were intended to be just hollowed out to accommodate circuit boards and batteries, but

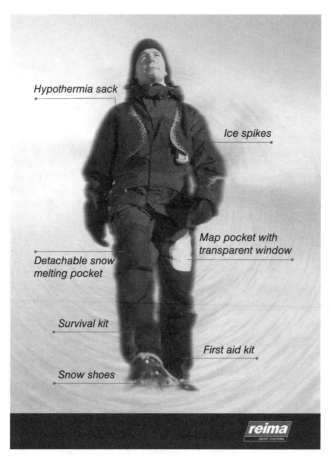

1.3 The Cyberia survival suit.

this soon proved to be impossible. Not only did the electronics eventually turn out to be much bigger in size and weight than everybody initially thought, but with the electronics built in, the pads would lose all their impact dampening capacity. Secondly, the mass of the electronics would have been completely impractical at joints, where the pads are, and would have restricted movement quite a bit. The weight would also be too much for any textile to carry. Even the tough Cordura™ used in the shell would pull and sag under the heavy blocks of batteries and components. The whole concept had to be changed.

The heavy system of electronics had to be divided into parts that could be spread out over a larger surface to distribute the weight. The early designs had the electronics shaped dynamically and embedded in the jacket shell in neat little pockets, customized for each part, but this would have to go. The new configuration placed the electronics inside an elastic harness vest, which wrapped tightly around the torso of the wearer. The vest was made adjustable with elastic bands, so that it would fit just as well on a small woman as on a big man. The electronics were spread out to allow for arm movement by leaving the shoulder area free, and instead, making as much use of the static back by placing most blocks along the spine and around the lower sides of the back. Even if the electronics were designed to be water-proof and shock-proof, they could now be removed for machine washing and so spare the garment from unnecessary wear in the washing cycles. The wearing comfort was increased tremendously as the jacket itself did not seem to have any added weight and was flexible and easy to wear. The harness did a fantastic job as a back protector, with the plastic encased electronics acting as ergonomically placed bumpers.

The underwear

It was clear to all from the beginning that the prototype garment would follow the principles for cold weather layering systems. The concept also required measuring the heart rate of the user and, to get reliable results, the sensors would have to have direct skin contact and so custom underwear was necessary. An all-extremities sensor layout was scrapped as it was considered too much hassle to deal with the uncomfortable wiring and cable connectors between the two-piece underwear. For measurement accuracy, the upper-body data proved quite sufficient and even the planned temperature sensors were moved into the jacket liner to rid the underwear top from unnecessary wiring.

Project outcome

The biggest obstacle during the work was clearing up the various misconceptions between the electronics people and the textile/clothing people. It was sometimes frustratingly difficult to explain things to one another due to the differing backgrounds; one holds many things as obvious after having studied them for years.

Once the team got to know each other better, the misunderstandings seemed to become less frequent, but around that time outside help with certain tasks became necessary. Although the team now knew to explain everything as clearly and completely as possible, the problem persisted. Most of the times, one or both of the following phenomena was encountered when introducing someone new to the work: (i) When shortly or thoroughly explaining the project and what in particular we would need from the person in question, they would stop listening after hearing about merging electronics with clothing. The whole subject was at that time so mind-boggling that accepting any further information proved too much for many to handle. (ii) After hearing about merging electronics with clothing, the person would block out all information except that related to their field, no matter how important it was to see the whole and things had to be explained over and over again.

When the suit was ready and taken to the media, the journalists and reporters were always first given a short introduction to the subject. Some details were always either missed, misinterpreted or sometimes replaced with incorrect assumptions. Even if those involved in the Cyberia project are still very happy with the concept, particularly how it addresses the idea of wearability and how the different parts are integrated, the concept is a little too complex. There is just too much stuff in it to be presented in a quick conversation.

1.4.3 The Georgia Institute of Technology's Wearable Motherboard

The Georgia Institute of Technology (Gatech) is well connected with DARPA, the research department of the United States (US) Army. Starting in 1996, Gatech developed the Wearable Motherboard,[30] a shirt with integrated conductive fiber circuitry monitoring the wearer's body and detecting a soldier's injuries from bullets and shrapnel. The Wearable Motherboard also has potential for civilian use. The simple, comfortable and non-intrusive way of measuring the human body with a familiar garment may well be applicable to healthcare from infants to the elderly.

The Georgia Institute of Technology's Wearable Motherboard was part of a U.S. Army research program called Land Warrior (1994–2007) The Land Warrior aimed at higher lethality, higher survivability and better communication by equipping an infantry soldier with a wearable computer. Although the system was successfully tested in practice, the Land Warrior was subsequently cancelled in 2007.

1.4.4 The MP3 jacket craze

In the first years of the new millennium, the world had just recovered from Y2K and was ready to indulge in all the inventions that had been predicted to happen in the 21st century. By 2000–2001, progress in wireless technologies and electronics was huge and gave the impression that everything was possible in small devices and that they would really make our lives easier. Wearables research was being published at an ever-growing rate and money men sniffed an opportunity to cash in. There were significant problems in mass producing wearable technology (see also Section 1.5), but there was a demand for *something*.

Wearable technology was usually connected to youthful, action-packed and cool themes in the media, to make the new scene more attractive. This can partly be attributed to the media itself, but many of the first applications were equally guilty of squeezing into this frame of reference. Therefore, the first commercial applications were to find their market in the young, urban, 'active lifestyle' people. Music was identified as one important keyword. Digital music players had not yet made their way into mobile phones, but already the Sony Walkman had shown how cool mobile music was. The Philips/Levi's ICD+ jacket made it first, but was too clumsy and was not really as usable as a commercial product should be.

The German electronics manufacturer Infineon unveiled a textile-integrated Bluetooth/MP3 system at the 2002 Avantex in Frankfurt, which would come to inspire many. The American snowboarding equipment manufacturer Burton had been eyeing wearable technology for a while and they too were shopping around for ideas. Like Infineon, they had made the connection between music and active lifestyle. Many snowboarders like to listen to music while riding, and around 2002–2003 the coolest professional riders wore big headphones over their hats connected to an iPod. The Apple iPod had established itself as the hottest MP3 player within a year of its launch in 2001. Burton approached Apple to create an MP3 jacket and successfully launched the Amp, the world's first iPod jacket, in January 2003. The Amp had integrated connectors for an MP3 player, headphones, a Bluetooth gateway for a mobile phone and textile operating buttons in the sleeve. It would be followed by new versions in following years, even if the first one cost $499 and was surely no commercial success.

Infineon had launched their technology in 2002 and made a courageous move to license their creation as a module and even allowed buyers to customize the user interface and other specifications. Several garment man-ufacturers launched a jacket with the same features as the Burton Amp. Among the first were sportswear giant O'Neill, and Rosner, a German casual fashion manufacturer. Rosner presented their MP3Blue jacket in July 2004: O'Neill got ahead with an announcement on The Hub jacket earlier, in January. Both products were launched for the 2004–2005 winter season.

Although Infineon were among the very first to offer integrated, textile-ready custom modules for mass production, the company decided to pull out of textile electronics in 2005. Their heir would become Eleksen, a London based company specializing in electro-textile systems.[31] Eleksen's patented ElekTex fabric is touch-sensitive and can be machine washed. After patenting ElekTex, Eleksen went on to compose a modular package for garment manufacturers to license and integrate to their product. In 2007, Eleksen systems have been licensed by among others O'Neill, Norrøna, Kenpo and Spyder in outdoor garments, and Microsoft, Orange and Mobis for textile input devices.

1.5   Conclusion: 'Where can I buy this?'

Although wearables were 'hot' and research prototypes were popping up everywhere, few commercial products appeared in the first few years of the new millennium. Mass production of almost every kind of conceivable goods had by then already shifted to the Far East, and production plants in the western world were downsizing and struggling to stay alive. For this reason, in the West, factories could not take on the challenge of investing into a new kind of production of goods which had no existing market to begin with and were at times difficult to describe to anyone outside the inner circle. Asian factories were growing at maximum rates and were running at full capacity all the time. Any excess resources were allocated to increasing capacity for their existing type of products; they did not need wearables and nobody seemed to have the time or will to take on such a challenge.

Old habits die hard and established routines can be difficult to change. Textile production and electronics production are worlds apart when it comes to production cycles, quality tolerances, markets, etc. Bringing these two closer to each other is a huge task, again as both are doing just fine without the other. Money can, of course, change everything, but without a sure-hit target market, even the big players seemed reluctant to invest. Large sporting goods manufacturers such as Nike and Adidas don't even own their production facilities and mostly have their designs produced at contract manufacturers who, in turn, must compete with others for each deal.

So, one major obstacle in marketing wearables is production, but more difficult problems are right around the corner. Personal electronics such as mobile devices and home appliances are traditionally sold in different stores from textiles and garments. Common sales channels can often be found only at large department stores, but these are, as such, too limited for a mass launch. Smart clothes and textile electronics would also require the sales staff to be educated in both textiles and electronics. Further problems arise in warranty issues and customer service, which are also linked to the product's life cycle. A wearable device's textile parts and electronic parts can have very different theoretical life spans. In the days of increasing awareness of environmental issues, an expensive, high-tech product in which one part outlasts the other sounds dubious and would undoubtedly face high recycling costs in many markets. An ecologically sound, and maybe the only, solution would be for the manufacturer itself to arrange for both warranty and disposal of their product. This is a very costly operation and often an unpleasant one. Especially for a smaller company eager to launch their product and to turn cash flow into profits after production start-up and marketing costs, the temptation to axe recycling costs may be pressing.

1.6   References

1 Manfred Clynes and co-author Nathan Kline coined the word 'Cyborg' in a story called 'Cyborgs and Space', describing a human augmented with technological 'attachments', published in *Astronautics*; September issue, 1960.

2 See Steve Mann's homepage; http://wearcam.org; 2007.

3 Thad Starner's non-technical biography; http://www-static.cc.gatech.edu/fac/Thad.Starner/01_cv.htm; 2007.

4 EDWARD O. THORP & ASSOCIATES, '*Second International Symposium on Wearable Computers: Digest of Papers*', The Invention of the First Wearable Computer; The Institute of Electrical and Electronics Engineers (IEEE), Inc. 1998.

5 E.O. THORP, '*Beat The Dealer*', 2nd Edition, Vintage, New York, 1966.

6 E.O. THORP, '*Physical Prediction of Roulette I, II, III, IV*', Gambling Times, May, July, August, October 1979.

7 E.O. THORP, '*The Mathematics of Gambling*', Lyle Stuart, Secaucus, New Jersey, 1984.

8 http://wearcam.org

9 http://wearcam.org/sousveillance.htm

10 The MIT lists following students and researchers as the 'Wearables v. 1.0': Rehmi Post, Jennifer Healey, Lenny Foner, Dana Kirsch, Bradley Rhodes, Travell Perkins, Tony Jebara, Richard W. DeVaul, Nitin Sawhney, Maggie Orth, Steve Schwartz, Chris Metcalfe, Kevin Pipe, Joshua Weaver, Pamela Mukerji and alumni Thad Starner and Steve Mann.

11 ISAAC ASIMOV, '*I, Robot*', Gnome Press, New York, 1951.

12 WILLIAM GIBSON, '*Neuromancer*', Ace Books, New York, 1984.

13 http://www.xybernaut.com

14 The Defense Advanced Research Projects Agency (DARPA) is the central research and development organization for the Department of Defense of the US Army, see http://www.darpa.mil/; 2007.

15 See http://wearcam.org/computing/workshop/; 2007.

16 http://www.iswc.net

17 A. PARASURAMAN, CHARLES L. COLBY; 'Techno-Ready Marketing: How and Why Your Customers Adopt Technology', *Case Study: Consumer Electronics*: Sony Walkman; Rockbridge Associates, Inc. 2002.

18 http://www.polar.fi/polar/channels/eng/polar/about_polar.html

19 DICK TRACY (Chester Gould, Chicago Tribune Syndicate, 1931–77), the cartoon detective and Michael Knight ('Knight Rider', NBC, 1982–86), the TV series private mercenary with his intelligent and computerized car both used their wrist watches for wireless communication.

20 F. GEMPERLE, *et al.*; 'Proceedings of the Second International Symposium on Wearable Computers', Design for Wearability; The Institute of Electrical and Electronics Engineers (IEEE), Inc. 1998.

21 J. HERSTAD and D. VAN THANH; *The Proceedings of The Third International Symposium on Wearable Computers* (ISWC '99); 'Wearing Bike Components'; The Institute of Electrical and Electronics Engineers (IEEE), Inc.; October 1999, pp. 58–63.

22 J. MIKKONEN, J. VANHALA, A. REHO and J. IMPIO; *The Proceedings of The Fifth International Symposium on Wearable Computers* (ISWC '01); 'Reima Smart

Shout Concept and Prototype'; The Institute of Electrical and Electronics Engineers (IEEE), Inc.; October 2001, pp. 174–175.

23 Polar Electro WearLink textile heart rate monitoring sensor belt; http://www.polar.fi/polar/channels/eng/segments/Fitness/F11/accessories/WearLink_transmitter/open.html, 2007.

24 POLAR ELECTRO; http://www.polar.fi/polar/channels/eng/polar/about_polar.html, 2007.

25 RHODES B., et. al.; *The Proceedings of The First International Symposium on Wearable Computers* (ISWC '97); 'The Wearable Remembrance Agent'; The Institute of Electrical and Electronics Engineers (IEEE), Inc.; October 1997, pp. 123–128; http://www.bradleyrhodes.com/Papers/wear-ra.html, 1997, http://www.remem.org/, 2007.

26 Launched in Europe in September 2000.

27 Media preview in Jämijärvi, Finland in March, 2000.

28 http://www.exposeeum.de/expo2000/index_e.htm#

29 Electro smog is a term used to describe the microwave radiation emitted by radio waves, power lines, transformers, wall sockets and all electronic appliances. According to http://www.electrosmog.com.au 3–5% of the population develops health problems when exposed to electro smog.

30 Georgia Tech Wearable Motherboard, see http://www.gtwm.gatech.edu/gtwm.html; 2002.

31 http://www.eleksen.com

2

Types of smart clothes and wearable technology

R. D. HURFORD, University of Wales Newport, UK

Abstract: This chapter aims to give the reader an overview of commercial smart clothes and wearable technology products. It will take a look at some of the important commercial and quirkier smart clothes products, from companies such as O'Neill, Burton, Adidas, Polar and Textronics, to name a few. The chapter will also look at market forecast and current emerging trends.

Key words: smart clothes, wearable technology, commercial, products.

2.1   Introduction

This chapter aims to give the reader an overview of commercial smart clothes and wearable technology products. It is not meant to be an exhaustive list of products, but to give examples that best illustrate the industry as it stands. Due to the fast moving nature of technology, there will more than likely be many new products in the market place shortly after this book has been published.

From a slow start, this new industry sector now seems to be growing at a good pace, with many new products being released and new companies getting involved in the development of smart clothes and wearable technology. There are two distinct types of companies within the sector: those that have their core business based around smart clothes and wearable technology; and those that produce a few products that sit within a larger mainstream product range. The first set of companies is usually supplying the technology to the second set.

Clothing and electronics have traditionally been separate industry sectors. They now have to work together to develop products and this cross-disciplinary working has caused problems. The collaboration between these sectors is sometimes difficult, with differences in language, working practices, development time-frames and marketing strategies. There are now several companies that specialise in the development of smart clothes and wearable technologies. Some of the products that have being produced do not always fit neatly into existing product categories. This was very noticeable for very early products: were they clothing or electronics?

Wearable computing has developed in parallel, being more focused on the development of advanced highly portable computing technology, but it has had an important influence on smart clothes and wearable technology.

In the past, large-scale uptake of smart clothes and wearable technologies seems to have been inhibited by the lack of sufficiently advanced technologies. This would often lead to products not meeting the required needs and expectations of potential users. The technology is now starting to mature and the products that are being introduced are beginning to live up to consumer expectations. The industry is still in its infancy and has a long way to go, but the journey is going to be very exciting. We will take a look at some of the important commercial and quirkier smart clothes and wearable technology products. There are many medical and military wearable products that are not discussed in this chapter.

2.2   A brief history

2.2.1   Personal portable technology

Some of the earliest personal/wearable technology was the pocket watch and later the wrist watch. The popularity and familiarity of the wrist watch may be the reason for the placement there of many current wearable control systems.

Personal or portable electronics really took off in the 1980s with the popularity of the Sony Walkman. During the 1990s, portable Compact Disk (CD) and MiniDisk became more popular. By 2004, the focus moved onto digital music formats, with the introduction of hard disk recorders. The Apple iPod was launched in 2001, and has paved the way for the digital music revolution. Portable devices now include video play-back functionality. In 2005 it was suggested that the iPod was one of the main drivers for the wearable technology industry (Marks, 2005).

The mobile phone is another example of personal/portable electronics. The earliest mobile phones were introduced around the 1950s, but they were very different from the devices that we use today. The first modern cellular networks were rolled out in the late 1970s and early 1980s. Handsets at this time used analogue signals and were very large. SMS (Short Message Service) was not introduced onto handsets until the early 1990s, but has had unexpected success as a communication medium. Mobile handsets are now capable of many advanced functions.

The power of mobile handsets is advancing rapidly, with current top-of-the-range handsets, such as the Nokia N95, Motorola MOTORAZR and Samsung F110, having the equivalent processing power of a computer from the early 1990s. As part of a 2007 advertising campaign for their N-Series smart phones, Nokia state that they no longer make mobile phones but make mobile computers or mobile communicators. Today, the United Kingdom and most of Europe has 100% mobile saturation (Budde, 2008).

2.2.2 Fabric technology

Fabric technology also plays a major role in smart clothes and wearable technology. The clothing industry was changed dramatically by the introduction of man-made materials such as nylon and polyester. Development of man-made materials has continued at an astounding pace. Current developments include nano-fibres and nano-coatings, which can provide us with a host of useful and unusual characteristics that enhance earlier technologies or that have been unavailable to us before. These include heat absorption and cooling, colour changing characteristics, stain and water resistance, abrasion and impact protection, and electro-conductivity. All of these characteristics can play useful roles in the development of smart clothes and wearable technology. Electro-textiles and other electro-conductive materials are currently proving to be very popular. There are three main companies developing commercial conductive switching and control systems. Softswitch (Peratech), Fibretronic and Eleksen. Eleksen has released evaluation and development kits for its fabric switches, allowing easier access to the technology.

2.2.3 Early commercial examples of 'wearables'

Philips and Levi's ICD+

The Philips and Levi's ICD+ Mooring Jacket was released in 2000 to a mixed reception. The jacket used a unified connector/controller to integrate a mobile phone, MiniDisc player, earphone and microphone. The electronic components were not fully integrated into the jacket, but were enclosed in pockets, with the wires held down using Velcro tabs. All of the components needed to be removed from the jacket for it to be washed. (Another company, ScotteVest, has developed a whole product line that utilises a similar system of enclosures and pockets.)

The jacket was not a commercial success (partly due to its £800 price tag). However, it has provided valuable insights for the development of smart clothes and wearable technology, highlighting problems of fully integrating clothing and electronics, washability and durability, and also the complexity of the design and manufacturing of such garments.

The ICD+ Jacket, was very much ahead of its time in 2000. However, the mobile phone and MiniDisc player were not swappable for newer models and they quickly became obsolete. The ICD+ Jacket was the pre-cursor to many of the current iPod jackets, from companies such as Burton and O'Neill.

France Telecom – CreateWear

France Telecom worked with French fashion designer Elisabeth de Senneville on a range of 'wearable communication' garments. The garments had flexible colour screens built-in, using a small matrix of Liquid Crystal Displays (LCD), providing visual communication when using a mobile phone. The mobile phone connects the garment to a central communications server where images, animations and messages can be uploaded. The wearer can then choose what should be displayed. It was rumoured that the prototype would be developed into a product, but this has not occurred. Recently, Philips have developed a flexible display technology called Lumalive.

2.2.4 More recent commercial products

Burton – AMP (now Audex)

In 2003, Burton and Apple unveiled the limited-edition Burton Amp Jacket (and later the AMP Backpack), the first and (at the time) only wearable electronic jacket with an integrated iPod control system. The jackets used integrated fabric connectors and fabric controls (which were mounted on the sleeve). The jacket was available exclusively through the Online Apple Store at a price of $499. Burton continues to produce a broad range of iPod Jackets under the Audex brand.

Rosner – mp3blue jacket

In 2004, Rosner brought the mp3blue jacket to market. It showcased a fully integrated mp3 player within a jacket, using a fabric control system. The Moving Picture Experts Group Layer-3 (mp3) player still has to be removed before washing. The electronics technology was provided by Interactive-wear AG (which, at the time, was part of Infineon Technologies AG). The mp3blue jacket was available for $599 through a dedicated website. The jacket can now (2007) be found on some third-party retailer websites for around $199.

O'Neill – Hub and the h.Series

Like the Rosner mp3blue jacket, the O'Neill Hub Jacket contained an integrated mp3 player, with a fabric control-system placed on the sleeve.

The Hub jacket also included Bluetooth, allowing the wearer to connect and talk on their phone via the microphone and earphones integrated into the jacket. The Hub jacket also came on the market in 2004. O'Neill worked closely with Interactive Wear to develop the Hub Jacket, and have continued to use their expertise developing the h.Series ranges.

In 2005, the Hub jacket turned into the h2 range of products which included two jackets and two back packs. There was no longer an integrated proprietary mp3 player, but a connector for an Apple iPod. One of the back packs included solar panel technology for charging electronic devices.

O'Neill has continued to develop the h.Series of products, releasing the h3 range in late 2006, which included a backpack with integrated camera for recording a trip down a ski slope. O'Neill released the latest refinements, the h4 range in late 2007. The range now comprises an Audio Beanie, Campack, Comm Ent Jacket, Ent Jacket, FatController (Glove), iPod Backpack and Walkie Talkie Jacket. All use integrated electronics and control systems.

2.2.5 Fashion focus

Studio 5050 (http://www.5050ltd.com)

Studio 5050 was founded in 1995 by Despina Popadopoulos and is based in New York. Studio 5050 have worked on a range of projects investigating the design and development of smart clothes and wearable technology. The technology used in each project is relatively simple and the focus of the designs tends to be around emotional and social aspects.

The HugJacket has an intricate quilted pattern conductive fabric sewn on the front of the jacket. When two people wearing a HugJacket embrace, power is transferred through the quilted pattern. This in turn lights up the integrated Light Emitting Diodes (LED) and activates the speakers. Once the embrace has ended, the power is disconnected and the HugJacket becomes inactive once again.

Studio 5050's first commercially available product is called Moi. It is a piece of jewellery (a necklace or bracelet; however, it can be configured into many different shapes) that lights up. Moi is essentially an LED on a fine malleable wire which is connected to a small battery.

CuteCircuit (http://www.cutecircuit.com)

CuteCircuit is a 'fashionable technology company that creates design excellence and beauty in the fields of wearable technology and interaction design. CuteCircuit was founded in 2004 by Francesca Rosella and Ryan Genz. CuteCircuit products are wearable technologies and smart textile-based intelligent garments' (CuteCircuit, 2008). CuteCircuit's most acclaimed project is the Hug Shirt. The Hug Shirt uses several inbuilt sensors and actuators to sense a hug and then send that hug (via a connected mobile phone) to a corresponding hug shirt. The Hug Shirt has won many awards, including a nomination for Time Magazine's best innovation of 2006.

XS-Labs (http://www.xslabs.net)

XS-Labs was founded by Joanna Berzowska and is based in Canada. It is a design research studio focusing on innovation in the field of electronic textiles and wearable computing. XS-Labs have undertaken many projects, exploring the technology, but also exploring the social and emotional considerations that need to be made when designing electronic textiles and wearable computing. Several projects focused on developing reactive garments that were used to display the history of their use. Different sensor and outputs were used to obtain different effects.

LilyPad (http://www.cs.colorado.edu/~buechley)

Leah Buechley has developed an e-textile construction kit, to allow people with little or no electronics experience to construct basic electronic textiles. The LilyPad components and kits are available for purchase via SparkFun. com. The kit consists of several components. First, the LilyPad, the main controller, functions as an easy to use Arduino (an open-source electronics prototyping platform). The flexible patch is roughly 10 cm in diameter, with a small hard section at the centre, which encases the electronic components. Second, there are several attachable peripheral components, which include light sensors, buzzers, LED, vibrator, accelerometers and battery holder. The components can be connected to the LilyPad by sewing with conductive thread. The LilyPad can then be programmed to perform various functions depending on the connected peripherals.

Electronic sewing kit (http://www.aniomagic.com)

Nwanua Elumeze's do-it-yourself (DIY) electronic sewing kit is stated as having 'everything you need to build soft, washable, interactive clothing' (Elumeze, 2008). The kit provides a set of materials that allows the user to stitch electric circuits in fabric. The kits are very basic but are intended for beginners, and highlight the basic principles of sewing and electronics. The electronic sewing kits were produced in collaboration with Leah Buechley (see above).

2.2.6 Very recent commercial products

The rise of the iPod jacket

Through 2006 and 2007, as popularity for the Apple iPod was growing faster than ever, the number of iPod clothing products increased rapidly. Clothing incorporating fabric control systems are now being produced by Burton, O'Neill, Levi's, Bagir, Ermenedildo Zegna, Koyono, Kempo, Jansport, Woolrich, Quicksilver and Craghoppers. The control systems are usually provided by Eleksen, Fibretronic, Softswitch or Interactive Wear. The trend for incorporating iPod controls is now moving into lighter softshell jackets, fleeces and hooded jumpers.

Bags and backpacks with integrated control for iPods have also been gaining popularity: for example, Burton, Spyder, Belkin, Jansport, Nike and Quicksilver. These use the same technology as is used in the jackets. Sometimes, flexible solar panels are integrated into the bags to provide a trickle charge function for iPods and other portable electronic products.

Belkin (http://www.belkin.com)

Belkin have developed an iPod control patch, the SportCommand, that can be used to wirelessly control an iPod. The patch, which uses Eleksen's fabric switches for flexibility, can be attached to the exterior of a jacket or bag with Velcro or a clip. These control patches are priced very competitively at around £50, which, compared with the high prices of most iPod Jackets, is very desirable as it can be used on any or all of a user's existing clothing.

ToBe Technology (http://www.tobe.nu)

ToBe Technology have developed a detachable iPod controller, using Fibretronic's Fiddler controller, which can be used in many of their garments. This gives greater freedom to the clothing manufacturer, to the designer and to the consumer, who may like to design their own clothing or upgrade their existing jackets or bags. The ToBe controller is currently only available from the ToBe website, at a price of $75.

Marks and Spencer (Bagir) (http://www.bagir.com)

Marks and Spencer has been distributing Bagir's MusicGIR™ iPod tailored suit. The suit is especially designed for an iPod and comes complete with a strategically placed pocket for the iPod and its supporting devices located so as not to bulge. Discreet loops are used to hide the earphone wires, and fabric controls, provided by Eleksen, are incorporated either on the left sleeve or on the inside of the front breast lapel, allowing the wearer easy and discreet access to the controls. The cut of the suit follows the dictates of the latest fashion; it looks like a regular, well-made suit. The suit is also reasonably priced at £150, making it comparable with other normal suits. The jacket can be bought on its own for £90.

Levi's (http://www.levi.com)

Levi's returned to the wearable technology scene in late 2006 with the introduction of the Red Wire DLX range of jeans, with an integrated iPod controller using a plastic mini joystick controller from Fibretronic mounted onto a patch of denim, which fixes onto the front pocket. The patch has to be removed for washing. The DLX range is no longer available via the Levi's US and UK websites, but the range is being actively promoted on their Asian website.

2.2.7 Training technology

Adidas (http://www.adidas.com)

The Adidas 1 (Frank, 2004) and 1.1 training shoes, provide adaptive cushioning under the heel. A small micro-processor is located under the arch of the foot. As the wearer runs, the magnetic sensor in the heel detects compression (therefore detecting the force at which the wearers foot is meeting the ground). By the next stride, the trainer has adjusted the cushioning in the heel to make the shoe more comfortable. The electronics are fully integrated and the battery can be replaced; it usually lasts for about 100 hours. The first version of the running shoe came to market in 2005, priced at about £175. Adidas has made several enhancements since, adding a basketball version to the range.

Textronics – Numetrex (http://www.textronicsinc.com/)

Textronics have developed an electro-textile sensor that can be incorporated into clothing; the sensor captures the wearer's heart rate. Textronics use the philosophy that the garment must work well as a garment as well as a piece of technology, so they have used an advanced seamless bra design that provides a high level of comfort, along with support and freedom of movement. The nylon/LYCRA® fabric gives a second-skin feel and improves moisture management and airflow. The sensors in the fabric pick up the wearer's heartbeat and relay it to a transmitter in the front of the bra. The transmitter captures the heart rate information and sends it to the heart rate monitoring watch, which displays the heart rate information instantaneously. In addition, the transmitter can communicate with some fitness machines such as treadmills and elliptical trainers that have integrated monitoring devices. The sensors are comfortable and cannot be felt by the body. The transmitter is removable, so that the garment can be washed without damaging the electronics. The sensors are, however, fully integrated and can withstand washing. Textronics has spent 2007 expanding its product line, to include a man's vest and a woman's vest version of their garment.

Adidas and Polar (http://www.adidas-polar.com/)

The Fusion range collaboration, a high-end consumer training product for runners, has two parts (a vest and a pair of shoes), each with a clothing component and an electronics component. The vest monitors the user's heart-rate to calculate optimal training regimes for that individual. The garment consists of a sports bra for women or close fitting vest for men, bonded into which are two fabric based electrodes (sensors). The pulse rate is captured via a Polar WearLink transmitter, which clips onto the front of the garment near to the electrodes. The signal is captured and wirelessly transmitted to a sports watch. The running trainers (shoes), have a Polar stride sensor incorporated. This also transmits data to the watch. Data from both transmitters is captured by the watch and is used to access the users' training and to calculate the most appropriate training regimes. The electronic components are removable, enabling the garments to be washed without damaging the electronics.

Nike and Apple (http://www.apple.com/ipod/nike/)

Nike Inc. and Apple Computers' collaboration has resulted in a system that connects a stride sensor embedded in a running shoe, wirelessly to an iPod mp3 player. The iPod can then selects songs based upon the user's running rhythm and music tempo. The iPod also calculates and stores the distance travelled and calories used during the workout. When the iPod is plugged into a computer for re-charging, this information is synchronised with Nike's website via the iTunes software. The iTunes software will also recommend songs to purchase, with recommendations of play lists from top athletes. The nikeplus.com website allows runners to create training programs and build teams that can compete against each other. This is a clear example of how many previously separate technologies, such as wearable electronics, personal computers, portable media players and Internet technologies, can be combined to create a 'system' of interlinked products and services. Early sales were strong Apple stated that over 450 000 units were sold within the first 90 days of release. Nike+ iPod is a good example of systems integration at all levels – electronic and garment technology, computer integration and web-based services; and marketing through co-branding.

2.2.8 Fabric switches and sensors

Fibretronic (http://www.fibretronic.com)

Fibretronic, producers of a 'unique' five function mini joystick controller called the Fiddler (used by Levi Strauss and O'Neill) which gives 'click' feedback when pushed, making the joystick much easier to use, have more recently developed the Connected-wear System. The Connected-wear range has standardised components, which enables any of the controllers to be integrated into any Connected-wear enabled garment, bag, or glove. This system allows the manufacturer to incorporate only the input part, the keypad, or the control system. The consumer will then have to purchase the connection components separately, allowing them more choice of potential functions (http://fibretronic.com/connectedwear).

Peratech/Softswitch (http://www.softswitch.co.uk)

Peratech was established to exploit a novel technology called Quantum Tunnelling Compound (QTC) technology. QTC is the underlying technology for the fabric-based Softswitch. Peratech provide wearable solutions for Bluetooth, Freeform Switches, Stroke Switches, Lights, Beep Switches and Heated Integrated Technology. Softswitch freeform switches were used in the Burton AMP jacket and backpack, O'Neill's Hub Jacket and Nike's ACG CommJacket. These were all early smart clothes and wearable technology products. More recently, SoftSwitch technology was used in the Ermenegildo Zegna iJacket.

Eleksen (http://www.eleksen.com)

Eleksen's core technology ElekTex is a touch sensitive interactive textile which is flexible and durable. It has been widely used in applications ranging from consumer electronics, such as iPod controllers and fabric keyboards, to the military. ElekTex switches have been complimented with a range of controllers and communications technology, such as Control electronics for iPod, Bluetooth, universal volume and FM radio.

Recently Eleksen developed a display solution for a laptop case. A small screen was integrated in the exterior of a case to allow for a mini Windows Vista display. Eleksen also provide industrial and electronics design services for customers, to develop products and help the integration process. Eleksen have been acquired by Peratech, after Eleksen had financial difficulties.

Subsequently, QIO Systems have gained a worldwide licence for Eleksen's consumer electronics business.

Interactivewear AG (http://interactive-wear.de)

Interactivewear (previously Infinieon Technologies AG) produce the P-Series entertainment and communications platform. The P100 is the company's most advanced device, providing integrated textile keypad operation of an embedded mp3 player. The device also has integrated Bluetooth, allowing for connection to mobile phones. Interactivewear also provide an iPod solution.

Smartlife Technology (http://www.smartlifetech.com)

Smartlife Technology specialises in the development of continuous body monitoring systems. Smartlife Technology have designed and developed garment-based sensor systems, allowing for personal vital sign monitoring, such as electrocardiogram (ECG), respiration and temperature. Smartlife Technology have developed garments for application in three main areas: sports, dangerous and critical situations and healthcare. Smartlife Technology's garment system is based on knitted sensor structures integral to the garment's manufacture.

Zephyr Technology (http://www.zephyrtech.co.nz)

Zephyr Technology specialise in the integration of physiological monitoring technology into garments in order to monitor high activity exercise and then communicate physiological data to the web and portable electronic devices such as mobile phones and watches. Zephyr Technology produces a range of fabric sensor products, including body and foot monitors, used in conjunction with an advanced software solution that allows for the detailed analysis of the information collected.

2.2.9 Finding the way – Global Positioning System (GPS) integration

Interactive Wear AG – Know Where Jacket

The Know Where Jacket has integral earphones and microphone, fabric controls, GPS, Global System for Mobile (GSM), Bluetooth and mp3 capabilities. There is no indication of how the wearer will interface with the GPS systems. The Know Where Jacket was first demonstrated as part of the 'Future Market' special exhibition at CeBIT 2006 (presseagentur.com,

2006). Interactive Wear AG are primarily a technology company that provides electronics, but they have developed a design and collaboration strategy for working with clothing and fashion companies, including striving towards full integration of electronics components.

O'Neill – NavJacket

Wearers will interface with the GPS either via the small sleeve mounted Organic Light Emitting Diode (OLED) display, or via the audio earphone interface. It is likely that the NavJacket is using Interactive Wear AG technology. O'Neill and Interactive Wear AG worked closely on the earlier O'Neill h.Series ranges. The NavJacket is expected to become available for purchase in the Winter 2008/2009 season.

FRWD – Outdoor Sports² Computer

FRWD claim to make the most advanced sports computer in the world. The arm mounted device records all essential information about the activity that is being carried out. This includes GPS positioning, altitude, as well as heart rate. The readings are processed and can be displayed in real-time on either a wristwatch or mobile phone. The accompanying software package can be used to download stored data, and can display and compare against previous activities. The mapping information can also be exported and used in applications such as Google Earth.

2.2.10 Displays and lighting systems

Philips – Lumalive (http://www.lumalive.com/)

Philips have demonstrated a lighting/display technology that can be integrated into clothing. It allows for the display of moving images, using a grid of multi-colour LEDs.

Bogner – 2036 (http://www.bogner.com)

Bogner, in collaboration with OSRAM, have produced a concept range of futuristic skiwear, integrating an LED lighting system that enhances the garments and provides valuable safety features to the wearer. The concept also uses solar panels as the power source for the LEDs. There do not appear to be any plans to produce and sell this product concept.

Luminex (http://www.luminex.it/)

Luminex is a non-reflective fabric that emits light. The technology has been adapted from electronic equipment that is used in nuclear physics experiments. The fabric can be integrated into garments, with the light controlled by an electronic microcontroller.

Electroluminescent

Electroluminescent (EL) technology can be used to provide small, flexible displays, which have a fixed output capability. EL panels are paper thin and flexible. They can often be found in gimmick T-shirts, as graphic equalisers or as dashboard displays. EL technology does have the potential to be used in other more serious applications. Current developments in EL technology may allow for EL inks to be printed onto garments.

2.2.11 Keeping warm

WarmX (http://www.warmx.de)

WarmX clothing uses a mixture of active and passive technologies to keep the body or specific areas of the body warm. The active parts of the garment use silver-coated polyamide thread, which is knitted into the fabric. This allows electric current to flow through the garment, causing heat to be transferred to the body. Due to the fact that insulated wires are not used, there is a sophisticated anti-short circuit control system that allows the heating circuits to function effectively. The passive elements of the garment use seamless knitting technology to zone the garment with different knit structures, allowing heat to be retained or to be dispersed more easily.

Zanier – Heat-GX (http://www.zanier.com)

Zanier have produced a processor-controlled heated glove with three levels of adjustment for the wearer. The glove was initially launched in 1999, but has been enhanced several times since. The glove uses powerful rechargeable batteries to power the heating elements. There are several other manufactures of heated gloves, including Blazewear, Resusch, Thinsulate, Hunters and Therm-ic. Heated gloves are predominantly targeted at winter sports and motor biking, however there are some specialist gloves for deep sea diving.

Reusch – Solaris (http://www.reusch.com)

Reusch are using the new ThermoTec system from Interactivewear in their new ski glove. It uses a specialized, highly-flexible and lightweight heating element, which is powered by two low-profile rechargeable batteries. The system also incorporates a temperature sensor so that the temperature can be automatically controlled, and an LED to give the wearer feedback on which mode the heating is set to.

Therm-ic (http://www.therm-ic.com)

Therm-ic use custom insoles to heat ski boots: these insoles can be used in any boot. The battery pack is clipped to the outside of the boot and can provide up to 18 hours of heating. The temperature can be controlled wirelessly. Therm-ic and X-Technology Swiss have developed (for the 2008/2009 season) a new heat-transfer sock for skiers and snowboarders, to meet the particular requirements of boot heaters. The design of the sock allows the heat produced by boot heaters to be conducted directly to where it is needed most around the feet.

2.2.12 Power

Generation of power has always been and will continue to be a problem for smart clothes and wearable technology. There are several potential solutions for generating power; however, until recently none have been feasible. Possible power generation solutions include solar heat and movement. The desirable features for battery technology include small, lightweight, flexible and rechargeable with high capacity and output. However, at the moment it is not possible to get all of these characteristics in the same battery.

Ermenegildo Zegna (http://www.zegna.com/)

The Zegna solar powered jacket uses a removable flexible solar panel that is mounted on the collar. The solar panel gives a trickle-charge solution for devices such as mobile phones and mp3 players, providing a small amount of power to keep the devices 'topped up'. The solar jacket was developed in collaboration with Interactivewear.

ScotteVest (http://www.scottevest.com)

ScotteVest produce a range of jackets that incorporate removable, flexible solar panels. The panels can also be bought separately as a slip on vest. The solar panels pack has a capacity of only 750 mA/hr. This is a relatively small capacity and may be insufficient for charging some of the more power-hungry personal electronic devices.

Solar bags and backpacks

The number of bags with integrated flexible solar panels increased dramatically during 2007. Bags and backpacks are now available from well-known sports manufactures such as O'Neill, Noon Solar, V-Dimension, Sakku, Sunload, Reware, Eclipse Solar Gear, Earth Tech and G-Tech. Voltaic

Systems have recently produced the 'Generator' laptop bag, which includes a solar panel capable of charging a laptop battery in a day. This is currently one of the fastest charge rates for a battery of this capacity. The bag is priced at $599 (http://www.voltaicsystems.com/).

2.3   Industry sectors' overview

The Industry can be broken into four loose areas: Sports, Healthcare/ Medical, Fashion/Entertainment and Military/Public sector. The areas do overlap in some places, often with similar or even the same technology being repackaged for a different end-use.

2.3.1 Sports

The sports sector has historically been an early adopter of technical textiles and 'high tech' solutions, and this early adopter status has not changed for commercial smart clothes and wearable technology products. This category can also be broadly divided into training/professional sports and casual sports. Each category uses different types of technology. Training and professional sports use bio-physical monitoring technologies, overlapping with healthcare. Casual sports have largely been incorporating entertainment and communication technologies into clothing. Adventure sports such as snowboarding and mountaineering have recently been leading this trend.

2.3.2 Healthcare

There are many wearable technology solutions for medical monitoring. They are usually designed with just one thing in mind, getting the readings, which is usually accomplished very effectively. However, these devices are often uncomfortable, unsightly and awkward to use, leaving patients feeling self-conscious and lacking the desire to use the devices.

Medical healthcare products have a greater emphasis for constant monitoring of biophysical data; for example, ECG, respiration rate, blood pressure, temperature and movement. Medical testing of these products is essential, with each country having its own set of clinical testing requirements that need to be achieved before a product can be used in a clinical environment. This is often an important factor for companies developing bio-monitoring technologies, as it can cost large amounts of money for product testing. Several companies have targeted the sports industry initially and have later moved onto the medical industry as their products have been developed further, gaining the levels of accuracy and consistency that are required for clinical use.

There are two main types of healthcare products: those that are prescribed by a physician (medical healthcare) or those that are purchased by individuals who are concerned with their general heath (well-being). Examples of the latter products and developments includes the Wealthy project (Wealthy, 2005), LifeShirt (Vivometrics, 2007), MyHeart Intelligent Biomedical Clothes (MyHeart, 2008) and HeartCycle (Phillips Research 2008). As there is a greater awareness of the need to keep fit and personal well-being, continuous monitoring and keep-fit solutions overlap with sports products. However, the design of the sports products may be inappropriate for this emerging market which has different demographic and end-user needs.

2.3.3 Fashion and entertainment

A fascination with technology means that the incorporation of cutting edge technology is common in the fashion industry. Designers such as Alexander McQueen, Erina Kasihara, Diana Drew and Hussein Chalayan have all produced ranges that explore the integration of technology into clothing. Suzanne Lee's book *Fashioning the Future: Tomorrow's Wardrobe* (Lee, 2005), takes an in-depth look at the relationship between fashion and technology.

The incorporation of electronic technologies into everyday clothing remains relatively uncommon. Entertainment in the form of music and communications is currently leading the way, with the integration of iPod and mobile phone control systems being examples of the few attempts at promoting smart clothes and wearable technology for everyday use. This is likely to expand in the future to include more sophisticated devices such as portable media and games players.

2.3.4 Military, public sector and safety

The military has been one of the largest funders of wearable computing and wearable technologies. Communication and battlefield command systems use a combination of personal, vehicle (ground, air and water), static and satellite technologies that all work together. The US military in particular has been very active in the development of smart clothing and wearable technology solutions. The Naval Air Technical Training Center's (NATICK) Future Force Warrior programme is developing fully bionic warrior suits that it hopes to introduce by 2020. This suit would include intelligent armour, biomonitoring, weaponry, communications and exoskeleton (NATICK, 2006).

First responder services, such as the Police, Fire and Ambulance, are now also beginning to use smart clothes and wearable technologies. In some parts of the UK, police forces are now using head mounted cameras and Personal Data Assistants (PDA) for storage and transmission systems to record information as they patrol and interview suspects and witnesses. This will act very much like the in-car video recording systems that are widely used. Safety products for monitoring personnel in hazardous or remote environments are also highly desirable, such as monitors of chemical exposure levels and biophysical status.

2.4   Current trends

There is a current trend towards mp3 player controls integration, the actual player being rarely fully integrated. The uptake of this is largely due to the fact that the suppliers of the fabric control systems are now producing Original Equipment Manufacturer (OEM) solutions for clothing manufacturers. Another contributing factor has been the massive uptake of the Apple iPod by consumers (see Fig. 2.1), and the growing iPod accessories industry. The continuation of this trend is likely and will probably evolve to incorporate other devices, such as mobile phones and smart phones, in more appropriate ways.

Over the next few years, commercial development of plastic electronics will become more widespread, offering many interesting opportunities for developing smart clothes and wearable technology systems, using potentially more robust, waterproof electronics. Recently, Plastic Logic, a spin-off company from Cambridge University, has started commercial development of plastic electronics (BBC, 2007). The firm is working on 'control circuits' that sit behind screens on electronic displays. In particular, it is working on the electronic circuitry for 'electronic paper' displays.

Power generation and storage will continue to be a problem for smart clothes and wearable technology products and is an important area for development. The recent surge in 'solar powered' bags and backpacks is

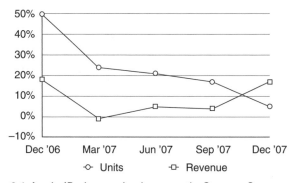

2.1 Apple iPod annual sales growth. Source: Company reports.

likely to continue and will hopefully drive the solar panel manufacturers to develop more appropriate solutions for the smart clothing and wearable technology market.

The durability and washability of smart clothes and wearable technology products still remains a problem and will continue to inhibit their uptake by the general public. The development of more advanced core technologies that are capable of withstanding standard washing techniques, or the development of appropriate flexible encapsulation solutions, are going to be essential for large-scale uptake.

2.5   Market forecast

In 2003, The Venture Development Corporation wrote one of the first market forecasts for Smart Fabrics/Interactive Textiles (SFIT), stating that the global market for SFIT-enabled solutions in 2003 had a value of around $300 million and was expected to reach $720 million by 2008 (VDC, 2003).

In 2005 VDC published another report breaking SFIT into specific sectors, giving a more detailed forecast. VDC estimated that in 2005 the market for general-purpose computing/communications wearable systems would exceed $170 million, and was expected to reach $270 million by 2007. This represents a 24% Compound Annual Growth Rate (CAGR). The biophysical monitoring wearable systems market was expected to have exceeded $190 million by 2005 and was expected to reach $285 million by 2007. Also they mentioned a small number of infotainment-based products such as Oakley's 'Thump' mp3 sunglasses and Bluetooth headsets. VDC felt that this market was nascent, and it was unclear what the dominant business model would be (VDC 2005).

In VDC's 2007 SFIT report (VDC, 2007) they estimated that global demand for smart fabrics and interactive textiles products and solutions totalled $369.2 million for 2006 and will reach $1129 million by 2010, representing a four-year CAGR of 32%. As we can see, the actual 2006 figures have already exceeded the forecast for 2007 (which was made in 2005) and the estimated CAGR rate has also risen by 10%.

North West Textile Network's smart and interactive textiles market report identifies 'Product/Person Tracking' and 'Smart Medical and Hygiene Products' as the most significant markets segments within the smart product market (North West Textile Network, 2005).

2.6   Conclusions

This chapter has intended to give an overview of the smart clothes and wearable technology industry as it stands today (and in its recent past), highlighting some of the key companies, products and technologies. Unfortunately, there is not enough space or time to give details of every product on the market, and it would probably need somebody in a full-time capacity to keep up with the new products emerging onto the market.

The chapter has shown the diversity of products currently available and the growth within the sector over the past few years, especially where iPod integration is concerned. However, it is not just all about iPod jackets! There is a wide range of bio-physical monitoring products hitting the market place, with capabilities suiting professional athletes to keep-fit beginners.

Hopefully, this chapter has highlighted some of the issues involved with the development of smart clothes and wearable technology and has given the reader a good overview of the industry.

2.7   References

BBC NEWS (2007), 'UK in plastic electronics drive' [URI http://news.bbc.co.uk/2/hi/business/6227575.stm, accessed 21st April 2008]

BUDDE, P (2008), 'Europe – Mobile Market: Overview & Statistics'. *Paul Budde Communications*. [URI http://www.budde.com.au/buddereports/3808/Europe_-_Mobile_Market_-_Overview__Statistics.aspx, accessed 25th March 2008]

CUTECIRCUIT (2008), About CuteCircuit, *CuteCircuit*. [URI http://www.cutecircuit.com/about-cutecircuit, accessed 23rd March 2008]

ELUMEZE, M (2008), Electronic Sewing Kit. *Aniomagic.com* [URI http://www.aniomagic.com/sewing.html, accessed 21st April 2008]

FRANK, R (2004), Engineering Feat. *DesignNews.com* [URI http://www.designnews.com/article/CA452888.html, accessed 21st April 2008]

LEE, S (2005), *Fashioning the Future: Tomorrow's Wardrobe*, London: Thames & Hudson.

MARKS, P (2005), Fashion Industry Covets 'iPod Factor'. *NewScientist.com* [URI http://technology.newscientist.com/channel/tech/mg18625025.600-fashion-industry-covets-ipod-factor.html, accessed 21st April 2008]

MYHEART (2008), Technical Objectives: Intelligent Biomedical Clothes for monitoring, diagnosing and treatment. IST 507816 [URI http://www.hitech-projects.com/euprojects/myheart/objectives.html, accessed 21st April 2008]

NATICK (2006), Future Force Warrior. [URI http://nsc.natick.army.mil/media/print/FFW_Trifold.pdf, accessed 21st April 2008]

NORTH WEST TEXTILE NETWORK (2005), *Smart and Interactive Textiles: A Market Survey*, Droitwich: International Newsletters Ltd.

PHILIPS RESEARCH (2008), 'HeartCycle Project'. Philips Research Technology Backgrounder. [URI http://www.research.philips.com/technologies/healthcare/homehc/heartcycle/heartcycle-gen.html, accessed 21st April 2008]

PRESSEAGENTUR.COM (2006), Electronic Clothing. Presseagentur.com [URI http://www.presseagentur.com/interactivewear/detail.php?pr_id=832&lang=en, accessed 21st April 2008]

VDC (2003), *Smart Fabrics & Interactive Textiles*, Venture Development Corporation: Natick, MA.

VDC (2005), *Wearable Systems: Global Market Demand Analysis (2nd)*, Venture Development Corporation: Natick, MA.

VDC (2006), *Smart Fabrics & Interactive Textiles (2nd)*, Venture Development Corporation: Natick, MA.

VDC (2007), *Smart Fabrics & Interactive Textiles (3rd)*, Venture Development Corporation: Natick, MA.

VIVOMETRICS (2007), *About Us*, VivoMetrics. [URI http://www.vivometrics.com/corporate/about_us/, accessed 21st April 2008]

WEALTHY (2005), '*Wealthy – Wearable Health Care System*'. Information Society Technologies IST-2001-37778 [URI http://www.wealthy-ist.com, accessed 21st April 2008]

3

End-user based design of innovative smart clothing

J. McCANN, University of Wales Newport, UK

Abstract: A hybrid methodology is required to guide the design research and development process of a smart clothing layering system from user-needs analysis to the potential research and testing of near market prototypes. The breadth of potential combinations of materials, design features and wearable technologies is daunting without a clear profile of the real requirements of the wearer or end-user. A design tool is introduced that identifies key considerations may guide the designer or product development team, new to this hybrid area of design.

Key words: hybrid design process, fit for purpose, smart clothes, wearable technology, end-user needs.

3.1   Introduction

This chapter investigates major elements for consideration in the design research process for those seeking to develop products that are attractive and fit for purpose in the relatively new design discipline of smart clothes and wearable technology. Clothing is a major contributor to how people define and perceive themselves and is a necessary part of their everyday lives. Clothing can promote a feeling of wellbeing and now has the potential for emerging wearable technologies to become embedded in the clothing system. To be acceptable and comfortable, as a vehicle for self-expression, products must look stylish and attractive and function reliably in relation to the technical and aesthetic concerns of the wearer, as well as from social, cultural and health perspectives. Good aesthetic and technical design, driven by meaningful end-user research, can help exploit niche markets where form and function work in harmony in the research and development of comfortable and attractive products that can assist us in many aspects of our daily lives. Now there is the opportunity for the application of a range of additional smart textiles and wearable electronics to enhance the functionality of clothing. While some technologies may be generic, their application and usability will demand different design solutions for different user-groups. This chapter looks at an identification of end-user needs that may be presented in a hierarchy of design requirements to guide the application of smart technical textiles and wearable technologies in a clothing 'layering system' that is comfortable, attractive and fit for purpose for the identified end-user.

We are at the beginning of a new industrial revolution with the merging of textiles and electronics. The current markets for clothing and electronics have been separate, with contrasting cultures, sources of production, product development critical paths, trade standards and routes to market. There have been quite different requirements for eventual disposal at the end of life of products within these disparate sectors. To date, the incorporation of technology into clothing has been crude at best, and has consisted of marrying existing consumer products into branded clothing lines, primarily for sports such as snowboarding, running and fitness. In the medical sector, devices have generally been developed for 'ill people', with little aesthetic appeal. Technological advances are not readily accepted by some of their intended markets due to badly designed user-interfaces, which often have controls or displays that are unattractive and difficult to read by a breadth of users. Innovation includes the concept of early adoption and the implementation of significant new services that may represent changes that are organizational, managerial and technological in the clothing and electronics product development chain. There is a challenge to bring technologies to their intended customers in an acceptable and desirable format.

There is also a need for a shared 'language' and vision that is easily communicated between practitioners coming from the relevant areas of expertise. Emerging electronic and textile related technologies are confusing to traditional clothing and textile development teams, while electronics and medical experts are not normally conversant with textile and clothing technology. The design development of new hybrid products that bridge traditional market sectors, demands a cross-disciplinary team, from a mix of disparate study backgrounds, to embrace a collective awareness of new technologies, research methods and design techniques. The development of a hybrid methodology is required to guide the design research and development process of a smart clothing layering system from user-needs analysis to the potential research and testing of near-market prototypes. The breadth of potential combinations of materials, design features and wearable technologies is daunting without a clear profile of the real requirements of the wearer or end-user. A design tool that identifies key considerations may guide the designer or product development team, new to this hybrid area of design. This chapter gives an overview of user-driven design requirements that may be addressed through the application of smart textiles and wearable technologies that is discussed in more detail throughout this book.

3.2   The garment layering system

The concept of the tried and tested military type 'layering system', commonly adopted in the performance sports and corporate wear areas, provides a reference point for the identification of design requirements for functional clothing development. This layering system is described in greater depth in Chapter 12, 'Garment construction: cutting and placing of materials'. In brief, this system normally comprises a moisture management 'base-layer' or 'second skin', a mid insulation layer, and a protective outer layer. Elements of personal protection, or body armour, may be incorporated into the system. The base layer is normally of knitted construction and, most recently, seam-free garments have become prevalent. Varied knit structures, often with elastomeric content, may be placed around the body to aid wicking and offer increased support and protection. Mid layers incorporate textiles such as knit structure fleeces, woven fibre or down-filled garments and sliver knit constructions, known as fibre pile or fake fur, primarily to provide insulation by means of trapping still air. The outer layer, or 'shell garment', provides protection for the clothing system microclimate from the ambient conditions by adopting a range of variations on woven or knit structure protective textile assemblies. Personal protective inserts, within the system, consist of knitted or woven spacer fabrics, wadding and foams or non-woven composite structures in varying degrees of flexibility or rigidity. To function effectively, the garments and components within the layering system must co-ordinate in terms of style, fit, silhouette, movement and closures.

3.3   Identification of design requirements

A user-needs driven design methodology is proposed that promotes collaborative design with users. It addresses a breadth of technical, functional, physiological, social, cultural and aesthetic considerations that impinge on the design of clothing with embedded technologies, that is intended to be attractive, comfortable and 'fit for purpose' for the identified 'customer'. A hierarchal process 'tree' revisits and adds a new layer of topics to the author's previous work in her *Identification of Requirements for the Design Development of Performance Sportswear* (McCann, 2000). This initial tree of requirements uncovers a breadth of topics and sub issues to guide performance sportswear designers in the merging of technical textiles, garment technology, human physiology, and cultural requirements. The major topics have been organised under the areas of Form and Function. If a product does not look good or work, the customer will not be satisfied. 'Form' embraces aesthetic concerns and the importance of respecting the culture of the end-user, and 'Function' embraces the generic demands of the human body and the particular demands of the end use or activity (see Fig. 3.1).

In working with designers in the academic environment, as well as in her involvement with commercial projects, the author has proven that this is a useful grounding for the identification of design requirements for a clothing system for a breadth of functional end-uses. In order to aid decision-making, the design process requires an overview of the profile of the target customer in terms of gender, age group, fitness level, and an indication of

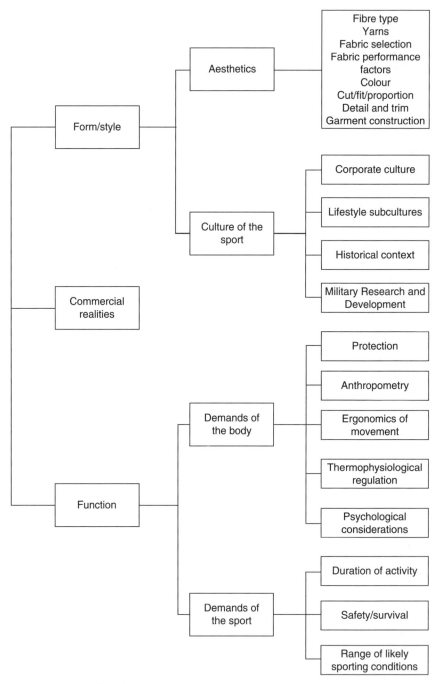

3.1 Tree of requirements for the design development of performance sportswear (McCann, 2000).

the proposed category of smart textile product to be developed. For example, design features that constitute a wearable, acceptable and useable garment with embedded smart wearable attributes for a child playing sport will be different from the needs of a fireman subjected to extreme hazardous environments or from the demands of everyday clothing for an older wearer. While the areas of investigation are organised under the main headings of 'Form' and 'Function', 'Commercial realities' is recognised as a topic with major impact on bringing innovative products to market, especially in the wearable electronics sector. This is a topic for further elaboration.

3.3.1   Form

Aesthetics of design

This chapter argues that aesthetic considerations remain key with regard to the acceptability and wearability of the final product. An appropriate balance of aesthetic concerns such as colour, fabrication, cut, proportion and detail will contribute to the psychological 'feel good' factor of the wearer. When viewed from a distance, the colour and silhouette of clothing gives a first impression. The colour selection for components that constitute the individual item of clothing, or a co-ordinating layering system, may be influenced by designer fashion, peer group trends, codes of culture, and tradition and individual preference (McCann, 2000). Colour choice may also be dictated by corporate image or health and safety requirements. At closer proximity, clothing provides an indication of the personality of the wearer, as the cut, fit, proportion, styling and embellishment may be seen to range from the flamboyant to the discreet. The fabric design and its drape, or 'handle', and the more subtle detail of construction methods and trim will indicate whether the garment is formal or casual, whether it is a corporate or team uniform, dictated by safety standards, or an individual's own choice.

Other considerations will include seasonal requirements and whether the garment is intended for an indoor or outdoor environment, or as a multipurpose clothing system for a range of conditions. The outdoor trade has always been driven by textile innovation, often as a direct result of military preparedness or exploration into space. For example, Teflon® and Gore-Tex®, developed for protecting astronauts and their spaceships from extremes of temperature, are now in common use. 'Kevlar®, Teflon® and Gore-Tex® all started at NASA and were later (from 1971) marketed by DuPont. Although we may think of these materials as ordinary and familiar, they continue to be used and developed for their remarkable properties in space exploration' (McCarty, 2005). The advent of synthetic fibres, such as nylon and polyester, in the 1940s led to the development of the lightweight waterproof textiles with waterproof coatings that emerged in 1960s. These were initially crude in design and had to be voluminous to enable air circulation due to their lack of 'breathability' (McCann, 2000). Subsequent developments have led to enhanced moisture management in hydrophilic laminates, such as Sympatex (Sympatex, 2008) and microporous waterproof membranes, such as Gore-Tex (Gore, 2008a). Bonnington's ascent of Annapurna in the 1970s coincided with the introduction of colour television, giving prominence to outdoor activities (Lack, 1992). The development of package deal ski holidays led to fashion merging with snow sport in the 1980s and, subsequently, to the snowboard image impacting on street fashion. The design of outer layer garments has since become more fitted and stylish as a result of enhanced comfort, through textile 'breathability' and stretch attributes, with the usual requirement to be super lightweight with minimum bulk for easy storage.

In textile-driven clothing design, aesthetic choices, in tandem with technical design considerations, begin with fibre and yarn selection that impacts directly on the properties of the fabrics and trims. Stretch fibres have revolutionised the aesthetic design of clothing since the launch of DuPont's Spandex® fibre in 1959 (Apparel Search, 2008). John Heathcoat Ltd, which continues to be an innovative producer of technical textiles, was the first mill in the UK to produce textiles with lycra for intimate apparel, as seen in Heathcoat's in-house magazines from 1963 and 1967. Stretch may be incorporated, in varying percentages, in warp and weft knitted structures, laces and nets, in woven constructions, including narrow fabrics and shock cords, and in some 'hook and loop' closures. Stretch enhances comfort and fit in providing engineered areas of support, or provides movement that also enables the wearer to put on and take off the garment with greater ease. The direction and degree of stretch varies within the warp and/or weft of a woven cloth; it may be referred to as 'one-way' or 'two-way' stretch. Stretch yarns are desirable in knit structures, for closely fitting styles and to provide extra ease and comfort in more tailored woven styles. Stretch is a characteristic of neoprene and of certain waterproof breathable laminates.

The notion of 'form and function' seldom embraces the adding of embellishment that is superficial to the identified design requirements. In creating functional design 'in the round', in relation to the contours of the body, both design lines and ergonomic cutting lines often work in harmony and may merge into clean, minimal styling. An increased awareness of human movement, the support of muscles, protection, moisture management, in combination with the application smart textile innovation, results in garments with design lines that maximise efficiency coupled with meaningful aesthetics. The positioning of textile structures, with the necessary functional attributes, by means of novel seam-free garment engineering techniques, enhances the aesthetic of the design and invariably provides the marketing

focus. Falke's Ergonomic Sport System is a prime example of fit and material specified to address the anatomical, supporting and climate control
demands of the body (Falke, 2008). Functional clothing, with intimate
apparel and performance sportswear in particular, has invariably been an
early adopter of fibre and textile innovation and novel construction techniques, often led by, or in collaborative design practice with, the end-user.
Falke promotes the principle of aesthetics and functionality blending in
perfect harmony.

The culture of the user

Successful functional clothing design is the result of designers becoming
thoroughly conversant with the culture, history and tradition associated
with the particular end-use or range of activities. The subtleties of identified
particular life-style trends will affect the style and mood of the clothing.
Certain cultures and their traditions may specify codes of dress. A design
that is considered attractive for a wearer from one community or age group
may be totally unacceptable for another. What is perceived to be a 'cool'
design for the youth market will not be appropriate and attractive to an
elderly user in a wheel chair. Concerns such as social and cultural issues,
historic context and tradition, corporate and work culture, participation
patterns and levels, status, demographics, and the general health and fitness
of the wearer will impinge on the design of smart clothes and wearable
technology. An investigation of the lifestyle demands of the wearer, in terms
of behaviour, environment and peer group pressure, is needed to provide
an awareness of both clothing requirements and the application of emerging wearable technologies that have appropriate functionality and true
usability for the identified user.

The development of activity specific technical clothing, in the cultures of
performance sport and corporate work wear, is often based on feedback
from wearers dissatisfied with ranges available. Expert practitioners are
often consulted by manufacturers in the design development of new products where cut, detail and fabrication respond directly to the wearer's perception of what is required for a particular activity. In the area of performance
sport, past sporting heroes such as Douglas Gill and Musto (sailing), Jean
Claude Killy (ski), and Ron Hill (athletics), have initiated the development
of their own clothing brands with heritage that continues today. Traditional
sports are evolving as modern practitioners become more aware of fashion
and as technical textiles enhance more stylish cut, fit and proportion. Many
practitioners at the peak of their performance have grown up at a time of
growing awareness of design in all areas of clothing and lifestyle products.
Charismatic sporting personalities, such as the golfer Tiger Woods, and
tennis players, such as the Williams sisters, are endorsed as leaders of style

to promote the image of major clothing brands for both team and individual sports in order to create product that is attractive to peer and fan culture. Woods already has endorsement deals with Nike, Buick, Tag Heuer, Accenture, Gillette, and Electronic Arts (Sports Business Digest, 2007). In contrast to the culture of mainstream fashion, serious sports practitioners are invariably knowledgeable in their understanding of generic textile terminology and the claims of the brands with regard to fibres, fabric constructions, coatings, laminates and finishes, prominent in performance ranges.

The designer must be aware of the commercial culture of major events and the specialist press pertinent to the target market, as well as the international trade fairs where leading fibre and fabric producers promote new developments. Trend forecasting with regard to colour, styling and mood may be consulted for fabrics and apparel design direction within certain sectors. The impact of the media, and especially colour television, may influence the use of colour and graphics for corporate uniforms, while sponsorship logos may be subject to given restrictions. Designers must communicate with sales teams, retailers and practitioners to achieve aesthetically strong design statements that also promote maximum performance characteristics in the product. Point-of-sale material must be produced in a language that is accurate and appropriate to the culture of the target market. Is the purchase influenced by the claims of branding and marketing and where is the purchase made? The culture of the target community may be associated with the high street, shopping malls, specialist outlets, catalogue and/or Internet purchases.

3.3.2 Function

Demands of the body

The designer will benefit from gaining an overview of human physiological issues that impinge on the design of the functional garment layering system. It is necessary to consider practical issues to do with the demands of the body that may be addressed in everyday clothing in terms comfort. The comfort of clothing will be affected by appropriate styling in relation to measurement, shape and fit, predominant posture and the ergonomics of movement for a particular community or target market. What is considered as 'small', 'medium' and 'large' sizing for one activity will be inappropriate for different age ranges and end-uses. For extreme requirements, a complete measurement chart may be needed for the wearer. Predominant posture will vary for different activities and different age groups. A visual record of the subject(s) in action will support verbal feedback and observation. Clothing should, ideally, enhance support for the body where needed without restricting movement. Sensory considerations are key to the effectiveness of the layering system where textile selection and garment features and trims may be designed to address issues such as handle and grip, protection for the eyes and the avoidance of impeding hearing (McCann, 2000).

Key aspects of the maintenance of comfort directly related to textiles are moisture management and thermal regulation. For effective thermal regulation in clothing and accessories, the design of the system must address the importance of sweating through the application of special fibres and fabric constructions and the provision of ventilation within the garment(s). The measurement of performance, in relation to 'workload', indicates a direct relationship between physical work rate and body heat production and the production of perspiration for a given activity or task. This informs appropriate fabric selection in terms of moisture wicking and, where necessary, spacer fabrics for ventilation. Wearers will not be comfortable and may be at risk in relation to survival in extreme cold if a high level of moisture is absorbed within the garment base layer; for example, cotton instead of fibres that wick away moisture (Turner, 2000). In terms of heat retention, the system design must combat the wind-chill factor and address prevention of wetting in conserving appropriate levels of still air space. The ambient conditions and climatic variability impact on all other aspects of thermo-physiological regulation.

Textiles and clothing may be designed to offer varying levels of protection, dependant on the end-use. Consideration must be given to vulnerable areas of body in anticipating injurious hazards and commonly occurring accidental incidents particular to the activity. Textile structures, such as spacer fabrics, provide light-weight personal protection in corporate body armour, for performance sport, and as impact protection for the elderly. Protection from the ambient environmental conditions may be provided by waterproof breathable textiles, with additional properties and finishes such as abrasion resistance and chemical protection for more extreme requirement. The psychological 'feel good factor' is directly related to appearance and style as well as to the reliability, or the perception of reliability, of the garment system (McCann, 2000).

Demands of the activity

Designers should carry out primary research in observing and obtaining feedback from wearers to identify their needs for the chosen activity or task. In respect of patterns of use or participation levels, for example, the clothing may be worn regularly for work or everyday wear, for leisure, at weekends, or once a year on holiday. The activity may be performed quickly or be of medium or long duration. Active wear fabrics and styling are increasingly being adopted for corporate workwear. In the performance sportswear sector, a balance of both technical and aesthetic design considerations has become the norm. For sport, the 'rules of the game' may include a dress code or restrictions to do with safety or fair competition that impinge on apparel design. The contrasting demands of sports training kit, team or competition uniform and post-activity clothing are clearly defined. The activity could be a sprint in a stadium, a sequence of contrasting sports, such as triathletics, trekking for a day, or an extended polar expedition. Sport warm-up kit may be worn for longer than the corporate strip for competition. An appreciation of the impact of the environment in which the activity takes place is required. The sport, job or leisure activity may be seasonal, carried out indoors in a controlled environment, or outdoors in predictable conditions or in rough terrain.

The clothing system may be required for protection from an unknown hostile environment or for a known predictable environment. The wearer may be subject to contact or non-contact activity with regard to aspects such as body protection and abrasion resistance. There may be the demands of an extreme climate or a range of temperatures and degrees of humidity. Designing for very cold, dry conditions or hot humid conditions, is less challenging than catering for both extremes. Transportation is also a consideration in terms of bulk and weight. The wearer may have a vehicle to travel to a destination or may have to transport heavy equipment manually over long distances. In sectors such as workwear, there may be safety and/or commercial considerations to do with identification and hierarchy. An appreciation of the functional needs of the end-user will impact on a breadth of design considerations with regard to comfort, protection, durability, weight, ease of movement, identification and aftercare. The designer's challenge is to engage with the end-user(s) to uncover, understand and prioritise a range of issues that set the scene for the specified activity, or range of activities, and inform the development of products that are attractive and fit for purpose in relation to the culture and lifestyle of the intended wearer(s).

3.4   The technology layer: the impact of emerging smart technologies on the design process

An infinite range of textile constructions and finishes, and novel garment engineering techniques may now incorporate smart attributes and wearable electronics into increasingly sophisticated clothing system design. This chapter looks at requirements capture primarily for the development of a clothing layering system for functional end-use which may now be enhanced by the application of smart textiles and wearable technologies throughout the layers. Smart attributes may be adopted from the initial choice of the fibre and fabric structures to the dyeing and finishing of materials, the encapsulating of soft electronics, and the integration of powering devices and displays. Smart textiles may be defined as 'passive smart' (can sense the environmental stimuli), 'active smart' (can sense and react to the condition or stimuli), 'very smart' (can sense, react and adapt) and 'intelligent' (capable of responding in a pre-programmed manner) (Gonzales, 2003). 'Passive smart' properties include insulating and moisture wicking structures, 'active smart' include attributes such as anti microbial protection, and 'very smart' include phase change impact protection, for example 'd3o' (d3o, 2008) and thermal regulation, for example, 'Outlast' (Outlast, 2008). 'Intelligent' technological advances permit power and signal pathways to be integrated into garments and accessories to facilitate applications such as heart rate, temperature and respiration sensing, and location monitors.

3.4.1 Enhancing and changing the aesthetic

A new aesthetic 'look' has become prevalent, predominantly in the performance sportswear sector, which has been an early adopter of novel garment design engineering techniques for garment cutting and the joining of materials. Textile heat bonding and moulding techniques may be applied primarily to technical materials made from a high percentage of synthetic fibres. Functional design details, such as zip openings, garment edges, and perforations for ventilation, may be laser cut to provide a clean and non-fraying finish, with the Canadian outdoor clothing company Arcteryx, an early adopter of these techniques (Arc'teryx, 2008) now adopted by many outdoor companies including Patagonia. Seams may be joined by a range of heat-bonding techniques, promoted by machine makers, for example 'Sew Systems' (Sew Systems, 2008), working in collaboration with adhesive specialists such as 'Bemis'. Yvonne Chouinard, founder of Patagonia states: 'There won't be sewing machines in future. We're using a lot of seamless technology, coming out with a jacket that has no stitching at all – everything is glued and fused together for a lighter, stronger, more waterproof product' (Wilson, 2004, p. 6).

Advances are also being made in laser welding techniques with research collaboration between The World Centre of Joining Technology and its member companies (TWI, 2006). These joining methods may be used in combination with garment moulding to produce 'sewfree' garments with clean design lines. Waterproof zips, for main closures and pockets, sleeve tabs and hood reinforcements may be bonded without stitching. Seam-free knitting techniques are also prevalent in the structuring and shaping of intimate apparel, base layer garments, in medical hosiery, for example 'Scan to knit' (William Lee Innovation Centre, 2007), and in heavier gauge knit-wear, such as Falke's 'Bjorn' ski pullover (Falke, 2008). These new garment construction and joining methods offer enhanced potential for embedding or encapsulating wearable electronics within smart textile materials and finishes, in garments constructed from knitted, woven and non-woven textile assemblies (Kathleen Gasperini, 2007).

A functional clothing layering system may benefit from tried and tested concepts derived from the natural world. In studying biomimicry, innovators of smart textiles are adopting structures and finishes found in plants, insects, animal skin and furs. Fir cones have been studied in creating a balance of breathability and waterproofness in phase change textiles that expand and contract for mid and outer layer garments, commercialised by Schoeller as 'c-change' technology (c_change, 2008). The lotus leaf has been mimicked in nano scale technology for moisture and stain repellent finishes by Schoeller, branded as Nanosphere (Schoeller, 2008). Terminology such as 'second skin' is used for base layer garments where textiles are designed to enhance moisture wicking and ventilation in areas or 'zones' related to the demands of the human skin. Past references have been made to the mimicking of shark skin in relation to the development of knit structure textiles for competition swim wear in the Speedo 'Fast Skin' project (Speedo, 2008a). In spring 2008, Speedo's Aqualab research team has launched the Speedo LZR Racer Suit (2008), 'an engineered swimsuit whose 3D anatomical shape has a Core Stabiliser with a corset-like grip to support and hold the swimmer so they can maintain the best body position in the water for longer without losing freedom or flexibility of movement' (Speedo, 2008b, p. 19). The hydrodynamically advanced suit is a fully bonded garment cut from a three piece pattern, that is ultrasonically welded with LZR Pulse panels embedded into the base fabric to provide 'a Hydro Form Compression System' to compress the swimmer's body into a more streamlined shape.

Fleece pile fabrics are being engineered by companies such as Polartec to incorporate variable patterning in developing fabric structures replicating animal fur, to provide areas of added protection and/or ventilation, relative to the demands of the body (Kick Public Relations, 2005). The Patagonia BioMap fleece 'puts performance where your body needs it through the strategic placement of variable-knit zones that address: warmth, dryness, cooling, mobility and fit' (Patagonia, 2008a). 'Comfort Mapping' has been used as terminology, by W.L. Gore, to market outer layer garments where features such as breathability, abrasion resistance and thermal regulation have been addressed through fabric selection in combination with garment cut and construction. Gore 'Comfort Mapping Technology' divides the body into different zones which have different temperature requirements as a basis for the selection of various material combinations in the same product. Gore 'Airvantage' provides a personal climate control system with a lining that consists of air chambers which can be individually inflated and deflated to adjust the temperature. Gore has now integrated its Airvantage insulation technology and Gore Comfort Mapping Technology, 2006, into one garment. Gore claim that their lightweight waterproof running jacket 'combines individually adjustable thermal insulation with an anatomic cut, outstanding breathability and superior wear comfort' (Gore, 2008b,c). The development of such engineered knitted and woven structures, with their zones of contrasting fabric and yarns properties, provides garment functionality that is integral with aesthetic appearance.

Stretch fibres contribute directly to the embedding of technologies, as a key attribute in the fit of intimate apparel and base layer garments for sport and medical applications. Variable stretch textiles are used in compression garments that offer targeted support in maintaining muscle alignment and a reduction in the loss of energy in athletic performance. Additional reinforced taping, outlining and supporting muscle groups, contributes to the aesthetic appearance of the garment. The Adidas 'TechFit Powerweb', which uses thermoplastic polyurethane (TPU) elastomers, has been developed to harness the intrinsic power in the human body and improve athletes' performance. The body-suit is manufactured by US extrusion specialists Bemis from an elastic base fabric with a special two-layer TPU film coating. The technology works by wrapping strategically placed bands around the body. The optimised directional elasticity and energy return of the garment supports athletes' movements and maximises the use of the kinetic energy produced by the body during exercise. The TPU is claimed to provide increased joint power, better posture, improved feedback for more accurate movement and faster recovery times due to increased blood circulation (Anon., 2008).

Body scanning is being used to directly inform the design and fit of customised compression hosiery (William Lee Innovation Centre, 2007). Stretch is key to the placing of textile sensors and wearable electronics that must be held in appropriate locations within the garment. The relevance of the sizing and shape of garments is discussed in more detail in relation to the demands of the body, to follow. The potential for the mass customisation of garment design is directly related to individual fit, proportion and personal choices for embellishment, now becoming a reality as a result of the digital garment chain ([TC]², 2008).

The application of colour may be introduced through technologies that include digital printing and the potential for large-scale repeat jacquard weaves and engineered knits that enable designers to position specific colours and textures in relation to the desired garment style. Colour changes may be specified in combination with differentiation in fibre properties and fabric constructions. 'Thermachromic inks can demonstrate interactivity, via a textile display that changes colour in response to temperature fluctuations. Water and UV responsive dyes are also available. Plastic optical fibre carries pulses of light along its length. Textiles can incorporate changing text or pattern in soft digital displays' (Hibbert, 2004). Further embellishment in terms of colour and texture may be added by means of digital embroidery and laser finishes. Conductive fibres and yarns are used in targeted digital embroidery to create electronic connections. Conductive inks may be used in digital print to create electronic printed circuits. Camouflage effects, colour change and high visibility finishes may be enhanced by smart textile finishes. Colour may be used to attract attention or give warning in complying with safety standards. Certain darker shades can prevent or reduce UV penetration, with test rates available, and some colours may repel or attract insects and wild animals.

Performance sportswear, an early adopter of technical textile innovation, has promoted a design-led approach to the design of wearable products. This is a prime example of a market where clothing must work and look good. Novel textile and garment manufacturing techniques are being adopted to aid both the permanent embedding of textile-based wearable electronics in garments and the incorporation of specific enclosures for removable wearable technology. Laser finishes and textile bonding techniques enable the encapsulation of miniaturised traditional and textile-based electronics. Textile moulding is used to create textile control buttons for electronic switches. Conductive metallic fibres and polymers are being used in lightweight textile assemblies for soft keyboards and as controls for wearable devices (Eleksen, 2008). Flexible solar panels are harvesting power on outerwear jackets. A flexible solar cell unit based on thin-film technology has been integrated into the upper back of the SCOTTeVEST to supply the power to operate and charge mobile digital devices such as phones or PDAs (ScotteVest, 2008) The Zegna Sport solar jacket is based on the iSolarX wearable solar technology platform, a collaboration between Interactive Wear AG and their technology partner Solarc of Berlin. 'It comprises solar modules, the textile integration kit, and the charging electronics. The mounting and connection techniques were developed in close co-ordination with the Zegna, Sport designers.' This garment can recharge a cell phone, iPod or other device using solar modules mounted on a neoprene collar (Zegna, 2008). Fibre optic displays have been developed for flexible displays on sports backpacks, initially by France Telecom (Banks, 2006).

3.4.2 The culture of wearable technology

The use of mobile and wearable technology is already widespread across all levels of society. Spectacles that enhance vision or offer UV protection have much of their value associated with their design and in brand recognition. Designer hearing aids are now available. Most people have a method of time keeping, either in the form of a traditional or digital watch or, for the youth market, often by means of the cell phone. Mobile phones, as communication devices for work, family and within peer group sub-cultures, now incorporate extended functionality that includes music, digital photography, gaming and internet access. Many people carry a lap top computer for work and/or pleasure, with comparable technology now becoming miniaturised into mobile phones. Mobile technology provides 'infotainment' and entertainment. Mobile phone communication provides enhanced confidence and safety for those potentially at risk, such as children, single women, the elderly and their carers. Digital watches and cell phones provide interfaces for feedback from wearable devices that monitor vital signs, movement and positioning for health, wellness and sport. Within corporate culture, smart wearable technology aids identification, communication, safety and protection.

Leading sportswear brands are collaborating with technology providers in the co-branding of design led products, such as Nike and Apple (Apple, 2008) and Adidas and Polar (Adidas-Polar, 2008) This requires the merging of the previously separate industries of textiles, clothing and electronics, with their respective cultures, and so has significance with regard to disparate time scales for research and development, sourcing and production, routes to market, industry standards and the product end-of-life and disposal requirements. The introduction of wearable electronics also demands an understanding of the way the human body works and related aspects of health and safety. For designs to be successful, products that embrace this new hybrid mix of technologies must be understood by, and be appropriate to, the needs of the identified end-user or customer. A relevant strategy may be adopted, prevalent in the area of performance sportswear design, where there is a culture of products being supported by extensive point-of-sale material. In contrast to more transient fashion, performance sportswear ranges are promoted with explicit point-of-sale information on the attributes of fibre and fabrics, garment detail, functionality and aftercare. This provides a model for the promotion and explanation of the enhanced functionality of wearable technologies if devised in a format suitable for the target customer.

Design-led wearable electronics which have already come to market have been targeted primarily at athletes and the youth market. Current technological advances have the potential to assist the everyday lives of the rapidly growing older market, but have often failed to address their social and cultural aspirations. A new market, the 'Baby Boomers', has been accustomed to making choices in the design of their clothing and accessories throughout their lives. Transient fashion ranges are not generally geared to the physiological demands of the changing older body, resulting in clothing that is often uncomfortable due to inappropriate fit, styling, proportion and weight. It may be difficult to take off and put on, fasten, and have compromises in relation to quality and aftercare. Medical devices have been developed for 'ill people' with little aesthetic appeal and with data feedback that may be difficult for the wearer to read and understand. The design of products that promote health and wellbeing may not be readily accepted by some older users due to badly designed user interfaces that have small controls or displays that may prevent someone with a minor impairment from using them effectively. Little has been done to address physical and cognitive limitations when developing these new products and services to ensure that they are appropriate to the culture and real-world needs of the variously described rapidly growing 'Grey', 'Silver', 'Third Age' or 'Rainbow Youth' market. Twigg states that 'Clothes are central to the ways older bodies are experienced, presented and understood within culture, so that dress forms a significant, though neglected, element in the constitution and experience of old age' (Twigg, 2007, p. 285).

3.4.3 Revisiting the demands of the body

It is recognised that a rise in obesity, diabetes and heart disease is resulting in an ever-increasing financial burden on government and private organisations. The worldwide growth for Biophysical Monitoring Wearable Systems has grown from $192 milion in 2005 to $265 million in 2007, within three distinct market areas. These are Health/Fitness, Medical and Government/Military (Krebs and Shomka, 2005). In order to reduce the level of illness and death associated with obesity and heart disease, governments are recognising the link between sport and fitness and the health of the nation. These trends stress the need for smart wearable textile products that help to make self-monitoring more accessible and positive for those who wish to keep fit or for those who find themselves gradually or sometimes rather abruptly becoming unwell. If products are designed in an attractive, wearable and useable format, the appropriate application of 'passive', 'active' and 'very active' smart and 'intelligent' materials will contribute to the comfort and wellbeing of the wearer.

Wearable technologies may be embedded within the clothing system to monitor vital body signs, respiration, movement and positioning, aid thermophysical regulation and offer impact protection from hazardous environments. Clothing and textile designers will require an appreciation of the properties and features of a garment that are required to harvest vital signs from the surface of the body. Where should sensors be positioned and how close fitting should the garment be? The relationship of predominant movement patterns and posture to body surface distortion, and potential impediments imposed by clothing on agility in terms of range, speed and repetition of movement will influence garment design and the selection of smart textile characteristics. Design-engineered garment moulding and bonding techniques, as well as seam-free technology, may be implemented to enable the 'zoning' or 'body/comfort mapping' of appropriate textile properties and assemblies within the layering system. In order to function effectively, wearable technology must be compatible with the physiology of the wearer in terms of comfort, size, shape, fit and ease of movement and use.

Anthropometry, or the measurement of the body, is a key to the positioning of smart and intelligent textiles around the body. 'SizeUK' has been the first national bodysize survey of the UK population since the 1950s and the first time that the shape of the population has been captured and analysed by means of 3D scanning technology. As a result of this survey, clothing retailers are in the process or have already updated and amended their size charts and garment specifications to ensure that their size charts satisfy the size and shape of their target customers (Sizemic, 2007). It has also been anticipated that the data will attract interest and be of considerable value to sectors outside the clothing industry, e.g. health and medical, particularly with concerns over the increasing level of obesity. This digital scanning technology will inform designers, and, in turn, the general public who are aware of the inadequacies of standard garment sizing charts with inconsistencies across regions, market sectors and brands.

Smart phase-change textiles can help to regulate temperature, such as the Outlast technology that absorbs heat when the body is subject to heavy workload and releases heat when the body cools down. In more extreme situations, as precautions against hazards to health and safety, phase-change textiles inflate when exposed to extreme heat to provide an insulating barrier in situations such as fire fighting. Other inherently flexible phase-change materials harden on impact, such as products 'd3o' (d3o, 2008) and 'Active protection' (Dow Corning, 2007), and then revert to a flexible state once the impact has passed. Wearable technologies may provide electronic arrays to enhance visibility. Tracking and positioning devices may be used for avalanche detection, and to monitor activity, movement and posture in sports training and in practice. Sensory textiles can aid the detection and prevention of bedsores. Textile switches and touch sensitive displays may be used to aid dexterity, as alternatives to control buttons for wearable devices that may not be further reduced in scale. In respect of olfactory sensing, work has been done on colour for therapeutic scent delivery as a healing platform to address emotional wellbeing. 'Scentsory Design' research explores the mood-enhancing effects of olfactory substances that act on the brain to influence performance, behaviour, learning and mood (Tillotson, 2007). Scented textiles have the power to drive emotions and spur memories, and may release aroma when agitated or warmed. Micro-encapsulation traps scented particles in the fabric with elements that may be associated with aromatherapy (Hibbert, 2004). In addition to gesture and voice control of electronic devices, research in the USA has demonstrated that it is possible to use brain movement to control computer programs (Jackson, 2006).

3.4.4 Enhancing the functionality

The adoption of a new generation of smart textiles and wearable technologies has the potential to enhance the comfort and functionality of a multi-purpose garment layering system to suit the life-style needs of the wearer. This chapter has made reference to a range of different applications primarily concerned with functional clothing products such as corporate wear, performance sport, health and wellness and inclusive design. The end-user driven design approach is intended, however, to be applied to the identification and prioritisation of the particular needs of any given individual wearer, or target group, where smart clothing should both function and look attractive within the remit of the selected end-user. The successful design and launch of the smart clothing layering-systems, or individual products, will demand compatibility between the styling of the textile-based products and the design and usability of the technology interface, as well as the design of promotional point-of-sale material. It will also require the support of relevant technology providers and services.

Considerations relevant to enhanced garment functionality include the monitoring of the duration and regularity of the activity, the range of likely conditions and environments in which the activity takes place, and issues to do with transportation. The adoption of technologies that monitor and communicate information, such as vital signs, posture, speed and repetition of movements, have important implications for the wearer and others such as employer, coach, team members, peers, family and carers. In addition to technologies that impinge on health and wellness, safety and performance, further data may be mapped as relevant to do with identification and hierarchy, location, weather forecasting, transport scheduling, travel information and a range of information to do with leisure and entertainment. The garment layering system may be altered radically in style and fabrication to address varying degrees of functionality, depending on the demands of a specified single end-use or range of activities.

Cross-disciplinary expertise is needed to take advantage of potential applications for wearable technologies that are emerging and relevant for a breadth of activities. Already the general public may obtain personalised data on sizing and shape through visiting a scanning booth in a high street department store such as Selfridges (Piller, 2007). The development process will begin with the use of such digital laser body scanning data, linked to 3D pattern drafting, to locate smart materials and wearable technologies within the garment layering system or individual item that are appropriate for the intended activity. As discussed, in addition to passive smart attributes, base layer garments may incorporate smart textile-based sensors and/or pockets to carry technology to monitor vital signs such as heart rate, respiration and skin temperature. The team must be suitably informed to identify the appropriate smart attributes required and the optimal (personal) feedback mechanism to the wearer, to ensure that maximum personal benefit is enjoyed. Technical expertise, in terms of software analysis, must address required measurements in terms of frequency, signal/noise ratio, accuracy and bandwidth. A breadth of expertise is needed in respect of materials cutting, moulding and joining, and the incorporation of a range of soft wearable devices such as zoned heat-controlled textiles for base layers and hosiery, and for mid layer fleeces, the positioning of soft textile switches, flexible solar panels, and, for enhanced safety, illuminated fibres, glow-in-the-dark textiles, reflective textiles and tracking devices (Wilson, 2004). The choice of technologies may be daunting. The future product development team must have meaningful engagement with wearers in requirements capture and throughout the design development process to garment testing and aftercare.

3.5   Conclusion: a new hybrid design process

The new design area of smart textiles and wearable electronics demands a merging of methodologies across disparate disciplines to inform the application of wearable technologies in smart clothes that have the potential to enhance the quality of life of the target wearer. A hybrid design-led methodology is proposed to provide guidance to designers and the product development team, embarking on co-design practice with end-users for the design of functional clothing within this rapidly developing cross-sector market (see Fig. 3.2). Experts familiar with the disparate topics may further elaborate the detail of the process tree. It may be refined and tuned to identify the needs of an individual wearer or a specified group. The implementation of a survey may uncover further relevant issues grounded in publications, conference proceedings and journals. A comparative review of existing products may be conducted through market research, attendance at international trade events and in findings from industrial liaison and visits. An on-going 'technology watch' should be maintained with respect to developments in the area of smart textiles, garment engineering and wearable electronics.

Primary qualitative research methods may be employed, in semi-structured interviews with wearers, to verify and elaborate design topics from the process tree and any further issues uncovered. An appropriate balance will be sought in terms of the technical demands of the body related to the specified end-use and relevant style considerations to do with textile selection and garment design, dependent on the life-style needs of the wearer. Aspects may be explored in relation to the design and comfort of existing clothing, the wearer's degree of understanding of the attributes of existing smart textiles and the function and usability of a range of wearable

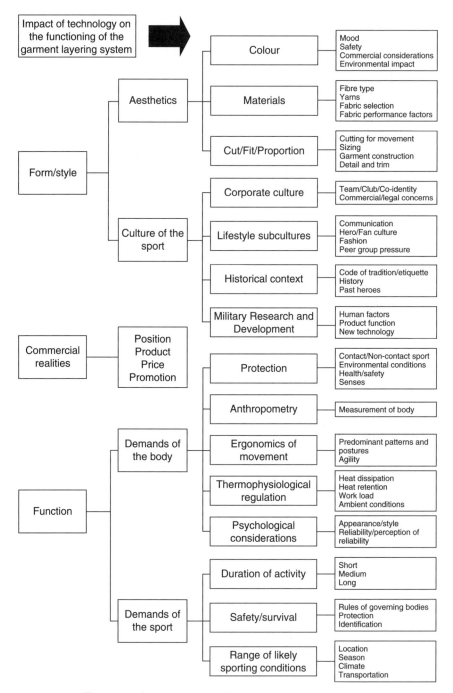

3.2 The start of a process tree for examining the co-design area of smart clothing and wearable technology.

technologies. Observation research techniques will support and verify the findings from the interviews. Expert input from academic collaborators and industry partners will aid the prioritisation of the issues uncovered. These findings will inform the drafting and refining of a comprehensive design brief. Verification of the findings will be carried out in the subsequent testing of initial to near-market prototypes. Final design specifications, with technical working drawings, explanatory text and a sequence of manufacture, will be drawn up and supported by a garment 'sealing sample'. The detail of the wearable technologies embedded within the near market prototype will be integrated within the final specification. This product specification must be in a new, shared, cross-disciplinary language.

3.6   Future trends: commercial realities

To date, there is a high element of waste in both the clothing and electronics markets. The design of fashionable clothing and electronic devices is generally transient and disposable, with an abundance of products available that are relatively cheap as a result of global sourcing, and primarily targeted at the youth market. With off-shore production, young designers have more of an understanding of styling and merchandising than traditional technical design development skills. Clothing and mobile technology may be bought on a whim, with the majority of an individual's wardrobe seldom worn and some electronic devices never taken out of the box (Bryson, 2007). A rapidly growing older community is generally neglected by fashion trends and by technology providers. A new breed of designer is needed to lead the cross-disciplinary product development team of the future. A new 'shared language' is needed to bridge traditional boundaries between the cultures of technical textiles and clothing, wearable electronics and ICT-based systems, social and health sectors. Designers need guidance in bringing smart customised products with added value to market, that are usable and attractive to new markets.

As the concept of wearable technology began to emerge, initially in somewhat crude prototypes, a leading technology provider recognised that soft electronics should be brought to near market in a format suitable for easy adoption by innovative textile and clothing brands. The UK company Eleksen bravely stated that 'the future is now' at international trade events and conferences over recent seasons. As a small company they created their Electex technology as ready-assembled technology kits for customers to integrate directly into textile-based products. The concept has created interest and been adopted by sports brands and the automotive industry. The company recently grew to PLC (Public Liability Company) status and, sadly, has now gone into administration. Design cannot be looked at in isolation from the impact of worldwide economic changes within the relevant industries. The design process tree should be further elaborated in the area of 'Commercial Realities'.

Designers need to be aware of commercial issues that have an impact on design considerations, such as product position, price and promotion. Where are current purchases made both for clothing and for electronics? What is the relevance of branding, or co-branding, or the need for the introduction of new brands to straddle traditional market sectors and attract new customers such as the ageing community? Consideration must be given to the merging of different time lines for materials and product sourcing and production. Cross-sector product testing and the recognition of health and safety standards will have implications for guarantees and instructions for maintenance and aftercare. The functionality of the product must be clearly described in effective visual and text based point-of-sale and promotional material. Finally, designers in this new century must be aware of sustainable design issues in relation to the whole product life cycle. The embedding of technologies in textiles and clothing, with controls, powering and communication devices, presents a significant design challenge.

Designers embarking on this new area of design must address sustainable design issues throughout the product life cycle chain, from design inception to garment disposal. These concerns are now becoming mainstream with a prominent high street chain, Marks and Spencer, launching 'Plan A' in tandem with the UK Government introducing a Sustainable Road Map for the clothing and textile industry (Defra, 2007; Hamnett, 2007). The outdoor sports industry has been conscious of the environment for more than a decade, with Vaude's Ecolog programme involving the recycling of polyester garments and the development of fleece fabrics from PET bottles (Eco-fi, 2008; Teijin, 2002). This community is already active in combating 'green washing' (Enviromedia, 2008) with companies such as Patagonia 'working within an environmental framework' having initiated the 'Common Threads' programme in 1995 and, most recently, 'The Footprint Chronicles', an interactive website tracking the carbon footprint of a range of their products (Patagonia, 2008b).

This introduction to a new hybrid design area is intended to set the scene and raise awareness of topics that will be revisited in more detail in subsequent chapters throughout this book.

3.7   References

ADIDAS-POLAR (2008), Part of your training: Part of you. [URI http://www.adidas-polar.com/phase5/index.html, accessed 10th June 2008]

ANON. (2008), Adidas Maximises Athletic Performance. *Industry News, Textiles: The Quarterly Magazine of the Textile Institute*. No. 1, p. 4.

APPAREL SEARCH (2008), Spandex fibres definition, history and characteristics. [URI http://www.apparelsearch.com/Definitions/Fiber/spandex_definition.htm, accessed 10th June 2008]

APPLE (2008), Nike + iPod Tune your run. [URI http://www.apple.com/ipod/nike/, acccessed May 12th 2008]

ARC'TERYX (2008), [URI http://www.arcteryx.com/, accessed May 19th 2008]

BANKS, R. (2006), Trends: All that matters is the display. [URI http://www.byz.org/~rbanks/movableType/webLog/trends/archives/cat_209_all_that_matters_is_the_display.html, accessed 10th June 2008]

BRYSON, D. (2007), Unwearables. *AI & Society* 22(1): 25–35.

C_CHANGE (2008), The bionic climate membrane, c_change a Schoeller™ technology. [URI http://www.c-change.ch/, accessed 10th June 2008]

D3O (2008), Intelligent shock absorption. [URI http://www.d3o.com/, accessed May 19th 2008]

DEFRA (2007), Starting on the road to sustainable clothing. News release, September 5th. [URI http://www.defra.gov.uk/news/2007/070905b.htm, accessed 11th June 2008]

DOW CORNING (2007), Active protection system [URI http://www.activeprotectionsystem.com/index2.html, accessed 10th June 2008]

ECO-FI (2008), Ecospun is now Ecofi. [URI http://www.fossmfg.com/bu_ecospun.cfm, accessed June 11th 2008]

ELEKSEN (2008), ElekTex textile touchpads. [URI http://www.eleksen.com/?page=products08/elektexproducts/index.asp, accessed May 19th 2008]

ENVIROMEDIA (2008), Greenwashing: It's not black and white. [URI http://www.greenwashingindex.com/, accessed June 11th 2008]

FALKE (2008), Ergonomic sports system. [URI http://www.falke.com, accessed 10th June 2008]

GONZALEZ, J.A. (2003), Advances in Technology: Smart and Engineered Textiles. Protective Clothing Research Group, Department of Human Ecology, University of Alberta. [URI http://www.ualberta.ca/~jag3/Smart_textiles.ppt, accessed 10th June 2008]

GORE (2008a), Our technologies – Gore-Tex Products. [URI http://www.gore-tex.com/remote/Satellite/content/fabric-technologies, accessed 10th June 2008]

GORE (2008b), Airvantage™ Adjustable insulation. [URI http://www.gore.com/en_xx/products/consumer/airvantage/index.html, accessed May 19th 2008]

GORE (2008c), Gore-Tex Fabrics – Gore Comfort mapping technology. [URI http://www.gore-tex.com.au/www/348/1001412/displayarticle/1024369.html, accessed May 27th 2008]

HAMNETT, K. (2007), Katherine Hamnett on sustainable clothing. [URI http://www.youtube.com/watch?v=3Y0NYGdfJQQ, accessed June 11th 2008]

HIBBERT, R. (2004), *Textile Innovation*. London: Line, 2nd ed. [URI http://www.textileinnovation.com, accessed May 27th 2008]

JACKSON M., MASON, S., and BIRCH, G. (2006), 'Analyzing Trends in Brain Interface Technology: A Method to Compare Studies'. *Annals of Biomedical Engineering*, 34(5): 859–878.

KATHLEEN GASPERINI (2007), Crossover of Winter Outerwear + Lifestyle, Wearable Electronics, + Skier/Snowboarder Styles Indicate Expansion Beyond Core to Lifestyle-Inspired 'Participants'. Label Networks: Global Youth Culture and

Street Fashion Marketing Intelligence. [http://www.labelnetworks.com/sports/sia_ overview_07.cfm), accessed 10th June 2008]

KICK PUBLIC RELATIONS (2005), What's new from Polartec®. [URI http://www.kickpr. com/polar_1105.html, accessed 10th June 2008]

KREBS, D. and SHOMKA, S. (2005), *Mobile and Wireless Practice. A White Paper on Wearable Systems, Global Market Analysis, Second Edition*. Venture Development Corporation, October. [URI http://www.vdc.corp.com]

LACK, T. (1992), Troll safety equipment report. *Mountain Ear*, 6(1), 50–55.

MCCANN, J. (2000), *Identification of Requirements for the Design Development of Performance Sportswear*. M.Phil Thesis. Derby: University of Derby.

MCCARTY, C. (2005), 'Nasa: Advanced ultra-performance' In: McQuaid, M and Beesley, P., eds. *Extreme Textiles: Designing for High Performance*. London: Thames & Hudson.

OUTLAST (2008), Outlast: Advantages. [URI http://www.outlast.com/index. php?id=71&L=0, accessed May 27th 2008]

PATAGONIA (2008a), Patagonia BioMap. [URI http://www.patagonia.com/web/us/ patagonia.go?assetid=2793, accessed 10th June 2008]

PATAGONIA (2008b), The footprint chronicles: Tracking the environmental and social impact of Patagonia clothing and apparel. [URI http://www.patagonia.com/web/ us/footprint/index.jsp, accessed June 11th 2008]

PILLER, F. (2007), IHT Reviews Bodymetrics' Mass Customization Program at Harrods and Selfridges in London [URI http://mass-customization.blogs.com/ mass_customization_open_i/2007/01/iht_reviews_bod.html, accessed 10th June 2008]

SCHOELLER (2008), Schoeller Textiels AG: Nanosphere. [URI http://www.schoeller-textiles.com/default.asp?cat1ID=128&cat2ID=134&pageID=269&emotionstate =0&emotionID=1&langID=2, accessed 10th June 2008]

SCOTTEVEST (2008), Scottevest/SeV® iPod and travel clothing solutions™. [URI http://www.scottevest.com/, accessed 10th June 2008]

SEW SYSTEMS (2008), Designers and developers of sewing systems and machinery for the textile industry. [URI http://www.sewsystems.com/, accessed May 19th 2008]

SIZEMIC (2007), SizeUK Data Set. [URI http://www.humanics-s.com/uk_natl_anthro_ sizing_info.pdf, accessed 8th Nov 2007]

SPEEDO (2008a), Aqualab [URI http://www.speedoaqualab.com/, accessed 10th June 2008]

SPEEDO (2008b), Now – Space Age Speedo. *Textiles: The Quarlerly Magazine of the Textile Institute*, (1): 19.

SPORTS BUSINESS DIGEST (2007), Gatorade . . . It's in Tiger Woods. [URI http:// sportsbusinessdigest.com/?p=46, accessed 10th June 2008]

SYMPATEX (2008), Sympatex: Navigator. [URI http://www.sympatex.com/index. php?id=20&L=2, accessed 10th June 2008]

[TC]² (2008), 3D Body Scanner and 3D scanning software. [URI http://www.tc2. com/what/bodyscan/index.html, accessed June 2nd 2008]

TEIJIN (2002), Teijin Close-up. Annual Report. [URI http://www.teijin.co.jp/japanese/ ir/doc/annual2002/ar2002_49-77.pdf, accessed June 11th 2008]

TILLOTSON, J. (2007), Sensory Clothing Concepts: Microfluidic Devices and their Fields of Application from Well-being to Care for the Elderly. *Avantex Symposium*, Messe Frankfurt, June.

TURNER, G.L. (2000), Hypothermia and the cave rescue environment: A review of treatment and advanced prehospital provider care. [URI http://www.wemsi.org/hypo_1.html, accessed 10th June 2008]

TWI (2006), 'ALTEX' – automated laser welding for textiles [URI http://www.twi.co.uk/j32k/protected/band_3/crks3.html, accessed June 2nd 2008]

TWIGG, J. (2007), Clothing, age and body: A critical review. *Ageing and Society* 27: 285–305.

WILLIAM LEE INNOVATION CENTRE (2007), Scan2Knit [URI http://www.k4i.org.uk/about/Scan2Knit_Case_Study_20_08_07.pdf, accessed June 2nd 2008]

WILSON, A. (2004), Sewfree – Bemis, 100 Innovations. *Future Materials*, Issue 6. [URI http://www.inteletex.com/adminfiles/PDF/fmBemis.pdf, accessed June 2nd 2008]

ZEGNA (2008), Solar JKT Ermenegildo Zegna. [URI http://www.zegna.com/solar?lang=en, accessed May 19th 2008]

4

The garment design process for smart clothing: from fibre selection through to product launch

J. McCANN, University of Wales Newport, UK

Abstract: This chapter provides guidance for a sequence of stages within the technical design development of a clothing system that supports the integration of wearable electronics. The development process is made up of disparate yet interdependent stages in a critical path that is driven by the identification of end-user needs. It focuses on the selection and application of technical textiles, and novel garment manufacturing methods, that enhance the functionality of technical oriented clothing that must look good and be fit for purpose as opposed to being influenced by more transient fashion.

Key words: design critical path, functional, layering system, end-user needs, technical textiles, wearable electronics, novel manufacturing.

4.1   Introduction

This chapter provides guidance for a proposed sequence of stages within the technical design development of a clothing system that supports the integration of wearable electronics. It looks at a design development process for smart clothes with wearable technology, from initial concept through to product launch. It focuses on the selection and application of technical textiles, and novel garment manufacturing methods, that enhance the functionality of technical oriented clothing as opposed to more transient fashion. A design development critical path is introduced for clothing that must look good and be fit for purpose. This represents garment sectors, such as performance sportswear, corporate work wear, intimate apparel and products that promote health and wellness, that are driven by fibre and technical textile innovation and a development process that incorporates product testing both in the lab and in the field. In particular, the performance sportswear sector has been relatively design-led as an early adopter of technical textiles advances, filtering down from military preparedness, and most recently wearable technology has emerged to enhance the functionality of both recreational and competitive sport. Many of the well-known sports and outdoor clothing brands are now employing technology and electronics specialists within their product development teams (Roepert, 2006).

A new industrial revolution is taking place as the technical textile-driven apparel industry and the electronics industry, with their previously quite different cultures in design development and production processes, come together. An initial contributor to the successful implementation of wearable technologies suggests that 'well thought out production directives save some surprises' (Roepert, 2006). Within the electronics industry, 'the cycle from the start of the development until production, takes at least nine to twelve months as a rule' when 'the delivery time for semi-conductor products alone can amount to between three and six months' (Roepert, 2006). The cycle within the fashion apparel industry is typically seasonal although, in the performance sportswear sector, the styling of components within a garment 'layering system' is less transient as these garments are designed to be adaptable to a range of climates and environments. In bringing wearable technology to market, 'a joint project plan and timely cooperation between "wearables" and textile partners helps to fashion the design successfully and to optimise costs' (Roepert, 2006).

A clear design brief is needed to drive the product-design development process. The garment-design development process is made up of disparate yet interdependent stages in a critical path that may begin with the identification of end-user needs. This informs progression, through a sequence of stages from fibre and textile selection and finishing to garment prototype development and construction and planning for product launch. The appearance, comfort and functionality of a garment is highly dependent on each stage of the sequence. The embedding and/or incorporation of wearable technology cannot be an add-on to the system but this 'technology layer' must be considered simultaneously throughout the design development process. The introduction of electronic functionality to 'the item of apparel should be achieved in a way that is imperceptible to the user.' (Roepert, 2006). The real need, ease of use, and reliability, as well as the perception of reliability of the product will be key to its acceptability. The functionality of the design of the 'clothing system' must be verified in both laboratory and in field tests. Design is an important consideration for the development of the interface between the technology and the user and for appropriate visual and text-based promotional material for product launch. The current and future product development team should collaborate in interrogating and addressing sustainable and ethical considerations throughout the design research and development process.

This chapter introduces topics for consideration within the cross-disciplinary design development process that will be addressed in greater detail within this book. See Fig. 4.1 for range of issues and Fig. 4.2 for an overview of the critical path.

The design brief
Setting an effective design brief: Choice of product/range
Function: activity/ range of activities, individual product or collection
Identification of end-user needs: Target group survey, consultation, observation, development of scenarios/personas.
Commercial realities: Comparative market research, price expectation, route to market, etc.

2D design development
Initial concepts/ideas: Preliminary style concept + more specific sketches + suggested materials
Technical working drawing: Suitable 2D format to effectively explain concept + text as necessary

Textile development
Textile sourcing: Textile and related trade events, agents, industrial liaison
Fibre and textile selection: Textile construction & finishing for knitted, woven, non-woven, nets, etc.

Initial 3D design development
Measurement/sizing
Traditional or new methods of capturing body size and shape.
Traditional pattern development
Flat pattern making, reverse engineering, working on the stand; etc., CAD.
Designing in the round
Pattern cutting for movement, pattern blocks with articulation, initial 3D toile
Modern construction and joining technology: Seam-free knit construction, bonding, laser welding, ultrasonic seaming/welding, moulding, etc.
Integration of wearable technology
Smart textile technology: Passive, Active, Very Active and Intelligent attributes.
Function, positioning, integration/encapsulation/compatability, interface, etc.
Interconnections and communications
Power
Design of technology interface

Final prototype development
Final 3D prototyping: Whole technical and aesthetic design evaluation, refitting, testing, amending in collaboration with 'customer' and technical team
Sealing sample: Sealing sample replication, size grade control, release for production
Product specification sheets: Drawing, materials, trims, pattern pieces, making up sequence/methods
Manufacture

Point-of-sale
Design of promotional material to effectively 'sell' concept to target audience

4.1 Diagram showing the overall garment design process.

The 'Critical Path'
The end-user needs have to be considered throughout

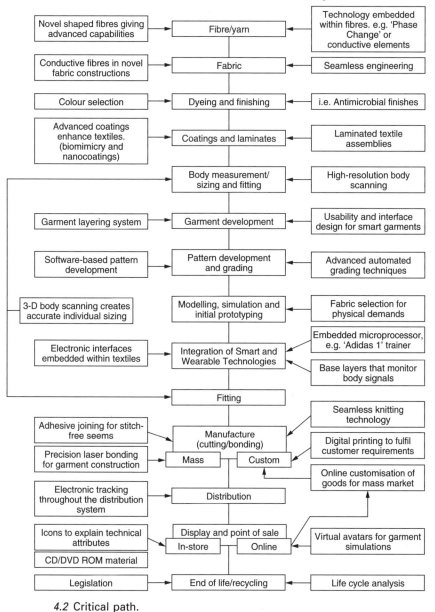

4.2 Critical path.

4.2   The design brief

4.2.1   Setting an effective design brief

Design outcomes will be dependent on the clarity of the brief. The design brief should inspire ideas, while providing an articulate written explanation of the aims, objectives and time scale of a design project. A comprehensive brief provides an essential point of reference for the design development process, for both the designer and the client. An effective brief ensures that any ambiguity is clarified, with differences in opinion resolved, before the design development begins. Sometimes clients have not made fundamental decisions about the objectives of their marketing budget. The designer may need to guide clients without patronising or victimising them. A creative brief should begin with the project title and a short introduction to the background of the organisation and the product sector or niche market. The target audience should be described with regard to occupation, gender ratio, average age, nationality, location, psychological demographic, and lifestyle preferences. The objectives must be clear, with aspects of the product or branding provided that may be used as a starting point for the design. Suggestions may be given of where to look for inspiration with style direction and product shots of the mood, imagery, diagrams, text, colour, fabrication and manufacturing methods. An explanation of what is considered both good and what is to be avoided will lead to more acceptable design solutions.

An effective design brief should direct progression through the design development process to successful product launch for the identified product category. Deliverables must be clear in relation to the development of an individual product or a coordinating collection of items that constitute a complete technical clothing system. Unique selling point(s) must be identified and addressed with a clear message, or design 'story', in relation to the marketing objectives. Detail should be agreed in relation to materials and trims sourcing and prototype development with any additional out-sourced specialist processing. The timeline must be scheduled for the whole project before a designer is briefed, incorporating appropriate feedback and incubation stages. The designer should be consulted and asked to inform in advance if deadlines are unrealistic. A realistic budget must be set and agreed with the designer for the design development in terms of fees, material costs and any sub-contracting and travel expenses. Approval requirements should be clarified in terms of responsibility for signing off the various stages of the process. A clear design brief should result in products that are attractive to, and with technical design attributes that are believed and understood by, the target customer.

4.2.2 Identification of end-user needs

'The right clothing can grant us access to the right places and the right people. Our acceptance of modernity, whether as designers, early adopters or consumers, serves as an indicator of our creativity, adjustment and preparedness for the future' (Jenkins Jones, 2005, p.28).

The profile of the end-user, in relation to their technical and aesthetic clothing requirements, has been discussed in detail in the previous chapter. A target group survey, through consultation and observation, will set the scene for decisions to be made throughout the design critical path. Visual, verbal and textual information may be collated to help the development of scenarios and personas to describe the wearer(s). Gender, age group and relevant human physiological concerns must be understood prior to commencing the design process. The demographic lifestyle demands will relate directly to relevance of clothing products that have appropriate smart functionality. Clothing may be issued for work, with a corporate identity, or selected through personal choice for either formal or recreational wear. It may have a specific intended end-use or demand multi-functional attributes for a range of activities. The product(s) may be for seasonal or non-seasonal use, to be worn indoors or outdoors and in predictable and comfortable conditions or in an unknown, potentially hostile, environment. The clothing may be worn for a short or extended duration with implications related to durability, disposal and aftercare. These types of consideration will be relevant in identifying both the category of clothing to be developed and the nature of the smart textiles and wearable electronics to be adopted.

4.2.3 Commercial realities

There will be commercial constraints that impinge on the setting of the design brief. Collaborative commitments, to partners or certain suppliers, may exist that may affect decisions to do with the sourcing of materials and manufacture. Product development within this new hybrid area of design will require new relationships between technical textiles, clothing and electronics technology providers with their previously separate cultures, time lines, routes to market and branding. The design critical path must accommodate a breadth of disparate procedures and processes, within a realistic time line, that promotes the opportunity for effective communication and collaboration across the range of required expertise. The application of emerging technologies for functional clothing applications raises quite different topics for consideration from those normally addressed within clothing design training. To date, academic design training for the clothing industry has been led by designer fashion, with many fashion design courses focusing almost exclusively on women's wear. Within the UK, the fashion market is said to represent approximately 30% of clothing companies but, overall, it represents less than 10% of overall clothing sales. Within this relatively exclusive market, women's designer fashion accounts for 57%, while men's designer wear is estimated at 24% (Jenkyn Jones, 2005).

Women's fashion may be considered the most fickle sector of the clothing industry, with high street chains putting increasing pressure on quick response manufacturing to follow designer trends in this market segment. Traditionally, the fashion trade has worked to seasonal cycles with ranges launched for spring/summer and autumn/winter. There may be additional party or holiday wear launches. Jenkyn Jones (2005) describes how the fashion designer and merchandiser can become skilled at reading the prevailing aesthetic, altering it and applying the imaginative result to the human body in a desirable and marketable way. The design process for designer fashion 'is born of inspiration, rarely for an individual item, but for a whole mood or look' (Jenkyn Jones, 2005, p.129). 'In their search for inspiration, designers must learn to keep their eyes and ears open, to visit shows, shops, clubs, cafes, galleries, and films; to read magazines, newspapers and novels; to party and listen to music; and, above all, to people-watch and absorb the subtle and incremental aesthetic changes that take place in society' (Jenkyn Jones, 2005, p.170). These high fashion trends are expected to wane after one season while classic ranges may continue to sell over more than one season. With greater market fragmentation, an international fashion mix provides a greater number of styles that may be acceptable concurrently and worn by individual users in different contexts.

Menswear design is characterised by more slowly changing silhouettes and fabric choices but with close attention to detail and marketing through logo and brand display. 'Now, although men individually are spending proportionately less on clothing and more on gadgets, sports and holidays, more of them are buying fashion more frequently and their preference is for comfortable, casual clothing ... blurring the boundaries between business and casual apparel' (Jenkyn Jones, 2005, p.59). Usage levels for sportswear in the UK are high. Items of sportswear are purchased by almost 90% of people under 35 years of age, and by 76% of the population as a whole. It is evident that sportswear is often worn for everyday use. The UK-based market research company Mintel (www.mintel.com) forecasts that the value of the UK sportswear market will remain static in 2007 and 2008 before rising gradually to £3.96 billion in 2011 in the lead-up to the 2012 London Olympics. The performance sportswear sector has traditionally been an early adopter of new textile technologies and, more recently, wearable electronics. As a new industrial revolution takes place, with technical textiles and electronics industries merging, training and support is required for a new breed of designer to approach clothing from a more product design user-driven perspective to practice within the future cross-disciplinary team.

The electronics industry has traditionally been technology driven, with performance gains in processing, power and memory at the forefront of development. The consumer electronics industry, which is closer to the culture of product design, has recently been taking more notice of what consumers/users want from their devices. More care needs to be taken when designing products that are part of complicated systems, which may include clothing, electronics and services that must all be integrated. The electronics market may have an initial conceptual design that they wish to explore prior to making an approach to a clothing brand. This may be in the form of a story board or initial mock-up for demonstration purposes. There is no industry standard for the merging of textiles with electronics. Companies may carry out some of their development in-house and some with external partners.

The designer should carry out a comparative market review prior to commencing design development. An overview of trends in relation to design concepts, range composition, technical garment detail, fabric usage and manufacturing techniques must be collated and, in particular, a report on innovation in wearable technologies prepared. An appraisal should be made with regard to the usability of the technology interface for the intended market, and clarity of information on function, care and aftercare. The designer must be aware of the prices set by the nearest competitors as well as the price expectations for any new functionality to be introduced. These price points will relate to the demands of the demographic target group and the perceived value in relation to the quality of the manufacture, availability, design content, durability, perceived reliability and the available budget. Brand recognition is usually built up over many years of satisfying customer expectations of consistency, and by advertising unique qualities. This new hybrid market may offer co-branding and, occasionally, new brand development.

4.3   2-D design development

4.3.1   Initial concepts/ideas

Traditionally, in the fashion business 'a collection is a range of garment styles (models) designed with respect to current fashion trends and economic realities. It is put together by a collaboration between creative, sales and technical resources. Individual development steps may be undertaken sequentially or in parallel. The time required depends on the quality level and the size of the collection' (Kilgus, 2004, p.212).

Preliminary design concepts may be explained in 'story boards' or in multimedia presentations that provide a visual explanation of design problems to be addressed. These may be informed by material collated in the initial end-user needs consultation and observation phase, and further elaborated through imagery gathered in design research and through the comparative review of existing products or related areas of the market. Initial concepts, or experimental 'roughs', will be further developed through the sourcing and selection of materials and trims at the fibre and fabric fairs, where swatches and samples may be ordered. These will enable the designer to produce more specific design sketches with cuttings and technical information on suggested materials. These materials must have their suitability for the desired range of manufacturing methods verified through laboratory tests and practical workroom experimental trials for garment details. Additional specialist trade events and conferences, with relevance to wearable computing, will include issues around new manufacturing practices, digital technology, medical topics, inclusive design, mobile communications and smart environments. The apparel and technical textile related story boards must be developed in collaboration with team members responsible for the development of technology concept diagrams and demonstration models.

Clothing designers require an understanding of anatomy with 'the ability to map and illustrate the human form and communicate that information to others' (Jenkyn Jones, 2005, p.78). In fashion design, the ideal human form has often been considered young, slim and elongated. Functional clothing design in areas such as corporate wear and performance sportswear is normally portrayed in a more graphic, diagrammatic style and in more realistic proportions. In the new area of smart clothes and wearable technology, both the textile and apparel designers, and the technology team, must develop a basic understanding of the physiology of the human body and its interaction with materials, and wearable electronics, to provide comfortable clothing that is safe and fit for purpose. Perceived comfort involves addressing the ergonomics of movement, measurement and fit, thermal regulation and moisture management as well as the psychological feel-good factor. The application of wearable technologies can enhance functionality throughout the clothing layering system. The cross-disciplinary design team must solicit medical advice about the optimal positioning of monitoring and actuating devices and take responsibility for addressing any health and safety issues. Two-dimensional drawings must be developed in a clear format, with the addition of three-dimensional models as necessary, to explain effectively the design/technology concepts to the cross-disciplinary team.

4.3.2   Technical working drawing

Traditionally, in fashion design 'finished illustrations' describe the highest quality specification sketches to be used in portfolio presentation. The design development of functional apparel relies on two-dimensional schematic 'working drawings' developed around the particular design requirements identified. The essence of this type of drawing style is to communicate, and it integrates well with the application of new technologies. A well-proportioned illustration may be added to put the design into context. Schematic drawings are normally used in workbook presentations and in catalogues. Design proportions and features should be explored methodically for front, back and side views, with an indication of garment articulation for movement. The positioning of materials in relation to functionality, and an indication of construction methods, should be clearly explained. Enlarged views should be inserted in a suitable format to explain garment details, such as trims, closures, and the embedding or incorporation of the wearable technology, and the technology interface. Drawings may be needed to depict the garment interior. Technical text should be added as necessary, with terminology explained in a shared language that may be understood by the cross-disciplinary team. A computerised, more precise flat specification drawing will have measurements and/or accompanying specification sheets with fabric swatches and technical specifications. The more accurate the information given to the prototype developer and/or supplier, the more successful will be the outcome.

4.4   Textile development

4.4.1   Textile sourcing

'Textiles are ideal as a flexible conduit, particularly where the product is designed to be worn' (Braddock Clarke and O'Mahony, 2007). In introducing 'Electronic Textiles', Braddock Clarke and O'Mahony state that 'the intention is for the wearer to communicate with people and interact with the environment with greater ease. Effectively such systems act as extensions of the senses and serve to create cyborgs. Though they did not consciously begin with any decorative element, this area has developed a strong design aesthetic that is readily recognisable' (Braddock Clarke and O'Mahony, 2007). The cyborg image is now somewhat dated as electronics are becoming increasingly miniaturised and incorporated directly into textile structures. Designers embarking on the development of products within the hybrid mix of smart textiles and wearable electronics must research fibre types and textile constructions and their applications, found beyond the limits of the traditional fashion sector. Wearable technology cannot be looked at in isolation from the immediate and wider environment. Interior, architectural and automotive design all have an important relationship to worn technology and, in particular, aspects of health and wellness have to be considered. The leading technical textile trade event, Techtextil, has categorised the breadth of advanced materials under sectors that include Sporttech, Meditech, Hometech, Clothtech, Protech and Oekotech (Janecke, 2004).

In on-going industrial liaison, the design team will become familiar with relevant brand leaders, specialist in fibre production, and material finishes, and the specialist textile mills and suppliers and their agents. Specialist suppliers of textiles and trims from Europe, USA and the Far East, are often found at sector-specific trade fairs such as the international sports trade fair, ISPO. Other events concentrate on areas such as intimate apparel and swimwear, work wear, military procurement and footwear. Premiere Vision in Paris leads the direction for fashion fabrics, run concurrently with the Indigo Fair where ideas for textile print and embellishment may be sourced for subsequent production. Designers must be sufficiently conversant with technical terminology to communicate with textile producers and their agents in the selection of 'swatches' for testing, and 'sample lengths' for prototype garment development, prior to the ordering of 'stock' fabrics or special-order 'minimum runs' for production. The team must also be conversant with related trade events such as trade fairs and conferences where textiles are currently less understood, such as medical- and computer-oriented events. The 'pHealth' annual conference now embraces research papers on textile advances in the medical arena and the 'Ambience' conferences put a high profile on textile-related research. The 'International Symposium of Wearable Computers (ISWC)' community could be identified as the original cyborg community. This group now recognises that 'wearables' are embracing textile-based applications. The author and colleagues have focused on the importance of user-needs-led design processes, at ISWC events, in a paper at Osaka 2005, a poster and workshop at Montreux 2006, and a workshop at Boston 2007.

4.4.2 Fibre and textile selection

The design of function clothing is driven by technical textile innovation. A combination of fibre content, fabric construction methods and textile assemblies and finishes dictate the suitability of garment construction processes. The garment layering system (Kilgus, 2004) may incorporate a range of textile constructions and finishing techniques within the categories of knitted, woven, non-woven and lace or net fabrics, as well as the potential application of composite materials for body protection. The layering concept is described in greater detail in Chapter 12. The use of a high percentage of man-made fibres in technical textiles enables the functional clothing sectors to adopt manufacturing techniques such as laser joining and embellishment, ultrasonic welding, heat-bonding and the moulding of components. Textile dyeing and finishing encompasses treatments such as antimicrobial odour control and moisture management wicking finishes for base-layer garments. Mid-layer garments embrace finishes such as down proofing and stain resistance, while outer-layer garments typically have attributes such as water repellency and finishes that address visibility for safety, through reflectivity, and prints for camouflage protection or fashion-led embellishment. Textiles may be coated or laminated as a single substrate or in multi-layer assemblies. The design brief may demand the production of specialist fabrics to suit the specific demands of the end-use in terms of properties such as fibre type, content, weight, construction, durability, colour and finish (McCann, 2005).

In addition to the use of relatively familiar natural, man-made and synthetic fibres in clothing and textiles, there is an additional demand for conductivity in fibres and finishes for smart textiles and wearable electronics. Metal fibres are used in knitted constructions for applications such as textile sensors for vital-signs monitoring, in wovens for low-current heating systems for car seats and in nets for soft keyboards. 'Carbon fibre provides strength, odour absorption, fatigue resistance, vibration absorption and electrical conductivity. Conductive yarns, with carbon fibre, are used for clothing for contamination-controlled clean room environments – for manufacturing processes such as the production of semiconductors' (Braddock Clarke and O'Mahony, 2007, p.138). Shape Memory Alloys, made up of two alloys that respond to heat differently such as nickel and titanium, can be deformed at room temperature but regain their original shape when heated to a few degrees higher (Braddock Clarke and O'Mahony, 2007). Ceramic fibres withstand very high temperatures without deformation and loss of tensile strength. Besides low thermal conductivity, ceramic fibres offer good insulation and chemical resistance. Powdered ceramics are used with polyester fibre by the Japanese company, Kuraray, to shield the wearer from heat while absorbing and neutralising ultra-violet rays (Braddock Clarke and O'Mahony, 2007).

4.5   Initial 3-D design development

4.5.1   Measurement/sizing

The process of three-dimensional garment design development begins with capturing the body size and shape by traditional or digital methods of measurement. A standard size chart for an upper garment will include measurements around the body, such as base of neck, bust or chest, waist and hip, top arm and wrist; horizontal measurements such as cross-chest, cross-back and shoulder length; and vertical measurements including back and front neck to waist, shoulder to waist and sleeve over-arm and under-arm. These measurements will normally represent a standing figure with relatively erect stance. They will not represent a more relaxed pose or changes to the figure across the life cycle. In the category of functional clothing there may be a typical figure type for a given end-use with a repetitive range of movements and predominant postures. Garments for active end-use may require an extended range of movements demanding data not normally given on a standard size chart. Enhanced ease will be needed for movement in areas such as cross-back, over-arm, trouser centre back and over the knee by comparison with formal wear. The designer must devise an appropriate measurement chart for the 'subject'.

4.5.2 Traditional pattern development

There are different methods of developing design concepts into three dimensions that may be categorised as flat pattern cutting, working on the stand and reverse engineering. There is no definitive way to translate designs into three dimensions, with fashion designers often adopting a mixture of flat pattern cutting and draping or modelling. Flat pattern cutting normally begins with the drafting of standard blocks that represent the silhouette appropriate to the garment category. Generic descriptors exist for recognized cuts that include raglan sleeves, princess line bodice, straight, circular or flared skirt; as well as more formal tailored jacket styles and casual wear. These blocks are made up of basic shapes such as front and back bodice and sleeve, back and front skirt or trouser. They have the desired standard fit introduced in darts or simple seaming. Blocks are kept on pattern card and/or computer to be adapted and manipulated into specific styles. The stand, tailor's dummy, or mannequin may be helpful for checking garment balance or determining the shape of a detail drafted by flat pattern making methods. The stand may be used in directly modelling the development of the initial concept, with fabric shapes subsequently transferred into flat pattern pieces. Reverse engineering involves the taking of appropriate measurements from an existing garment in order to draft a replica garment, or derivation of that style. This may require the designer to unpick the seaming, lay the garment pieces flat and accurately mark off the shapes. All these methods will require verifying in initial prototypes, or 'toiles', made of material that has similar properties in terms of weight and drape to that intended for the final design. Any amendments to the 'toile' must be transferred to the flat pattern. Final patterns may be plotted onto computer. In the industry, more predictable styles may be stored and amended by CAD methods, with little 'hands on' pattern work.

4.5.3 Designing in the round – pattern cutting for movement

While a choice of textbooks exists for the development and adaptation of standard flat pattern blocks for men's, women's and children's wear, with slight variations in their approach to traditional garment cutting, little guidance is available to inform the cutting of functional clothing. As fabrics have become more breathable and an element of stretch may be added for comfort, leading edge performance clothing for end-uses such as active sport and corporate work wear, is becoming more tailored and streamlined. A close fit that follows the silhouette of the body promotes agility, while larger looser shapes may impede movement. Terminology is adopted such as 'second skin' for base-layer garments and 'body mapping' for mid-layer insulation or external protective 'shell'. Articulated cutting is required for the elbow and knee, underarm, in hoods and sometimes within the overall silhouette to address the predominant posture for a specific activity. Specialist standard blocks may be refined by individual companies for garments to move with the body in relation to the demands of the particular end-use.

There is little text available to explain cutting for movement. Functional clothing design should be developed in the round, avoiding the conventions of front and back views. The body should be looked at in terms of 3-D shape, size and posture, with both style and cutting lines working in harmony to address the demands of movement. The designer-pattern cutter should blend flat pattern cutting with working on the stand during the technical and aesthetic design development process of 3-D prototypes. The cut should address the positioning of technical textiles for their functionality in terms of protection, moisture management and thermal regulation. This process becomes more complex with the introduction of wearable technology. Textile sensors and miniaturised electronic devices and their controls will demand accurate positioning and attaching, embedding or encapsulation. Fabric specification and directional positioning should be clarified in accurate working drawings. The overall comfort and functionality of the style must be evaluated and amended as necessary, and the detail of the wearable technology verified in collaboration with the design team and end-user. If design detail is rejected, new solutions may be discussed and this stage of the process revisited.

4.6   Modern construction and joining technology

In the functional garment sectors, traditional garment manufacturing methods are being overtaken by new construction and joining technologies. In the knitted textile sector, recent innovations in computer-aided 'integral knitting' have led to 3-D technologies with the placement of ergonomic details that are integral to the design, predominantly of intimate apparel, base-layer garments and hosiery. The producers of knit machinery have developed these innovative techniques, with their various descriptors: 'Seamless' (Santoni), 'WholeGarment' (Shima Seiki) and 'Knit and Wear' (Stoll). Modern joining technology that is compatible with textiles made from a high percentage of synthetic fibres, includes heat bonding, laser welding and ultrasonic joining and finishing. These processes may be used in combination with moulding and targeted stitch construction. Garment bonding (Sew Systems, 2008), and laser welding (TWI, 2006) stitch-free joining technologies are used for both woven textiles and for knitted constructions, primarily in high-end markets such as performance sportswear and intimate apparel. Bonding involves the use of adhesive tape to join two fabrics together, with the relationship of heat, pressure and time as key considerations (Sew Systems, 2008). For recent advances in laser welding, ultrasonic cutting of the seam edge is required in order to achieve a clean and durable finish. The ultrasonic seaming process involves increasing vibration, focused to the point where it welds the fabric. It may be used for disposable paper-type garments, such as forensic suits and clean room garments for industrial and medical end-use. Heat shape-moulding and engineering of garment components is commonplace in the intimate apparel market (Sew Systems, 2008; Bogart, 2006), with stretch fabric development that is lightweight yet strong with different types of localised support.

4.7   Integration of wearable technology

4.7.1   Smart textile technology

Smart textile attributes may be described as Passive, Active, Very active and Intelligent. 'A "wearable electronics" application combines intelligent micro-systems with both electronic and textile components, supplies them with power, and sets up interfaces for textile integration' (Roepert, 2006). Miniaturised microcomputer systems can be integrated into clothing in the form and size of a button or logo, or 'more complex systems ... usually connected to the textile through a docking station, or ... constructed to be flat and flexible with the aid of flexible substrates (Roepert, 2006). Product development spans simple applications, with integration of individual elements such as light emitting diodes or headphones, up to complex knitted textile based systems with textile sensors for the monitoring of vital signs. Designers need to consider the real need and ease of use in terms of functionality, positioning, compatibility, maintenance and aftercare, and the clarity of the interface between the technology and the user.

4.7.2   Interconnections and communications

'The connection technology is the key to the integration and is actually the essence of wearable electronics since it was from this that the unity of electronics and textiles arises. The connection implements the power supply and data exchange between the individual components' (Roepert, 2006). Today, for individual connections, as well as crimped connections and solder joints, sewing techniques are also being researched and implemented in prototypes. A plug connection, such as the classical snap-fastener, is durable and can be used inconspicuously directly in the textile. As an alternative, wireless radio connections can be deployed.

4.7.3 Power

The power supply is a critical consideration to enable wearable technology to operate autonomously. 'Depending on the application, the challenge lies in achieving as long a period of autonomous operation as possible or else a high energy density. The dream is of a wearable electronics application which generates the power itself in the way automatic watches do.' Roepert (2006) explains that 'within expert circles, local power generation from motion or temperature is referred as energy harvesting. The human body generates in, as it were, "stand-by operation" about 100 watts of waste heat. Normal activity results in about 400 watts of power dissipation, and top athletes can emit as much as several kilowatts.' The harvesting of mechanical energy at the soles of the feet or from breathing or arm movements has been investigated and the basic techniques ascertained with fundamental work carried out at MIT. Researchers at Michigan Technological University have designed a strap made from a piezoelectric material that can harvest the free energy generated by the up-and-down movement of a hiker's pack and turn it into enough voltage to power small electrical devices. Flexible solar cells are now a source of energy for integration into rucsacs and jackets, with organic solar cells emerging to provide much more cost-effective solutions (Reware, 2004). Micro-fuel cells will also come to market that may be recharged in an instant and have higher energy density than traditional lithium-ion batteries. In practice, it is not currently possible to dispense with the good old conventional or rechargeable battery.

4.7.4 Design of technology interface

A major player in bringing wearable technology to market has been the UK company Eleksen, who would present to international audiences 'The future is now!' (Eleksen, 2008). To solve challenges experienced by garment manufacturers, Eleksen introduced 'eSystem' control electronics for a range of consumer electronics devices. 'The enabling components are three new sensors, incorporating SensorID technology, which allow designers to use any of 16 different 6-button layouts in their design with any eSystem control electronics' (Wilson, 2004). Eleksen has also developed a 'computer auxiliary display product' suitable for integration into bags, backpacks and clothing. This Wearable Display Module (WDM), based on Microsoft Windows

SlideShow, has a six centimetre 'Active TFT LCD screen, with one GB of storage for data files, controlled by an ElekTex interactive fabric touchpad' (Eleksen, 2008). Badira's DAB jacket incorporates a TV screen in the sleeve that 'can receive DMB (Digital multimedia broadcasting) TV programmes and DAB radio stations, as well as being able to play MP3s and show pictures and video clips.' It incorporates a new receiver from a Korean portable device partner, iRiver, and circuit technology, Ensonido, developed by the Frauenhofer Institute. Kamerflage has launched a context-sensitive display based on technology whereby 'digital cameras see a broader spectrum of light than the human eye. By rendering content in these wavelengths, for example, by printing onto a garment or banner, it is possible to create displays that are invisible to the naked eye but can be captured on a digital camera' (Wilson, 2004).

'The integration of plastic electronics with electronic paper displays is poised to revolutionise publishing as energy-intensive paper-making comes under pressure' (Anon, 2007a, p.8). An EPD (electronic paper display) possesses a paper-like high contrast appearance, ultra-low power consumption, and a thin, light form. It gives the viewer the experience of reading from paper while having the power of updatable information.' Electronic ink is ideally suited for EPDs because it uses a reflective technology which requires no front or backlight, is viewable under a wide range of lighting conditions, including direct sunlight, and requires no power to maintain an image. Current technology resembles paper in terms of contrast, brightness and viewing angle but has the thickness of a paperback book. Future versions will integrate E Ink's flex-ready products with plastic electronics which will be much closer to paper in form, being light, flexible, and rollable, from companies such as Plastic Logic, the Philips spin-off company Polymer Vision, and Epson. 'Independent experts forecast that plastic electronics will be a $30 billion industry by 2015 and could reach as much as $250 billion by 2025. 'Ultimately, electronic ink will permit almost any surface to become a display, bringing information out of the confines of the traditional devices and into the world around us' (Anon, 2007a, p.11).

4.8   Final prototype development

4.8.1   Final 3-D prototyping

Once the technical and aesthetic detail of the cut and proportion of the initial garment prototype has been fitted and amended, the functionality of the attached, encapsulated or embedded wearable technology must be verified. The clothing experts and the technical team, in liaison with the target end-user, should be involved in the fittings and decision making with regard to the overall comfort, fit, ease of movement, general functionality and appearance of the final design. The usability of the technology and the suitability of the design of the interface must be fit for purpose. The styling and technical specification of the whole garment system should be appropriate for the culture of the end-user in promoting a psychological 'feel good factor'. Any amendments must be recorded in terms of manufacturing techniques, detail and fit; with relevant alterations made to the garment pattern pieces, technical specifications, materials selection and placement, and the order of the make-up sequence.

4.8.2 Sealing sample

A pre-production prototype contract 'sealing sample' will be made to an agreed central standard size for the particular garment category. This sealing sample will be produced in the correct fabric quality in terms of measurement in relation to shrinkage control, weave or knit structure and yarn count and, ideally, should be produced in the location where it is destined to be produced in bulk. At this stage the fabric colour is not a concern. The standard will be affected by factors such as moisture content in the air, hardness of the water, the impact of local plant machinery and equipment and the expertise of the operators. Bulk fabric will be ordered once the customer has approved lab dips. The sealing sample replication must be standardised in size grade control, prior to release for production. The garment grades should be produced across the size range to verify pre-production sampling with measurements carried out to check consistency in relation to shrinkage, colour fastness, abrasion resistance, durability, etc. Once the central standard size has been verified, appropriate size grades will be developed, tested and confirmed. 'Before production of an electronic device can begin, an extensive, complete test of the system and its components is necessary. In addition, for commercially oriented [products], particular certifications (e.g. CE, FCC) are required. Only after this is it possible to order all the components and manufacture the systems' (Roepert, 2006). The final agreed sealing sample, with integrated wearable technology, may be released for production.

4.8.3 Product specification sheets

The sealing sample must be accompanied by clear diagrammatic working drawings of front, back and side views, with enlarged details, and must have clear references to materials and trims, measurements and the positioning of wearable technology. A sequence of make-up, with the relevant methods and finishes, must be explained in detail. Garment pattern pieces should be clearly marked with appropriate references to style name and number, name of component, size, grain lines, seam allowances, notches, direction of predominant stretch and positioning of wearable devices. Patterns must indicate the final textile specification in relation to garment cut and the placement or zoning of different qualities in relation to their functionality. New construction techniques, such as bonding and moulding, will demand precise guidance. 'Whatever the market sector or category of clothing, it is essential that the designer and pattern-cutter work together within the framework of the capabilities of the manufacturing available to them. They must be aware of the limitations of their technological and human resources' (Jenkyn Jones, 2005, p.59).

4.8.4 Manufacture

Until this new century, there had been limited modernisation of production methods within the garment manufacturing industry since the introduction of the electric sewing machine in 1921. The UK industry had factories remaining after the production of uniforms and work wear during the war, as well as smaller contract suppliers, while France and Italy have had family-run companies (Jenkyn Jones, 2005). By comparison with robotic production found in the automotive industry, the clothing trade has relied on many of the skills being carried out predominantly by female machinists, within both large-scale factory production and smaller scale contract work, supplemented by skilled outworkers. 'Manufacturers' may be referred to as 'vertical producers' that handle all operations from buying the fabric, designing, or buying in design, and making the garments. Many fashion design companies use 'Wholesalers' who design, buy materials, plan cutting, selling and delivery but do not make. Some small and often family-run companies provide a 'CMT' service (Cut, Make and Trim), often subcontracted by larger manufacturers at busy times. They do not produce patterns or take on design or sales risks. 'Contractors' produce small groups or 'stories' designed around a silhouette, fabric or perceived market demand for a store's 'private label'. 'Outworkers' are usually skilled women working from home, usually for independent designers who give them a set of cut bundles, threads and trimmings (Jenkyn Jones 2005).

In recent decades, large-scale manufacture has increasingly moved offshore to Asia, to take advantage of lower labour costs, with smaller production runs in countries within Eastern Europe, the Baltics, Turkey and North Africa. The sourcing of off-shore production has become less time consuming as a result of developments in computer-operated systems. Computer Aided Design (CAD) systems may be used to aid pattern cutting, size grading, lay planning, garment cutting and the dye cutting of selected mass-produced components. CAD speeds up production stages for what is referred to as 'Just In Time' (JIT) manufacturing. Garments may be tracked through manufacture to distribution and sales. The Universal Product Code (UPC) tracks style, size, and colour, while Electronic Point of Sale (EPOS) technology tracks sales and orders replacement garments. However, within the emerging and highly complex wearable electronics market, clever new ideas and technological innovation are becoming more important than keeping prices low. Sourcing them locally instead of globally will not affect their success if the public is willing to pay the necessary premium. But new technology may be easily replicated, which presents another reason for establishing production close to the nerve centre of the operation. 'It's a recognised fact that global sourcing blunts most competitive edges' (Shih, 2007).

4.9   Point of sale

Ultimately, it is the consumer who decides whether or not an innovation is worth spending extra money on. The design of promotional material must be sympathetic to the target audience to effectively 'sell' the concept. The designer must 'inform our impulse to buy and satisfy our real, and subliminal, needs' (Jenkyn Jones, 2005, p.29). Clear instructions for use, a FAQ, and a good product description facilitate the introduction of the product for distribution. This primarily applies to more complex products and particularly for distribution channels which do not yet have experience in the sale of wearable technology. The emerging market brings together retail cultures that were previously separate. In terms of convenience and service, the end-user must find what they are looking for quickly and easily – 'styling, quality and price points must be finely tuned.' Garments that lack hanger-appeal may be better merchandised in catalogues and on the internet. The functionality of the electronic device, maintenance and aftercare must be user-friendly. 'New developments in computer-aided manufacturing are resulting in a return to offers of individually personalised designs and made-to-measure items' (Jenkyn Jones, 2005, p.31).

4.10   End-of-life cycle

There is a major environmental and social impact across the life cycle of clothing, from the acquisition and processing of raw materials, fibre production, fabric manufacture and finishing, garment production, packaging, distribution, retail, use and end-of-life management. The UK's minister for Climate Change, Waste and Biodiversity, Joan Ruddock MP, has criticised the trend for 'unsustainable "fast fashion" that comes at a great cost because cheap clothes are going in the bin, and from there to landfill or into an incinerator after they have been worn once or twice' (Anon, 2007b, p.3). To initiate formal engagement with the clothing and fashion stakeholders, she launched the sustainable clothing roadmap at an event at Chatham House in London on 5th September 2007 (Maxwell, 2007). This has been described as '. . . part of a broader "sustainable consumption" project the government was engaged in, saying it would highlight and share best practice in this arena and celebrate success.' She said that there was a need for new thinking, with more ethical trading and a greater focus on sustainability. 'Action is being taken across the European Union as a whole on sustainability and our roadmap is part of that.' The major European sports trade event, ISPO, in Munich, 2008, has raised the profile of these issues within the sporting goods industry by hosting a one-day event sponsored by Volvo.

4.11   Future trends

4.11.1 Cross-disciplinary design

For the successful design development of wearable technology, the design critical paths of technical textiles and functional clothing and that of electronics cannot be considered separately. Stages in the process introduced above must be integrated across the disparate disciplines. 'Advanced tools and materials are making the designer's task even more complex. As a consequence we are starting to see some changes in design practice. A wide range of disciplines are being included in design teams. Design is no longer regarded as the task of just one person' (Braddock Clarke and O'Mahony, 2007). This statement is particularly pertinent in relation to the emerging cross-disciplinary design research and development of smart clothes and wearable technology. The clothing trade is still in the early stages of adopting advanced technology by comparison with disciplines such as automotive design or animation for the film industry. Computer-aided design and the digital revolution are now rapidly speeding up the clothing and textile design process from initial concept through creative, technical and scientific phases to product launch.

4.11.2 Sustainable fibres

An increased commitment to sustainable fibre development is impacting on the textile driven performance clothing market. Biopolymers are becoming big business, with a range of recycled and reduced-impact products on the market such as Unifi's Repreve, and American Fibres and Yarns Company's (AF&Y) Innova. Compared to other synthetics, AF&Y claim that Innova yarns require less energy to produce, have the least waste generation of synthetic fibres with none of the harmful industrial waste, do not require potentially harmful topical treatments for performance, and are recyclable. Teijin, Japan, is forming a joint venture with the biopolymer fibre producer Cargill, USA. Cargill's NatureWorks polylactic acid biopolymer (PLA), branded as Ingeo fibre, is produced from corn and has applications that include textiles for clothing and the home. Teijin, with partners, has the 'ECO CIRCLE®' programme that embraces the recycling of polyester fibres in a complete clothing system. DuPont also has a corn fibre branded as Serona® and the newly launched Cerenol®, from a DuPont/Tate and Lyle Bio Products joint venture. For outerwear, the waterproof, breathable, laminate brand Sympatex® has introduced Ecocycle-SL technology, which bonds the original 100% recyclable membrane to an upper material without using solvents. The upper material is made of recycled polyester and 100% reusable. Ecocycle-SL is claimed to be 'the world's first 100% recyclable laminate that delivers 100% performance.'

4.11.3 The digital critical path

As the computer interface has become more intuitive, the design of surface pattern for fashion and furnishing fabrics that traditionally would have demanded considerable time and resources and labour to realise, may now be produced in small quantities to unique designs. Designers can use a graphics tablet to draw and paint directly onto the computer, adding images directly from a digital camera as required. Digital ink-jet printing is changing the way designers conceive their imagery with the potential for different scales of unique prints in an indefinite number of colours. No screens are necessary as the designer can work directly from the computer screen onto the fabric. Images such as sketches, fabric swatches and trims can be scanned in and altered. Original drawings, collages, photographs and even three-dimensional objects can be scanned, further manipulated as desired and then printed as one-off art work, all-over designs, or complex repeat patterns (Jenkyn Jones, 2005, Braddock Clarke, 2007).

Traditional and present-day techniques may be combined to create a new aesthetic. If something has been made by hand, it generally feels more special, luxurious and individual. Surface treatments may be for decorative embellishment and/or for functional end use. In addition to the breadth of scope for print techniques and thermoplastic finishes, the manipulation of textiles may involve overlaying a mix of digital or hand embroidery and appliqué, combined with dyeing and finishing effects. Textile finishes can be engineered to enhance comfort and safety through 'zoned' features that offer protection, grip, thermal regulation, freedom of movement, support, visibility or camouflage and identification. Spacer fabrics can also be 3-D knit constructions for cushioning and support in sports and medical applications. Industrial and medical applications for digital embroidery have been developed by Julian Ellis of Ellis Developments (UK). Synthetic textiles, which are inherently thermoplastic, can be transformed through heat into new configurations which, on cooling, are permanently stable. Resins or other coatings can be added to provide additional performance characteristics such as rigidity. Three-dimensional effects such as embossing, crushing, pleating and moulding are thermoplastic properties used both for embellishment and for garment structure. Diverse patterning may be achieved through laser or ultrasound treatments that can etch or cut single or multi-layered materials.

4.11.4 E-commerce

The internet is now used throughout the textile and clothing design research and development chain. Designers may carry out market research on competitors and on the sourcing of information on materials, manufacturing methods and services. A calendar of events may be compiled with relevant sector specific trade fairs and conferences. The internet has informed and speeded up textile and clothing design development, sourcing and distribution. A two-way flow of Business to Consumer (B2C) information helps the design room identify the tastes of the market. Design and Visualisation Software saves labour and money in enabling designers and merchandisers to work together earlier in the design and merchandising cycle. The software tools can be used in a variety of ways, from ensuring that the garments fit the customer size range and make a good design story, to checking that the brand image is consistent in stores. The computer-aided visualisation systems help designers to show their ranges to wholesale buyers. Virtual shop layouts allow design co-ordination and in-store mood to be created without the need to buy items or have sales people handling the merchandise. Shop layouts may be distributed electronically. The digital design portfolio is now being followed by the virtual catwalk (Jenkyn Jones, 2005). The Internet provides a means for direct marketing and promotion, as a new development of mail order. Electronic shopping enables the virtual trying on of styles with unrestricted access 24/7, to otherwise inaccessible customers.

The Business to Business (B2B) Internet has helped speed up the supply chain for both fashion and functional garment production by the instant communication of orders and pattern data through electronic data exchange (EDI). Patterns and specifications can be transmitted across the world and easily downloaded and zoomed up to size (Jenkyn Jones, 2005). B2B enables members of the supply chain or the wholesale customer to contribute ideas, feedback and stock figures to the core company (Jenkyn Jones, 2005). The USA company [TC]^2 is recognised for the development of innovative technology in revolutionising the apparel and related textile goods supply chain ([TC]^2, 2008). Their body scanning technology was adopted for the recent sizing survey SizeUK (Sizemic, 2007). [TC]^2 proposes that 'tomorrow's manufacturers, brands, and retailers are going to thrive through the use of digital technologies and processes that are integrated across continents. From product development to delivery and logistics, these companies will implement systems that enable the shortest cycle time from concept to market, provide the most rapid and efficient response to consumer demand, and offer the best value for the investment' ([TC]^2, 2008).

4.11.5 Competitive edge

Global garment sourcing has resulted in manufacture becoming geographically separated from design, product development and marketing. Specifications, materials and trims are sent to off-shore locations to have different construction processes applied to them. The logistics of production planning, manufacturing and delivery are a vital aspect of efficient order fulfilment and satisfied customers. Now that smart technologies are emerging and becoming embedded rather than attached to textile-based products, the logistics of production become more complex. The demands of advancements in intelligent textiles, where electronics and sensors are combined in 'a soft and unobtrusive format', may open up some possibilities for companies to make new products in the US and Europe again. The North Face Advanced Products Manager, Winston Shih (2007), states 'What we want to do in the future starts with ideas, ideas that come from all the research and developments folks, all the people who are living the brand every day.' Clever new ideas and technological innovation are becoming more important than keeping prices low. 'Sourcing them locally instead of globally will not affect their success if the public is willing to pay the necessary premium. It's a recognised fact that global sourcing blunts most competitive edges.'

4.11.6 Mass-customisation

The Japanese designer Issey Miyake's A-POC series of whole garment technology (Vance, 2008) provides an innovative example of engineered textile as 'a piece of cloth', whereby the customer can make a choice of cutting lines to determine their desired proportions for details such as garment length and neck line. Mass-customisation and e-tailoring are entirely new approaches, where clothing is created from individual customer measurement data (sometimes generated by three-dimensional body scanners) and e-mailed directly to the factory. In these cases the supply chain is reversed and both fabric and garment can be made after the customer has tried on a sample and placed an order. In the past a fabric print had to be made in hundreds of metres in order to be commercially viable. Today, ink-jet printed fabric can be made in very small and unique quantities, allowing for customized prints and fashions. Three-dimensional knitted fabrics may be produced as seamless forms for medical supports to customer size requirements for leg ulcers, informed by body scanning (William Lee Innovation Centre, 2007). In future, designers and customers will have to collaborate in the design process. This is already occurring in sports shoes and sportswear and with men's suits and shirts. As the production of transient fashion has been moved off-shore, the EU is funding the Leapfrog research project into digital automated systems for the manufacture of value-added textile and clothing products (see the website http://www. leapfrog-eu.org/LeapfrogIP/main.asp).

4.12 References

ANON. (2007a) The death of paper? *Future Materials*, November 2007, ITtech.

ANON. (2007b) Minister calls for the end of fast fashion, *On Track: Business News. World Sports Activewear: Performance & Sports Materials* 13 (6) November/December.

BOGART (2006) *Bogart Lingerie*. [URI http://www.bogartlingerie.com/, accessed June 27th 2008]

BRADDOCK CLARKE, S.E. and O'MAHONY, M. (2007) *Techno Textiles: Revolutionary Fabrics for Fashion and Design*, Bk 2. London: Thames and Hudson.

ELEKSEN (2008) *ElekTex textile touchpads*. [URI http://www.eleksen.com/ ?page=products08/elektexproducts/index.asp, accessed May 19th 2008]

JANECKE, M. (2004) Innovative and Intelligent Textiles: Challenges, Chances and Perspectives. Director and Brand Manager Techtextil and Avantex Messe Frankfurt Gmbh, p.29, 30 [URI http://www.iafnet.com/files/iaf_04_presentations/ 09.%20JANECKE.pdf, accessed June 27th 2008]

JENKYN JONES, S. (2005) *Fashion Design*. London: Lawrence King Publishing, 2nd ed.

KILGUS, R. (2004) *Clothing Technology from Fibre to Fashion*. Nourney, Vollmer: Europa-Lehrmittel, 4th ed.

MAXWELL, D. (2007) On the Road to Sustainability. *Textiles*, 34(4).

MCCANN, J. (2005) Material requirements for the design of performance sportswear. In: Shishoo, R. (ed). *Textiles in Sport*. Cambridge: Woodhead Publications Ltd.

REWARE PRODUCTS (2004) The original juice bag, solar powered backpack. [URI http://www.rewarestore.com/bags.html, accessed June 27th 2008]

ROEPERT, A. (2006) *Wearable Technologies*, Interactive Wear AG, WSA, November/ December p.16–17.

SEW SYSTEMS (2008) *Designers and developers of sewing systems and machinery for the textile industry*. [URI http://www.sewsystems.com/, accessed May 19th 2008]

SHIH, W. (2007) Why global sourcing will survive 'homecoming' reaction. *World Sports Activewear: Performance & Sports Materials* 13 (6) November/ December.

SIZEMIC. (2007) SizeUK Data Set. [URI http://www.humanics-s.com/uk_natl_anthro_ sizing_info.pdf, accessed 8th Nov 2007]

[TC]² (2008) *3D body scanner and 3D scanning software*. [URI http://www.tc2.com/ what/bodyscan/index.html, accessed June 2nd 2008]

TWI (2006) *'ALTEX' - automated laser welding for textiles* [URI http://www.twi.co. uk/j32k/protected/band_3/crks3.html, accessed June 2nd 2008]

VANCE, L. (2008) *Issey Miyake's A-POC: A piece of cloth* [URI http://findarticles. com/p/articles/mi_qa3992/is_200105/ai_n8936766, accessed June 27th 2008]

WILLIAM LEE INNOVATION CENTRE (2007) *Scan2Knit* [URI http://www.k4i.org.uk/ about/Scan2Knit_Case_Study_20_08_07.pdf, accessed June 2nd 2008]

WILSON, A. (2004) 100 Innovations. *Future Materials*, Issue 6. [URI http://www. inteletex.com/adminfiles/PDF/fmBemis.pdf, accessed June 2nd 2008]

5

Designing smart clothing for the body

D. BRYSON, University of Derby, UK

Abstract: This chapter looks at the relationship between design and the body, addressing the implications of anatomical features 'Structures and shape', physiology 'How the body works' and psychological 'Our love hate relationship with clothing' and the impact these have when designing for the end-user's needs. For a design methodology to claim that it is inclusive or universal it must address the real needs of the body from the outset not design for the technology then place it in a softer shell. Smart clothing and wearable technology must add to the functionality of the human body not decrease it through poor or ill-considered design.

Key words: designing for the body, anatomy, physiology, psychological demands, variability design.

5.1   Introduction

This chapter looks at the relationship between design and the body, examining its anatomical, physiological and psychological demands and the impact these have when designing for the end-user's needs. For a design methodology to claim that it is inclusive or universal it must address the real needs of the body from the outset, not design for the technology then place it in a softer shell. Smart clothing and wearable technology must add to the functionality of the human body, not decrease it through poor or ill-considered design. The designer can draw inspiration from the human body, as well as those of other species, in respect of the many ways that form follows function.

A recent paper addressed the issue of unwearables (Bryson, 2007), the many products that have been designed without consideration for the human form whether deliberately as 'Art' or through neglecting to consider the body's needs. The issues considered here provide an overview, building on those raised by McCann in her thesis (1999) and developed further in Fig. 5.1: they point to areas for further research. It is not possible to cover all the issues in detail but this chapter is designed to indicate key directions for research and to take an holistic view of body–clothing interactions. Trying to reproduce *Gray's Anatomy* (Ellis, 2004) or Marieb (2006) on Anatomy and Physiology is inappropriate.

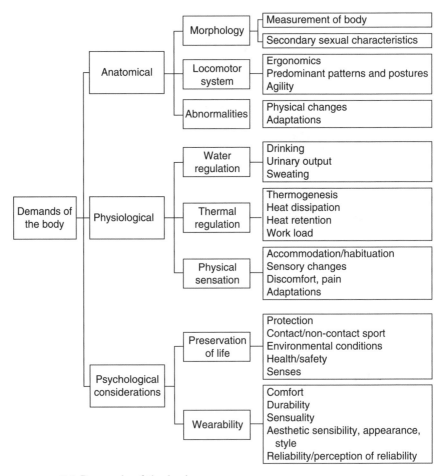

5.1 Demands of the body.

5.2 Anatomical, physiological and psychological considerations

5.2.1 What are the demands of the body?

The traditional tool for examining human needs is Abraham Maslow's hierarchy of human needs, see Fig. 5.2 (also page 376). As Huitt (2004) explains 'Maslow posited a hierarchy of human needs based on two groupings: deficiency needs and growth needs. Within the deficiency needs, each lower need must be met before moving to the next higher level. Once each of these needs has been satisfied, if at some future time a deficiency is detected, the individual will act to remove the deficiency' (Huitt, 2004).

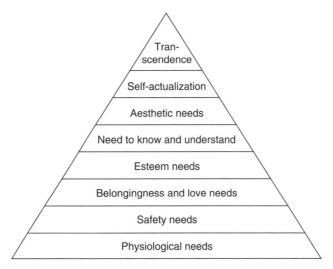

5.2 Maslow's Hierarchy of Needs.

In looking at body–clothing interactions, one of the issues not raised by Maslow's hierarchy of needs is anything to do with anatomy. This may be because, anatomically, we are what we are as defined by our DNA and how that is expressed. Anatomical factors have an impact but only in relation to other issues. If we wander around naked, anatomically this doesn't matter, but if we feel cold, a physiological phenomenon, we may decide to put on clothes.

It is the impact of our anatomy on our shape, more correctly termed morphology, and consequent physiological effects as a result of our environment (whether that is a building, the outdoors or our clothing) and what we do there, that affects us and our needs. Anything a designer does should take into account our anatomy and its impact on our physiological and safety needs.

Taking a simple example of a piece of wearable technology, many of us cannot do without, a pair of glasses. The relationships between the morphological and physiological considerations and the actual functional design are important. The width of the head, the position of the ears (normally we have one ear higher than the other), the distance between the ears and our nose which determines the length of the arms of the glasses, the position of the optical elements in relation to the position on the nose as we need to see through the glasses at the position of best visual acuity for the optics. The width of the nose will affect the position of the pads, the weight of the glass (or plastic) and frames will affect the skin over the nose and the bend of the arms or not by the ears will affect how the glasses stay on and their comfort.

Instantly, by looking at a piece of wearable technology that possibly goes back to early Rome (Ford, 1985), but definitely was around in the 13th century, we are presented with key issues related to the demands of the body.

Looking at Fig. 5.1 we can see the multifactorial implications of anatomy and physiology, as the design of glasses must address the following:

- Morphology – Measurement of the body
- Locomotor system – Do my glasses slip off when I look down or run?
- Abnormalities – Much more difficult if have only one ear; and adaptation – I have been wearing glasses so long, I have callouses behind my ear where the ends of the sidepieces go.
- Physical Sensation – Are they comfortable to wear for long periods or short periods? I may get used to them but might never really like wearing glasses.
- Discomfort – If I knock my glasses, I may tear the skin of my nose.
- Preservation of life – Glasses are important for me to see but also may be for protection against different types of light, welding, ultraviolet, for sports such as squash or racquet ball, where the ball is small enough to cause the loss of the eye as it fits into the orbit.
- Wearability – This is very much psychological as well as physiological and these have to be balanced. I may hate wearing spectacles but I know I need them. The more aesthetically pleasing, the more the psychological impact is overcome.

Already, by looking at a piece of technology that we take for granted, we are addressing six out of the eight demands of the body.

5.2.2 Getting to know your anatomy

This may seem a little strange but it can be useful, using an anatomy textbook, to feel where the various parts of the body are located. This can be done with muscles and bones, even some nerves and blood vessels. If you have ever had to take an examination, you know how annoying it is not being able to remember everything. The beauty about anatomy and human biology is that you take your own crib sheet into the exam each time, your body!

What you are doing by looking at the text and photographs and feeling where these part are is examining the 'surface anatomy' of the body. An easy example to try is to place your hands against the side of your face alongside your ears. Then, if you open and close your mouth you can feel the movement of your temperomandibular joint. In this way, with a human biology or anatomy textbook in front of you, you can feel a large number of parts of the body to get an idea of where they are and what they do.

5.2.3 Form follows function

This classic phrase 'form follows function' is very well known in design, but its origins are really in addressing the biological. If you look at the body, you can see that form follows function. The knee allows movement only in one direction. I can flex and extend, whereas a more mobile joint like the shoulder allows far greater movement; but with that comes a greater vulnerability – it is more likely to dislocate than the knee.

The range of movement possible at joints varies according to the formula stability = 1/mobility. The more mobility, the less stability; the more stability, the less mobility.

Looking at the inter-relationship between the body and smart clothing or wearable technology is an essential part of the design process, not just in terms of fit but how the clothes affect the body's ability to function. To fully understand the needs of the end-user, one should have some knowledge of human movement to inform garment cutting, and of human physiology to inform the types of material used in the design detail and material selection. It is also important to recognise that a sport or occupation can affect the body; indeed, that wearing anything can affect our function and, over a longer period, lead to physical changes in our body (see Table 5.1).

Table 5.1 Changes to the body as a result of a sport or occupation

Body change	Sport or occupation
Large shoulders	Canoeists and swimmers, especially for breaststroke and butterfly
Wiry/thin	Marathon runners, mountaineers and rock climbers. Female marathon runners have a tendency to be anorexic as it helps them cut time off their runs.
Large leg muscles	Rugby, squash, hockey, skiing. There can also be changes from novice to expert as the necessary muscles and bony attachments are developed with adapting to the rigours of the sport.
Lop-sided	Tennis players and javelin throwers tend to have one arm more developed than the other, their dominant arm.
Ears	Wearing glasses for long period, more spectacularly playing rugby, especially taking part in scrums.
Knees	Bursitis caused by being on knees, especially carpet-layers

5.2.4 Human variability

Until recent sizing surveys (National Size Survey, 2004), human variability and clothing design had been neglected. What is still neglected or forgotten is the extent of human variability. We readily acknowledge that each of us has a unique genetic coding through our genes, yet the connection of this to our actual form (what is termed phenotype) is glossed over. Taking a very simple example, my brother and I are both six feet tall, yet when we sit down my brother is taller than I am by about two inches. The difference is that I am longer in the leg; his legs are shorter but his back is longer. This classic difference between standing height and sitting height may not be visualised in a 3D scan but, like all things, the key is understanding and interpreting the data with an underlying knowledge of anatomy.

Measurement technology has been developed, which enables better design for basic human shape but these still rely on static models rather than on active measurement systems. These types of systems are being developed but will always rely on interpretation of figures and information into a design that can be constructed. Many measurement systems for product design take into account anthropometry, e.g. furniture design. Similarly, tailors producing what is now considered elite clothing rather than 'off the peg' take into account the personal requirements of the client.

Another example is looking at feet. Nine per cent of us have a squared foot where the big toe (Hallux) and second toe are identical in length; 22% have what is called a Morton's or Greek foot with the second toe larger than the hallux; leaving the majority, 69%, with a larger hallux, called an Egyptian foot (Magee, 1997). Yet rarely, unless they are hand-made, would this be accounted for in shoe design. There are different size and width fittings, but I could not go in and ask for a shoe for an Egyptian foot.

Sizing research has been examining more closely the mismatches between sizing systems and actual sizes. For example, findings from the Standards Australia sizing system by Honey and Olds (2007) showed that 'there is a significant mismatch between the real 3D shapes of young Australian women and the shapes assumed by the SA sizing system, and that there is a difference of more than two sizes when comparing reported dress sizes and best-fit clothing sizes.'

Stylios (2005) looks at how measurements and textile functionality can be integrated. '... the integration of four important research areas attempting to introduce new ways of designing, selling and producing garments; hence exploiting global internet retailing is an area of enormous interest because it challenges the conventional way of buying, selling, producing and distributing clothes. The four areas are as follows: (i) geometrical reconstruction of real humans; (ii) digital cloning of 3D face and body; (iii) virtual human locomotion; and (iv) cloth simulation' (p. 142).

There can be no better training for a clothing designer than to look at the human form when designing, whether it is through anatomical drawing or by gaining a knowledge of human biology. This allows designers to address 'real' people and their needs, rather than the stereotypical shape of the 'dummy'.

5.2.5 Accommodation and habituation

Physiologically, we can adapt to a piece of smart clothing or wearable technology over the years. Since the advent of the watch, we have become acclimatised to wearing one; so much so that many of us now walk around with multiple means of telling the time. For example, a wrist-watch, an iPod and a mobile phone. We have become used to wearing something on our wrist (although it is noticeable that watch wearing is decreasing).

Accommodation is the means by which our body becomes used to certain feelings or sensations. The attenuation of discomfort from wearing contact lenses is a classic example, as the wearer adapts so they can wear their lenses for longer and longer. To begin with, it might be recommended that they are worn only for two hours, then that is increased to three hours, and so on. The body adapts similarly to our clothing or the technology we wear; it may take time but we do adapt. Physiological comfort of biofunctional textiles has been addressed by Bartels (2006).

This process of accommodation and habituation is key to many items we wear. We break-in a new pair of shoes unless they are so ill-fitting that the blisters they cause mean that we cannot wear them. We feel good in a new shirt that fits perfectly and may wear it so often that it eventually wears out, but we also gain an awareness of the brands and labels that fit us. We buy a piece of wearable technology and get used to how it functions and how big a bulge it puts in our pocket or wherever we put it about us. However much the body adapts, it does have its limits and, whilst a certain amount of latitude is possible, there are often some things that limit uptake through lack of attention to the body's needs.

5.3   How smart is smart?

As we keep on covering it up, it is all too easy to forget what skin is capable of doing and how vital it is to our everyday lives. The textile industry has a long way to go before it can replace the skin; it is trying, but each material should be compared with the basic functionality of the human body and in particular 'the skin'. The skin's functions are detailed in Table 5.2.

In modern life, we almost always wear some form of second skin as we are expected to conform to the norm of wearing clothing. How smart that clothing is may well depend on our choice of clothes, our working

Table 5.2 Skin. The sales pitch as a design material and its functions

The sales pitch:	Functions:
I have the perfect material ready for you to use together with your design ideas. It has these properties:	Each part of the skin is involved in preserving the integrity of the body:
Perfect fit every time Stretches every way you want Non-slip surfaces where they are needed Waterproof and water permeable at the same time Helps control heat loss and heat gain	Protective coat – against friction, drying, ultraviolet light and bacterial invasion Minor metabolic role can produce vitamin D. Extensive sensory organ Excretes water and salts in the sweat Regulation of body temperature Stores materials such as glycogen and lipids

environment and how much we have to spend. However, that second skin can have either a negative or positive impact on our bodies.

The injuries that may occur to skin at the body–clothing interface come from:

- Contact with dyes in textiles – leading to allergic reactions. This has been noted over a long period of time and is not a new factor in design considerations (Farrell-Beck, 1998).
- Abrasion – from textiles or rubbing, especially if the fabric is coarse or the skin is abraded by mud, sand or grit. Once the skin is broken, infection is far more likely (Strauss, 1989).
- Continual rubbing from sporting activities – cycling and running may lead to sore areas more quickly than other sports.
- Sore areas are more likely to get infected, especially areas of thin skin in the groin and between digits, e.g. athlete's foot, a fungal infection, often between toes but can occur elsewhere on the body.
- The skin's reaction to continual rubbing may be to produce callouses which themselves may cause problems in fitting footwear and pain on being pressed.
- The deeper layers of muscle, tendon and bone may also be involved in the formation of callosities, which may require surgery.
- Sores may also arise under the breast through a combination of sweat and rubbing during activity.
- The pulp in the fingers may be injured, especially during rock climbing or occupations that involve climbing.
- Fingers may become infected or sore due to overwashing, prolonged periods in water or through long wearing of gloves that do not breathe, e.g. rubber or latex.

The modern smart materials with antimicrobials, antifungals, silver threads and other added functionalities will be equally as suitable or unsuitable for the skin to tolerate. The current literature, both medical (Hipler, 2006; Lee, 2006) and general (Elsner, 2008), is recognising the dangers these materials may lead to, but also their potential benefits in treatment (Senti, 2006; Wienforth, 2007) and means to analyse how the body functions (Priniotakis, 2005). 'Interaction of textiles with the specific immune system of skin is a rare event but may lead to allergic contact dermatitis. Electronic textiles and other smart textiles offer new areas of usage in health care and risk management but bear their own risks for allergies' (Wollina, 2006, p. 1).

5.4   Human–garment interaction

There is now wide understanding in the computer industry of what is termed Human–Computer Interactions (HCI), the design of interfaces, looking at the way humans interact with technology from using a coffee dispenser to working with an on-screen Graphical User Interface (GUI). This work now needs to move into the realm of garments to lead to an appreciation of what I am terming Human–Garment Interaction (HGI). In a traditional garment, this may include looking in more depth at the interaction between the user and garment in terms of sweating and the wicking properties of the garment. In a garment which includes technology, it would look at the usability of the technology for the end user, i.e. the user experience (Kuniavsky, 2003). This is already being examined in performance sportswear such as swimwear, where research into performance directly relates to garment technology, e.g. Speedo Fastskin.

An example of the first (traditional garment) is the rate of sweat production of the body (see Table 5.3). Humans have a natural system of thermoregulation. However, more and more smart clothing is being sold on the basis of its ability to cool the body, wick away sweat, and work with and support the body's natural functions. These garments do work but they too have their limits, especially in humid environments, and the way that the

Table 5.3 Heat produced and sweating with different types of activity

Activity level	Heat produced (watts)	Perspiration rate	
		(cm³/h)	(g/24 h)
Sleeping (cool, dry)	60–80	15–30	360–700
Walking 5 km/hr (comfort zone)	280–350	200–500	5–12 000
Hard physical work (hot, humid)	580–1045	400–1000	10–24 000
Maximum sweat rate (tolerated for a short time)	810–1160	1600	38 000 plus

user uses these garments and in what combinations with other garments can affect their function. A specifically designed layering system does function well, but all too often users will create their own layering system, not from one manufacturer's garments, but from several.

The second set of HGI, which will become more important, is the interaction of the user, not with traditional wearability, but with how the buttons/sensors of the garment work. How do we interact with the wearable technology? If there are buttons that link to technology, are there versions for right- and left-handed individuals? How do people use the technology in reality, not just what the manufacturer imagines will be its function? The roles that technology inhabit in our lives, are not going to be the same as they are now and the key to this is going to be integration into our lives, not as add-ons. So what is commonplace in the computer industry will need to be adopted by the clothing design industry.

5.5   Demands of the body and wearable technology

The relationship between the body and wearable technology needs to take into account all the factors listed in Fig. 5.1, as even the most basic wearable technology impacts on the body – from what we put in our pockets through to what we wear. The considerations again come down to issues of size and dexterity – how easily can we use an interface? Do we use it one-handed or two-handed? How large are the touch areas for numbers? What might be fine for me might not be for someone with Parkinson's disease or cerebral palsy, where fine motor control is difficult.

I tend to use my thumbs for my mobile phone as it is easier to hold that way, but I have to be very careful not to hit two or more pads at once. The iPod touch screen is bigger so I tend to hold it in one hand and use a single or two finger(s) to control and type. There is a need for creators of wearable technology to look beyond designing for themselves, to look at the implications for mass production and more importantly mass use.

I must admit that I have been solidly behind the thinking of wearable technology as something I carry as an extra or add-on. Something like the Burton jacket or Bristol Cyberjacket (Randell, 2002) doesn't grab my attention – how often would I actually be able to wear the jacket in an everyday setting going from home to work? For snowboarders, however, the idea worked and, in a trend-driven market, it was bought and now there is a far wider variety of jackets available with this device integration. 'My favorite thing about the Burton Amp is it's like having a portable DJ booth in your jacket,' said Keir Dillon, Burton Global Team Snowboarder. 'It makes changing songs before a pipe contest or on the lift so much easier – you don't have to take off your gloves or unzip anything' (Apple, 2003).

The possibilities in relation to wearable technology need to be examined more closely and attention paid to psychological considerations:

(i)   Is the technology attached to us or to what we are wearing?
(ii)   Do we carry the technology, e.g. in pockets or bag and look at it only when we want to?
(iii)   How do we view or get feedback from the technology, i.e. related to (i) and (ii), if carrying the hardwear, do we view the screen through glasses and a head up display?
(iv)   Is the technology embedded in an item of clothing or is the clothing the technology?
(v)   Is the hardwear embedded into our bodies surgically?

We need to willingly *want* to wear a piece of wearable technology, which means that it must be 'fit for purpose' and take into account the needs of the mass of people, not just the trend-driven younger market.

Prototypes that have sufficient functionality tend to look big and bulky (Baber, 2001; Rantanen, 2002), but the rapid development of entertainment technology and the functionality of their interfaces will progressively lead to smaller hardware configurations but more complex user interfaces and display methodologies. The entertainment world's perspective is very much pick-and-mix – 'I will have that portable device with this screen and these earphones, and when I want it in my car I will have this cable and at home I will link it up to my stereo speakers.' The direction is not just functionality but multifunctionality. The clearest direction of this is that electronics companies are moving away from cut-down versions of operating systems for mobile devices to full operating systems that are found on a laptop, with all the programmability inherent in that for linking into wireless networks.

How we use technology and the physical effects of wearing and using it have previously been addressed through fixed eye-track cameras that record eye movement while we use computer screens. The decrease in size of technology now means that we can attach the camera to the person and record what they do and how they do things, and this can then be analysed. Armed forces, police, firefighters and reporters are often now seen on television with wearable technology for recording what they see and feedback to them from live viewing (Nicol, 2006; Johnson, 2007).

Developments, for example, those by Tao (2002), looking at the integration into textiles of the equivalent of nerve fibres, is beginning to look more at the integration level of body-clothing and how the body may affect the clothing. This level of integration is similar to the body's use of proprioception. We know the position of our body, for example, leg, through feedback from mechanoreceptors in the muscles and tendons. Clothing is starting to develop these properties in that it will know where we are through GPS, it will know our body temperature and, should this become compromised due to hypothermia, it can transmit details of where we are and our condition to the emergency services.

5.6   References

APPLE (2003), *Burton and Apple Deliver the Burton Amp Jacket. World's First Electronic Jacket with Integrated iPod Control System.* [URI http://www.apple.com/pr/library/2003/jan/07burtonipod.html, accessed 15th April 2008]

BABER C. (2001), Wearable Computers: A Human Factors Review. *International Journal of Human–Computer Interaction*, 13(2), 123–145

BARTELS V.T. (2006), Physiological Comfort of Biofunctional Textiles. In: Hipler U.-C., Elsner P. (eds): *Biofunctional Textiles and the Skin. Curr Probl Dermatol.* Basel, Karger, 33: 51–66

BRYSON D. (2007), Unwearables. *AI & Society* 2007; 22(1): 25–35.

ELLIS H. (ed) (2004), *Gray's Anatomy: The Anatomical Basis of Clinical Practice* London: Churchill Livingstone, 39th rev edition.

ELSNEr P. (2008), *Textiles and the Skin.* [URI http://www.karger.com/gazette/67/Elsner/art_1.htm, accessed 14th April 2008]

FARRELL-BECK J. and CALLAN-NOBLE E. (1998), Textiles and apparel in the etiology of skin disease 1870–1914. *Int J Dermatol* 37: 309–314.

FORD B.J. (1985), *Single Lens. The Story of the Simple Microscope.* New York: Harper & Row Publishers.

HIPLER U.-C. and ELSNER P. (2006), Biofunctional textiles and the skin. *Current Problems in Dermatology*, Volume 33.

HONEY F. and OLDS T. (2007), The Standards Australia sizing system: Quantifying the mismatch. *Journal of Fashion Marketing and Management.* 11 (3): 320–331.

HUITT, W. (2004), *Maslow's hierarchy of needs. Educational Psychology Interactive.* Valdosta, GA: Valdosta State University. [URI http://chiron.valdosta.edu/whuitt/col/regsys/maslow.html, accessed 5th July 2005]

JOHNSON P. (2007), *Controlling the bandwidth of war* [URI http://news.bbc.co.uk/2/hi/americas/6657309.stm, accessed 15th April 2008].

KUNIAVSKY, M. (2003), *Observing the User Experience: A Practitioner's Guide to User Research.* San Francisco: Morgan Kaufmann Publishers.

LEE A. (ed) (2006), *Adverse Drug Reactions.* London: Pharmaceutical Press, 2Rev Ed edition.

MAGEE D.J. (1997), *Orthopedic Physical Assessment.* London: WB Saunders & Co. 3rd edition.

MARIEB E.N. (2006), *Human Anatomy and Physiology.* Addison Wesley.

MCCANN J. (1999), *Establishing the requirements for the design development performance sportswear.* MPhil Thesis, Derby: University of Derby.

NATIONAL SIZE SURVEY. (2004), *Size UK information* [URI http://www.size.org/SizeUKInformationV8.pdf accessed 15th April 2008].

NICOL M. (2006), *Storming towards the enemy . . . with their bayonets fixed and helmet-cameras rolling* [URI http://www.dailymail.co.uk/pages/live/articles/news/news.html?in_article_id=414531&in_page_id=1770, accessed 15th April 2008]

PRINIOTAKIS G., WESTBROEK P., VAN LAGENHOVE L. and KIEKENS P. (2005), An experimental simulation of human body behaviour during sweat production measured at textile electrodes, *International Journal of Clothing Science and Technology* 17 (3/4): 232–241.

RANDELL C. and MULLER H.L. (2002), The Well Mannered Wearable Computer. *Personal and Ubiquitous Computing* 6: 31–36.

RANTANEN J., IMPIO J., KARINSALO T. *et al.* (2002), Smart Clothing Prototype for the Arctic Environment. *Personal and Ubiquitous Computing* 6:3–16.

SENTI G., STEINMANN L.S., FISHCHER B. *et al.* (2006), Antimicrobial Silk Clothing in the Treatment of Atopic Dermatitis Proves Comparable to Topical Corticosteroid Treatment. *Dermatology* 213 (3): 224–227.

STRAUSS R.H., LEIZMAN D.J. LANESE R.R. and PARA M.F. (1989), Abrasive shirts may contribute to herpes gladiatorum among wrestlers. *New Eng J Med* 320(9): 589–9.

STYLIOS G. (2005), New measurement technologies for textiles and clothing. *International Journal of Clothing Science and Technology* 17 (3/4): 135–149.

TAO X. (2002), Nerves for smart clothing – optical fibre sensors and their responses. *International Journal of Clothing Science and Technology.* 14 (3/4): 157–168.

WIENFORTH F., LANDROCK A. SCHINDLER C. *et al.* (2007), Smart Textiles: A New Drug Delivery System for Symptomatic Treatment of a Common Cold. *J Clinical Pharmacology* 47: 653–659.

WOLLINA U., ABDEL-NASER M. and VERMA S. (2006), Skin Physiology and Textiles – Consideration of Basic Interactions. In: Hipler U.-C., Elsner P. (eds). *Biofunctional Textiles and the Skin, Curr Probl Dermatol.* 33: 1–16.

Part II

Materials and technologies for smart clothing

6

The influence of knitwear on smart wearables

F. SAIFEE, University of Wales Newport, UK

Abstract: This chapter conveys the application of principles from knitwear to the development of smart wearables. The two forms of knitting technology, weft and warp knitting, are considered in depth, from spacer fabrics to tubular and whole garment structures. Also reviewed is how traditional concepts of new and existing technologies in knitwear have evolved and how they have affected the design process, medical textiles, sportswear, inclusive design and use in industry.

Key words: weft, warp, spacer fabrics, whole garment, tubular.

6.1   Introduction

This chapter focuses on conveying the application of principles from knitwear to the development of smart wearables. Whilst the smart wearables industry is not new, it is one that is evolving at an incredible rate and one that is being influenced by a variety of disciplines. These disciplines come together to generate new thinking, yet traditional garment technology still holds strong. The traditional practices that encourage advancements in knitwear bear heavily on advancing the smart textiles industry, both aesthetically and strategically. Some of the key questions when addressing these issues include: What can principles of knitwear contribute to the smart textiles industry? What does smart fabric mean to knitwear? Who adds the smart – the designer, the product, the user – and is there a real need for it? Or are we just pandering to consumer trends? Looking at areas such as product marketing and brand innovation can help us to see what products are out on the market at present and how relevant they are to everyday needs and what they offer as innovation.

The chapter will also review the challenge of traditional concepts of new and existing technologies, how have they evolved and how they have affected the design process. How can knitwear products complement the design and manufacturing process? How does one arrive at the final garment? Has the thought process and the evolution of a product changed – for example, how can one arrive at the concept of a whole garment, which way round does it happen? Are traditional skills still the grounds to base new technologies on. One has to understand a number of issues outside of knitwear to innovate in this area. Traditional skills are required from a variety of people, which immediately calls for the need for a multi-disciplinary approach for both process and end artefact.

This chapter will address the background of knitwear and its unique qualities, giving an overview of the technical capabilities available in knitwear and its applications, what the current trends are and the user needs of smart fabrics, brand innovation and future trends.

6.2   Technical background of knitwear

'Technical textiles are determined by the type of fibres used, their finishing or the requirements arising from their intended use' (Stoll brochure, 2005).

This section will explore various knitwear structures. It is important to note that the end product of a garment is determined by all of the attributes mentioned in the quotation. Yarns will not be discussed in this chapter; there will be more of a focus on the structural properties of the knitted fabric. However, it would be impossible not to mention that, in particular relation to whole garment and seamless technology, stretch yarns play a huge role in the performance of the finished product. These yarns contribute to the stretch and the pressure of the garment, especially in sports and medical applications. This subject will be covered in a later chapter when woven structures are being discussed.

Table 6.1 shows the basic technical structures in knitwear and what these unique qualities allow, what challenges arise and the challenges that are faced because of the cross-over of technology and traditional methods. There are two basic forms of knitting technology: weft and warp knitting. Listed in Table 6.1 are the basic attributes of the two structures, which will be considered in more depth. Knitwear has many valuable attributes. It is a highly flexible structures that can be strengthened and will offer support

Table 6.1 Basic technical structures in knitwear, weft and warp knitting, and their unique qualities

	Machine	Structure	Attributes	Uses
Weft	Hand flat Circular	Fabric formed in loops in one continuous thread	Highly elastic and highly droppable, porous and comfortable	Ideal for apparel outer and under garments
Warp	Tricot Raschel	Structural threads of fabric running along the length of the fabric	Durable, run resistant, insulating properties, good surface resistance	Sportswear, automotive, industry, military

when various knit technologies and yarns are employed. The benefits of knitted fabrics are demonstrated in examples throughout this chapter.

6.2.1 Warp knitted fabrics

The main advantage of a warp knitted cloth is that, unlike weft knitted fabric, it is not easy to unravel. However, these fabrics are not as elastic as weft knitted fabrics. Both areas of knitwear have attributes that are applicable to many types of apparel and industrial uses. Warp fabrics are most often used in swimwear, intimate apparel and sports wear.

Spacer fabric is a type of warp knit (see Fig. 6.1). It is a 3D construction, which is a sandwich, constructed of two layers of fabrics with connecting yarns spaced from approximately 5–6 mm apart. These fabrics are widely used in the athletic shoe market and are being introduced as body cloth face fabrics in cars because of their durability. They have the potential to replace foam in seats and headliners but there are issues with weight and cost. These fabrics are knitted on warp knitting machines such as the double bar Rachel and Tricot machines and they can be yarn or piece dyed.

They have significant insulating properties, which make them ideal for use with electronics. Subash Anand at Bolton University described his recent research at the Technical Textiles Conference in Manchester (July 2005). He commented upon the insertion of cables through the centre of a warp knitted fabric, the cables then being attached to a power source which can carry electricity to provide heat or activate communication tools.

Warp knitwear is becoming increasingly popular in apparel and sportswear clothing due its high performance attributes and developments enabling the fabrics to be made much finer. Nike® use warp fabrics in their garments because of the excellent moisture wicking properties which are essential to high performing athletes. In all sports, the temperature of the body and muscles need to be kept at optimum level to operate to their full potential.

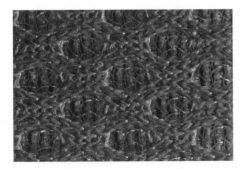

6.1 Warp knitted fabric (photograph – David Bryson).

6.2.2 Weft knitted fabrics

A weft knitted fabric has more stretch in the course (horizontal direction of the fabric) than in the wale (vertical direction) (see Fig. 6.2). The benefit of a weft knit is that, in garment construction, maximum stretch can be placed around the body; vertically, the garment has minimal stretch. When a fabric is constructed, it often appears bulkier than the yarn that it has been knitted from. This is because the fabric is composed mainly of air and space, which leads to its being air and water permeable. However, the bulkiness also imparts good insulating properties.

Plain fabric

Plain fabric is the most common weft knitted fabric and is produced by a wide variety of knitting machinery discussed in Section 6.3 concerning three-dimensional fabrics (see Fig. 6.3a).

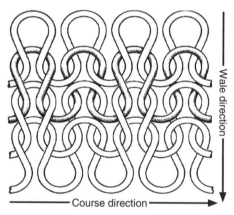

6.2 Weft direction of knitting.

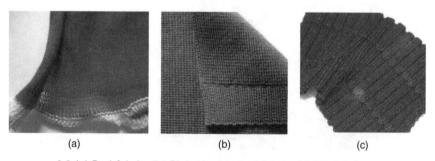

(a)   (b)   (c)

6.3 (a) Purl fabric, (b) Plain/double bed fabric, (c) Rib fabric.

Purl fabric (double bed)

As identified in Table 6.2, a purl fabric has notable insulation properties. It is much more stable than a plain fabric as it does not have a tendency to curl at the edges (see Fig. 6.3b).

Rib fabric

Rib fabrics are constructed with alternate face and back loops touching one another, as can be seen in Fig. 6.3c. This is a sample of a 4 × 4 rib fabric.

The above classifications are all absolute and all weft knitted fabrics can be categorised into one of these three classes. It is possible to modify and displace the loops but this does not alter the basis of the three main categories.

6.3   Three-dimensional fabrics

The following technologies all rely on the weft knitting principle and are used to create fabrics in a three-dimensional form:

Table 6.2 Sub-divisions in weft knitted fabrics

	Machine	Structure	Attributes	Uses
Plain	Uses one bed on a knitting machine.	Each loop is pulled through the previously knitted loop in the same direction, has a different appearance on each side.	Course direction extends twice the wale direction. Maximum stretch around the body. Excellent drapability.	Undergarments, intimate apparel, whole garment.
Purl (Double bed)	Uses both beds on a knitting machine.	Alternate courses knitting face and back loops. Looks the same on front and reverse of fabric.	More wale-wise stretch than a plain fabric. Bulkier than a plain fabric.	Sweaters, excellent thermal insulation properties.
Rib	Uses both beds on a knitting machine with a needle set up, 1 × 1, 2 × 2 etc.	Loops formed in the opposite direction of each other. Most common rib is 1 × 1.	Clings to the shape of the body yet is capable of stretch when required.	Waistbands, cuffs and collars.

6.3.1 Tubular fabrics

Tubular fabric is produced on a circular knitting machine. The threads run continuously around the fabric. Needles are arranged on the machine in a form of a circle and are knitted in the weft direction. There are four types of circular knitting – Run resistant circular knitting (applications include underwear, swimwear); Tuck stitch circular knit (used for underwear and outerwear); Ribbed circular knit (swimsuits, underwear and men's under shirts); and Double knits and inter lock. Many undergarments are made from tubular fabrics as it is fast and effective and requires very little finishing.

Traditionally, tubular fabrics have had a large application in the hosiery industry and still do. However, there has been a revolution in streamlined knitwear and there has been much innovation and re branding of this traditional fabric as 'seamless', which has helped to create a new demand. Figure 6.4 shows a seamless undergarment. It has no side seams and it is knitted on a circular Santoni machine. This type of product will increasingly replace cut-and-sew products as elasticity zones can be controlled, areas of single jersey can be built-in with three dimensions and ribbing can be incorporated. This can create shaping in the garment without any or with very little sewing required.

6.3.2 Whole garment

One stage on from circular seamless apparel is whole-garment production. The pattern is created using Computer Aided Design (CAD) and then sent to the electronic knitting machine, which reads the program and then manufactures the garment. This process allows a complete piece, including sleeves, to be knitted as one whole. This shortens the post-knitting process as all the shaping can be done without expensive tailoring and waste – the actual garment comes off the machine as a ready-made garment.

6.4 Seamless undergarment knitted on a Santoni machine (photograph – David Bryson).

6.3.3 Integral shaping

Integral shaping in combination with the whole garment process can add elements such as darting, pocket details, buttonholes and pleats (see the Shima Seiki website http://www.shimaseiki.coijp/collectione.html for examples of garments made using integral shaping techniques). If an integral section is knitted in Lycra® (covered elastic yarn), this naturally pulls in the section, simulating a dart on a conventional pattern. Here is an example of how traditional knowledge, technology and innovative thinking have come together (the knowledge of a stitch structure and yarn, coupled with forward thinking).

6.3.4 Variable knits/body mapping

Another type of integral shaping is called 'Variable knits' and this is used in conjunction with 'body mapping'. Patagonia has developed a garment using this technique (see http://www.expo.planetmountain.com/pages/imgs_prod/195.jpg). Body mapping is a technique that is able to detect areas on the body that require extra padding, moisture absorption or heat insulation. Certain knitwear techniques are employed to add various structures to maximise these requirements on the selected body parts. The Patagonia garment has a rib around the neck and cuffs for a snug fit, and extra insulation and jersey knit along the upper arms balances warmth and breathability. The seamless nature of the garment also reduces any chaffing where seams would originally be. This garment is ideal because of the moisture wicking properties and snug fit, which are essential in many outdoor sports. Moisture wicked fabrics draw moisture away from the body due to the micro-construction of the yarn. The fabric moves the perspiration away from the body to the outer surface where it can evaporate quickly. This type of knit also assists the wearer in maintaining a constant body temperature, which is vital in extreme sports due to varying climate.

6.3.5 Basic joining processes in knitwear

As important as the structure of a fabric is, the finishing of a garment can affect its performance greatly. Some needs that should to be addressed when choosing an appropriate finishing for a knitted fabric are suitability of fabric, durability, aesthetics, available technology and price point. Below are some examples of finishing methods in knitwear.

Unit method

The unit is knitted in one piece (as discussed above). The benefits of garments knitted whole or in units are reduced waste, as pieces can be knitted to shape, and shorter manufacturing processes, making it more cost-effective.

Linking/component method

Linking is a unique process to knitwear as it has been specifically designed for joining knitwear (see http://www.b-hague.co.uk/knitting/Ruth-Wood-3.jpg.). This method involves knitting the component pieces of a garment to shape, then linking them together. Fully fashioned knitwear is made this way. Linking is the most common method of seaming the pieces together without affecting the stretch of the fabric and it is achieved by joining two fabrics together, loop by loop, along the wale of the fabric. This method is commonly used in commercially high-end garments, e.g. sweaters. It is also used in side seams and is a common way of joining cuffs and collars to a garment.

Cut and sew method

This method can be used for weft, warp and circular knitted fabrics and it involves knitting the fabric, cutting out the shapes of the garment components and sewing them together. The method is commonly used in cheap mass production. The seams do not have as much stretch as a linked seam would have and the finish is not as refined.

Ultrasonic bonding

Ultrasonically bonded fabrics were originally used for intimate apparel. They are suitable for warp knitted fabrics used in sports wear. Specialised sports companies such as Nike® employ this method to bond garments together. Seams are joined with ultrasonic radio frequency. This technique can be used to cut, seal and sew synthetic materials, as long as they have at least 60% thermoplastic components. Woven/non-woven knits, coated materials and laminates can all be ultrasonically processed. The method provides a durable lightweight fabric and has many applications, especially in sportswear and outerwear. This is an advanced development in the joining process of knitwear. It could be compared to the linking technique, as the objective is to create an invisible seam that does not alter the stretch of the fabric.

6.4   Body scanning and texture mapping

Knitted garments are made by various methods of conventional pattern making; however, there have been major developments in the pattern making process.

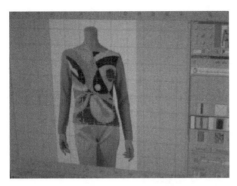

6.5 Body mapping with SDS-ONE® system.

Stoll (knitting machine manufacturer) have created a system called 'Knit and Wear'. Parts of the body can be scanned in; then the data is fed into a CAD system and transmitted to a computerised knitting machine, which can knit the 'bespoke' item.

Shima Seiki have another system for apparel design which was demonstrated at the 2006 Pitti Filati trade fair (see http://www.shimaseiki.co.jp/product.cge/sdsoneseriese.html). It is a CAD CAM system called SDS-ONE® (see http://www.shimaseiki.co.jp/product.cge/sdsoneseriese.html) whereby the designer can create patterns virtually using a method called texture mapping (see Fig. 6.5). The system works by scanning in a piece of yarn and creating a virtual fabric from this. The apparel design is created using the mesh making technique and loop (stitch) simulation. One can also use a yarn or fabric from the archive. Once the fabric has been designed, it is then 'mapped' onto a virtual 3D model. The pattern pieces are available to print out to the panel to view the actual proportions. The design process is quick and simple as all the designs are stored, so modifications such as pattern making, grading and marking can be easily made.

The method with which the garments are made is vastly different to the traditional process. Instead of feeding a yarn directly to the machine, a virtual simulation of it is scanned and sent to the computer. It could be argued that there is the loss of the tangible quality, which could lead to a misunderstanding of how a fabric works in terms of drapeability, which could lead to making ill-fitting clothes with little design imagination. However, the SDS-ONE® system can be enhanced by keeping some traditional methods in the process at the beginning rather than at the end; for example, the pattern can be printed out and a toile can be mocked up to understand the final fit and any adjustments can be made accordingly.

6.5   Application of knitwear trends to the smart wearable industry

So far we have understood the technical nature of knitwear, having described brief examples of how knitwear can be made in industry. It would be easy to become carried away with the promise of technology whilst forgetting the actual context. The next section provides examples of the applications of knitwear within the industry:

6.5.1 Medical applications

UMIST body scanning/one piece garment

Compression garments are normally provided in standard sizes rather than made to measure. Such garments are designed to apply pressure which forces lymphatic fluid towards a functioning lymph node, and they are the most effective form of prevention and treatment currently available. The William Lee centre based in Manchester (Knowledge for Innovation Conference, 2008) has created a system called Scan2Knit® (see Fig. 6.6a). This

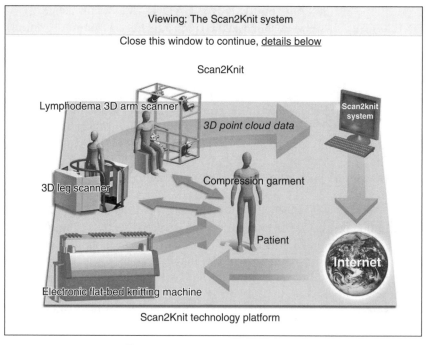

6.6a Scan2knit® – Knowledge for Innovation. Case study: Innovation – Scan2Knit Technology. [http://www.k4i.org.uk/about/Scan2Knit_Case_Study_20_08_07.pdf, accessed May 12th 2008].

system can knit bespoke medical compression garments for people who have developed severe ulcers through age, infection or injury. The aim of this project is to improve the effectiveness of compression bands by using a specialised computer-assisted manufacturing process. The Triform® scanner (see Fig. 6.6b) scans the leg using a normal white light to scan. It takes approximately 6 seconds per scan and is completely safe for patients as no special equipment or protection is required. The technology is very easy to use and apply with some training. A 3D model is constructed from scans of the affected leg and this model is fed into a knitting machine to produce a seamless garment. Much of this process occurs online, making easy communication possible between the hospital (where scans are conducted) and the garment manufacturer. The whole garment technique is ideal in this instance as there are no seams to create any interference to the ailment.

When the garment is implemented commercially, its first impact will be to save the NHS 1 to 1.2 million pounds a year in nursing time alone within the UK Venous Ulcer units and bring great long-term benefits of comfort and compliance for individual end users (see Fig. 6.7).

6.6b Triform® Leg scanner (courtesy of William Lee Innovation Centre, UMIST, Manchester, Advanced Therapeutic Materials, Coventry).

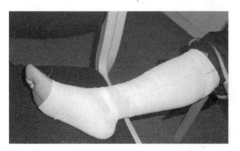

6.7 Compression sock courtesy of William Lee Innovation Centre, UMIST, Manchester. [http://www.fibrestructures.com/profiles/tilak_res. pdf].

Design for disability

The term 'Smart Wearable' seems to embrace many aspects of design, textiles and applications. Design for disability is an area that has not been so easily embraced by the fashion-conscious design world. Even though a certain amount of knowledge and attention is needed to design garments for specific needs, there is no reason why there cannot be a cross-over into the main stream. Smart wearables are 'smart' for the reason that the thought processes encourage consideration of user needs, which results in an easy transition (overlap) from mainstream fashion to inclusive design. Someone who has explored this concept is Caterina Radvan. Caterina is a knitwear designer by profession, currently studying for a PhD at the London College of Fashion. Her research is concerned with using the inherent unique qualities of seamless knitting to design and create fashionable womenswear within the inclusive design discipline (see Fig. 6.8). This enables disabled

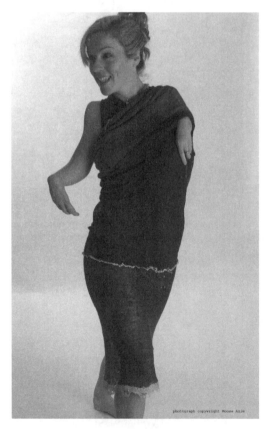

6.8 Seamless clothing for inclusive design (courtesy of Caterina Radvan).

women access to the full retail therapy experience that is available to non-disabled women and allows them equal access utilising clothing as a means of self-expression. The use of seamlessness allows for the subversion of normal ideas of back, front and left–right symmetry, and for the distorting of the garment shape; the rules governing the placing of seams is completely avoided. When worn, the garments produce drape in unexpected areas on the body, depending on the shape of the body wearing them (see Fig. 6.9). When this is applied to the disabled or unconventional body shape, an interesting and attractive look is created using the body form underneath. In this way disabled women are able to dress without the need or desire to camouflage or hide their unconventional shape, thus allowing for freedom of self-expression.

photograph copywright Moose Azim

6.9 Seamless clothing for inclusive design (courtesy of Caterina Radvan).

6.10 Clothing for disability (Daijeero and Saifee) [http://assets.aarp.org/ www.aarp.org_/articles/international/CassimPPT.pdf] page 35.

6.5.2 Easy wearability

Saifee and Daijeero have designed a prototype of a garment addressing the issue of the ease of putting it on (see Fig. 6.10). Research with user groups revealed that traditional garments do not allow easy access, especially for people with limited dexterity. For disparate body shapes, this often means having to have expensive bespoke clothing made. This technology addresses both these problems. The integral shaping in the garment has allowed expandable gussets to be created, allowing the wearer easy access with minimal friction and effort. All of these garments are adaptable and can be worn by people without any disability, thus challenging the concept of the 'conventional' shape which could benefit the image-obsessed world we live in.

6.5.3 Well being

There are textiles that encourage well being; for example, through aroma-therapies, vitamin delivery systems and skin calming properties. Diana Irani is a designer who completed her M Phil in Constructed Textiles at the Royal College of Art (London). Whilst studying she explored the above-mentioned aspects. Since then she has an established her own fashion label and will soon be launching a new range of clothing called Re_Medi, which will be made from specially developed fabrics that release a controlled dose of a complementary or homeopathic medicine into the bloodstream through the skin while they are worn. Irani likens the system to a nicotine patch, a widely used and effective means of administering an active ingredient by absorption. The simplicity of the system ('trans-dermal drug delivery') belies the imposing sound of its name. The wearer of the garment just chooses what type and dosage of natural remedy they want, and then the medicine releases itself from their clothing as soon as it comes into contact with their skin, by responding to the natural pH of their bodies. With NESTA's Creative Pioneer award, Diana will be able to devote the launch of her business to developing a working prototype of the Re_Medi range (see http://www.nesta.org.uk/diana-irani-2C-re-medi/). At the heart of her development and design will be a belief in the value of herbal remedies to treat a whole array of stress-related illnesses, including insomnia, eczema, headaches, and fatigue.

From the projects outlined above, the research found that it was important to the wearer that they 'fit in' – They did not want to draw attention to themselves. However, if the garment had designer appeal, then the user was not afraid to make a statement. In this area it seems that careful marketing plays a key role in taking the product to market.

6.5.4 Sportswear

Sportswear is an area that has been able to exploit the nature of knitwear technology to create high performance fabrics and garments. Seamless garments offer streamlined silhouettes which make the garment seem like a second skin. An added feature to the example below is the insertion of a sensor to measure heart rate.

Figures 6.11 a and b show the heart sensing bra recently launched by Textronics. This is an example of a seamless, fully integrated garment. It is knitted on a Santoni machine. The sensors are fully integrated into the fabric of the garment and the transmitter snaps onto the special compartment in the bra. The sensors in the fabric worn next to the skin pick up the heart rate of the wearer from the surface of the skin and relay it to the transmitter in the front of the bra.

The seamless nature of the bra ensures maximum comfort. The integration of the sensors knitted into the fabric enable it to pick up a better signal as it is worn so close to the surface of the skin, which reduces interference. This technology also allows elasticity to be employed within areas of the garment, to make sure the sensors are tight for a good fit. The bra is made from a quick-drying nylon and Lycra®, which also helps to keep the shape due to its stretch qualities. This example demonstrates how an intelligent

(a)

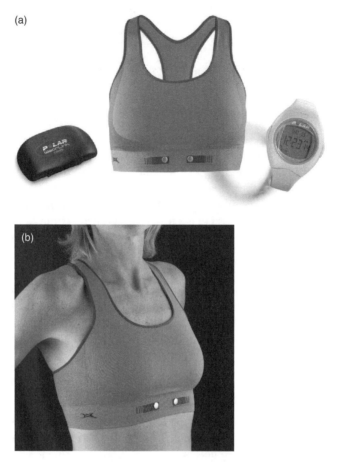

(b)

6.11 (a) Numetrex bra watch and link (courtesy of Textronics),
(b) Numetrex bra (courtesy of Textronics).

function has been added to the garment by using technology outside of
knitwear. There is a huge potential in applying this type of technology
to people who have a heart condition. They can be monitored throughout
the day in the comfort of their home. The technology can be linked to
a database which can be monitored remotely by a healthcare professional and alert emergency services if the heart rate drops below a certain
level.

Another advantage of the seamless technology is that it is possible to
have more than one layer in the process, which is ideal for taking moisture
away from the body and adding insulating properties.

6.5.5 Industrial

The machinery that has been developed to date for knitwear can offer unlimited possibilities. Some machines are able to process heavy duty and metal yarns; others create an open structure surface, considerable flexibility due to their adaptability of stitch parameters, and a high level of compressibility. This makes them ideal to create fabrics for use in industry.

An example of a high strength yarn is the aramid fibre Kevlar®. The characteristics of aramid fibres are high strength, stiffness, low moisture regain and low density. They are commonly used in fireman's turnout coats and other industry and military clothing that requires strength, stiffness and flame resistance.

6.6   Relevance of knitwear to smart wearables, lifestyle trends and brand innovation

6.6.1 Smart wearables

Clothing as a medium for technology is ideal because it is versatile, mobile, universal and adaptable. From research it is apparent that smart wearables are a mixture of innovations that challenge traditional concepts by use of a multi-disciplinary approach. As well as being an innovation in technology, smart wearables also reflect changing lifestyle. Knitwear is an ideal medium for smart wearables as the technology can be integrated within the structure of the fabric, which make it ideal for travelling. Current lifestyle trends demand products that are versatile as well as mobile, and knitwear as a medium can suit such requirements. It is suitable to carry these type of concepts because it is a fabric that lends itself to comfort. For example, seamless technology offers different structures that can hug and stretch where need be and have increased breathability when worn next to the skin because of the yarn and stitch structure of the knitted fabrics.

6.6.2 Lifestyle trends

The company Invista are currently developing a product with scent capsules imbedded in the fibre. The finished garment is made to smell of freshly washed clothes. The scent lasts for up to 20 washes and there are ways to revive the smell by giving a special wash treatment. Charlotte Mills of product development at Invista commented that knitted garments are ideal for this type of product as 'knitwear tends to be worn as an outer garment so it requires less washing, which keep the capsules inert and the fact that they are seasonal – sweaters tend to be folded away during a period of time during the year which also keeps the scent fresh.' Another issue that is open for discussion is the environmental aspect: if the garment is perceived to need less washing, this has a beneficial impact on the environment in saving water and electricity.

6.6.3 Brand innovation in knitwear

'Consumers respond positively to a range of benefits from aromatherapy to skin toning and cellulite reduction and are open to receiving these benefits through their clothing' (Charlotte Mills, Invista, July 2006).

From researching into the market it seems that any new product that comes to the market needs to have addressed certain criteria as discussed earlier in this chapter, user needs being key. However, sometimes the user may need a little coaxing to understand his needs and this is where brand innovation plays a very important part. In conversation with Diana Irani, she stated that a product with 'special' qualities needs to be introduced slowly; the consumer needs to build a trust with the brand before buying into it. Diana has already established her knitted accessories brand 'BLANK' and since then has launched her Re Medi line which is being received with a positive response (Section 6.5.3). This reiterates how careful study of what the consumer wants is vital when launching new technologies.

At this point in the chapter we acknowledge that the transition from a technology concept to the product is a culmination of the user need, the lifestyle, what practicalities are involved and how the product is marketed. Using the case studies of Invista and Dian Irani, it seems that knitwear is being seen as more than a fabric with conventional applications but, as discussed previously in the chapter, it has become a medium, a communication tool, communicating what kind of lifestyle one leads. Knitwear, in conjunction with other disciplines, has spectacularly enhanced many concepts with the aid of technologies seen in the examples that have been described.

6.7   Future trends

Looking at technology that is already available could help us to see a glimpse into the future. The Scan2Knit® technology could be a way to buy clothes in the future. One could have a scanner set up in the bedroom. Once the appropriate part of the body has been scanned and sent to a virtual stylist in a virtual boutique who can take you through an extensive catalogue of fabrics, colours, shapes and finishes to choose from, the information can be translated through the web to a receiving manufacturing unit, which then delivers to your doorstep.

Designing bespoke clothing from one's bedroom is already happening, with brands such as Nike® offering a range of options such as colour, fabric and style, to design a completely individual shoe over the internet. One may find in the future that mass production will increasingly be replaced due to the desire to create an individual statement.

From various field researches it has been found that many knitwear manufacturers in the UK are looking towards developing specialist products. For example, Stoll are developing products and technology for the medical field as described in Section 6.4. As the mass-market industry is becoming increasingly competitive, looking towards innovative products is one way forward. Companies have the knowledge and technology to innovate new thinking.

6.8   Conclusion

In the past, the perception of knitwear has been reliable, comfortable and old fashioned. As we have seen in many of the applications exemplified, this is not the case in contemporary culture. However, this notion has helped to bring new products into the market as the consumer is already aware of the steadfast qualities that knitwear can offer and can pave the way for brand innovation. It also shows that, as a discipline, knitting is capable of carrying new technologies that sit well with the old. An example of this is in the factories and design rooms alike; sampling is still done on a manual machine and then advanced using new technology.

As lifestyles are changing rapidly, user needs have to be addressed. The design process in garment design is already changing as consumers have more control of what is available to them. As has been described, there is the possibility of creating made-to-measure clothing using mass manufacturing tools. Eventually there could be a new trend for upmarket garments where whole garment techniques could provide high trend, good-quality garments at low cost for high-end retail prices.

Having addressed the ever-changing population and their role in product development, it is seen that the designers are still an important factor in the design process. The cross-over and collaboration from other disciplines will ensure that future products will embody functions as well as aesthetics. New materials will cross-over with traditional technologies, which promises exciting wearables to suit needs of a wider cross-section of the population.

6.9   Other areas of interest

Areas of interest that have not been studied in this chapter are the ageing population – as the average life-span increases, so will our demands for a better quality of life Products will need to offer good design and fulfil desires of what our minds want but are restricted from having by physical ability.

Environmental issues have also not been discussed in depth – for example, the life-span of a smart piece of clothing and its end use. Product after-life is an area which has room for discussion as new technologies require specialist yarns and fabrics for high performance. How is the environment accounted for and what effect will this have on our ever-depleting energy sources?

6.10 Bibliography

SARAH E BRADOCK CLARKE & MARIE O'MAHONEY, (2002). 'Sports tech'. Thames and Hudson.

BRADLEY QUINN, (2002). 'Techno Fashion'. Berg Publishers.

SARAH E BRADOCK CLARKE & MARIE O'MAHONEY, (2007). 'Techno Fashion'. Thames and Hudson.

KNOWLEDGE FOR INNOVATION. Case study: Innovation – Scan2Knit Technology. [URI http://www.k4i.org.uk/about/Scan2Knit_Case_Study_20_08_07.pdf, accessed May 12th 2008].

William Lee Centre. Manchester. 20th July 2005. 'Technical Textiles: The Innovative approach'.

7

Woven structures and their impact on the function and performance of smart clothing

L. THOMAS, Woven Textiles Consultant, UK

Abstract: This chapter discusses how a woven structure has great impact on the functionality of a textile. The importance of understanding an end-user's needs is highlighted as an important starting point to developing a new woven fabric. Key characteristics of the primary weave structures are discussed and applied to different performance needs. The chapter concludes by discussing some thoughts around future trends inspired by craft weaving.

Key words: woven structure, weaving, textiles.

7.1   Introduction

Smart Clothes and Wearable Technology are becoming ever more common as we demand more of our attire. Fabrics are the raw material, but are not passive components of garment construction. They need to be fit for purpose and of integral design to the garment. Whether they are traditional or technical, they must perform. Fabrics for Smart Clothes and Wearable Technology are invariably described as 'technical' or 'intelligent'; textiles that can be described as having 'added value'. They are designed and manufactured to perform a particular function, rather than for aesthetic appeal.

Most of the characteristics of technical textiles are as a result of the technicalities of the fibre, yarn or finishing process. These fabrics, whether they be knitted, printed, stitched or woven, are often a vehicle purely for these characteristics. However, the chosen fabric construction naturally has a great impact on the functionality of the textile. There are numerous methods of constructing a cloth, each with its own merits. At its most basic, a knit structure usually has an inherent lateral and longitudinal stretch whilst a woven fabric is far more stable and robust. Of all the textile construction techniques involving the interlacing of yarns, a woven structure offers the greatest strength and stability.

The art of weaving is one of our most ancient crafts, with examples of the evolution of woven textiles in nearly all cultures. The basic premise of weaving a fabric remains unchanged to this day. Two sets of biaxial threads are interlaced together at right angles to each other; the vertical, or length-wise taut warp, and the width-wise horizontal weft (woof). Warp threads are referred to as 'ends' and weft threads as 'picks'. Most cloths are biaxial and two-dimensional. However there are examples within the technical textile field that challenge this assumption such as triaxial fabrics, which are formed by the interlacement of three sets of threads running at 60° angles to each other.

The development of the automated and then later, computer-controlled loom, has allowed us to create complex fabrics at ever-increasing speeds. Nevertheless, there are still many intricate techniques that remain in the craft realm and offer specific useful properties that have yet to be fully exploited by commercial industry. There is still undoubtedly room for further innovation in woven textile structure and design.

This chapter assumes a basic understanding of woven textiles, and its notation, and thus does not go into depth describing the workings of the loom or the mechanics of weaving. For an introduction to weaving please refer to: J.J. Pizzuto's *Fabric Science* Eighth Edition, 2005; *The Handbook of Technical Textiles* edited by A R Horrocks and S C Anand; or G.H. Oelsner's *Handbook of Weaves*, to name but a few of the many books published on this subject.

7.2   Analysing a fabric's needs

When designing a 'performance' or 'technical' fabric to fulfil certain needs, there are many factors that need to be considered. These needs should be fully explored and identified by starting with an analysis of the intended end use. Questions that would need to be answered to help formulate a criterion for the desired apparel fabric might include:

* What is the garment?
* What is the range of activities it will be worn for? Is it multifunctional?
* What will be the environmental conditions it will be worn in?
* How long will it be worn for?
* What does it need to protect the body from?
* Will it have direct contact with the skin?
* Is this garment part of a layering system? What are the other layers?
* How will the garment be made? Will the chosen joining technology influence the fibre content and fabric construction?
* What are the ergonomic requirements of the garment?
* What is the garments projected lifespan?
* What are the 'end of life' requirements of the garment?

The answers will then lead on to being able to specify detailed fabric characteristics as exemplified in Table 7.1.

Table 7.1 Fabric characteristics

Fabric characteristic	Characteristic variables
Weight	Light weight Mid weight Heavy weight
Temperature and moisture regulation	Keep cool Keep warm Respond to changing temperatures Wicking
Protection	UV Water repellent Waterproof Waterproof – breathable Wind resistant Chemical resistant Stab proof Anti-ballistic Abrasion/tear Anti static Anti electromagnetic radiation
Movement	Stretch – elastomeric yarns Pleats
Easy care	Washability Quick dry Durability Non-iron
Electronic	Monitoring vital signs Tracking device Communication
Seamless	3D construction Circular weave Moulded
Light	Reflective Emitting
Health and well-being	Injury support Trans-dermal medicine delivery Healing Ion therapy Moisturising Anti-odour Anti-microbial Aromatherapy/scent delivery

Table 7.2 Fabric performance factors

Some of the many factors that influence fabric performance:			
Fibre	Fibre harvesting/processing	Fibre blend	Fibre finish
Yarn	Yarn count	Plied or filament	Twist
Weave structure	Weave density	Structure combinations	Warp/weft tension
Finish	Wet finish	Dyeing and printing	Coating or Laminate

The performance and aesthetic characteristics of a cloth are achieved by multiple interrelating factors; the fibre, the yarn, the weave structure, the thread density, the finish and any further surface treatment such as a laminate or coating. Each factor has to be considered carefully, both individually and as an interrelating group (Table 7.2).

Once a brief has been fully formed and the specific fabric characteristics decided upon, decisions can then start to be made as to the most appropriate method of construction, with reference to the fibre, yarn and finishing choices.

7.3   Choosing an appropriate woven structure

7.3.1 Plain weave

Plain weave or 'tabby' is the simplest and most commonly used of all weave structures. Weft yarns interlace in an 'over one, under one' fashion across the whole width of the fabric and the structure looks exactly the same on the face or the reverse of the cloth (Fig. 7.1).

Key characteristics:

- Firm
- Stable
- Tears easily
- Frays less
- Creases easily
- Thin
- Lightweight
- Minimum yarn slippage
- Less absorbent
- Poor drape.

7.1 Plain weave.

7.3.2 Hopsack and basket weave

These are variations on plain weave, whereby instead of the weft yarn interlacing over and under each warp yarn, it interlaces over two adjacent warp threads, and under the next two. With the hopsack weave structure there are two shots of the weft yarn for each lifting sequence. Like the plain weave, it looks exactly the same on both the face and reverse of the cloth (Figs 7.2 and 7.3).

Key characteristics:

- Frays easily
- Flexible
- More crease resistant
- Not durable, unless woven in at a much higher density – then very durable
- High yarn slippage
- More bias stretch
- Good tear resistance
- More open, breathable
- More likely to snag.

7.2 Basket weave.

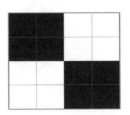

7.3 Hopsack.

7.3.3 Twill

A twill structure is easily recognisable by the distinctive diagonal lines on the fabric surface. A traditional denim is the most common and well-known example of a twill fabric. There are many variations on the twill structure. The variations are referred to as 2/1, 1/3, 2/2 and so on. These numbers refer to how many warp threads the weft yarn is floating over and under. So a 1/3 twill (Fig. 7.4) is 'under one, over three'. A 2/2 twill (Fig. 7.5) is 'under two, over two' and is known as a 'balanced twill'. The repeat size of the structure can start from three warp ends, going up to eight or even more. However, it should be remembered that the longer the floats, the more unstable the fabric. Unless it is a 'balanced' twill structure, such as 2/2 or 3/3 or 4/4 etc., the fabric will have a different appearance on the face and reverse of the fabric. A balanced structure will look the same on either side.

There are fewer interlacements in a twill weave in comparison with a plain weave, thus necessitating comparatively higher warp ends per cm and weft picks per cm. This increased yarn density makes a stronger and more robust fabric, with greater tear resistance.

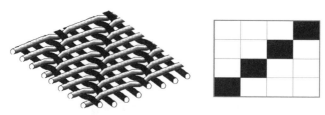

7.4 1/3 twill.

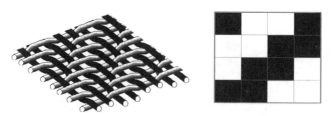

7.5 2/2 twill.

Key characteristics:

- Distinctive diagonal line
- More densely woven, thus heavier and with higher yarn content
- Strong and stable
- Durable
- Good abrasion resistance
- Medium tear strength
- Creases less
- More dirt resistant
- Pliable and flexible
- Good drape
- Snag resistant.

7.3.4 Satin and sateen

In a true satin weave there is only one interlacement of the weft yarn in the structure repeat size. In an 8-end satin, the weft will float under one warp end and over seven to make the repeat. It is referred to as 'warp-faced', as you will predominantly see the warp yarns on the face of the fabric (Fig. 7.6).

A sateen weave is the exact opposite of a satin: it is what you would see on the reverse of a satin fabric. This is a weft-faced weave whereby in an 8-end repeat the weft floats over seven warp ends and under one to make the repeat (Fig. 7.7).

With both satin and sateen weaves, the smallest structural repeat is over five warp ends. It can also repeat over six, seven, eight warp ends, and so on. However, it is important to remember that the long yarn floats which give the fabric its characteristic lustrous surface, also make the fabric more unstable. To counter this, the fabric needs to be woven very densely and so necessitates much higher ends per cm and picks per cm. This has the effect of making the fabric much more compact, thus using a great deal more yarn. A satin/sateen weave is therefore more expensive and time-consuming to produce, but the high density does provide some very positive attributes,

 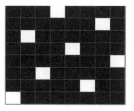

7.6 8-end satin.

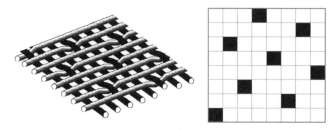

7.7 8-end sateen.

such as strength and very good tear resistance. Also, depending on the yarns used, it can be highly durable.

Key characteristics:

- Needs to be very densely woven
- Long floats – snagging
- Stable and strong
- Can be very durable (yarn)
- Wind repellent
- Helps water repellency
- Smooth with good drape
- Very pliable
- Frays easily
- Yarn slippage
- Very good tear resistance
- Lustrous, flat surface.

7.3.5 Honeycomb/waffle weave

The honeycomb structure is commonly used to make towels and blankets. Its distinctive and easily recognisable three-dimensional cellular structure makes it both highly absorbent and very insulating. The dimensional structure traps air but is not a dense heavy weave. This is a particularly useful trait in blankets as the fabric does not become overly heavy yet is very thermal.

The minimum repeat size for the honeycomb structure is eight warp ends, increasing up to sixteen; any larger than that and the fabric starts to become unstable. There are long warp and weft floats in the honeycomb structure which makes it susceptible to snagging; however, it is these long floats that make the fabric so absorbent, hence its common use in towels (Figs 7.8 and 7.9).

Key characteristics:

- Three-dimensional
- Good insulation – traps air

- Poor abrasion
- Poor durability
- Breathable
- Very absorbent
- Textured surface
- High bias stretch.

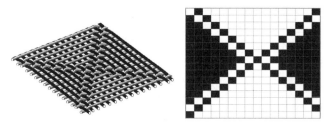

7.8 16-end honeycomb.

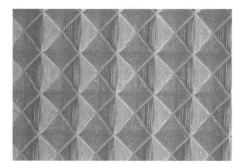

7.9 Schoeller honeycomb fabric (Photograph – David Bryson).

7.3.6 Leno

There are many variations on the leno weave structure, each with its own particular characteristics. However, the common factor in all these structures is the twisting of adjacent warp threads around each other. At its most simple, the 2-end leno, as shown in Fig. 7.10, sees two adjacent warp threads twist around themselves, held in place with each weft pick in a figure-of-eight pattern. This has the effect of firmly locking the weft yarn into place, thus eliminating yarn slippage. It is possible to weave extremely open, breathable fabrics that are very stable and lightweight. The reduction in the quantity of required warp yarn makes it a very economical fabric structure.

Like honeycomb, it is also a common weave structure to make blankets. Some leno constructions have a very cellular structure that traps air

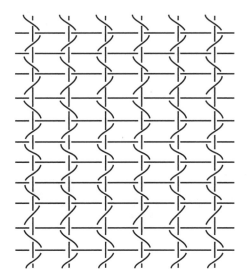

7.10 2-end leno.

and acts as insulation. In addition, the very low thread count makes it extremely lightweight. This combination of thermal properties and light weight makes it a popular fabric structure choice to make blankets for hospitals.

Key characteristics:

* Very stable
* Lightweight
* Open
* Breathable
* Strong
* Reduced yarn slippage
* Reduced warp yarn content
* Firm
* Reduced distortion
* Good shear resistance.

7.3.7 Layered weave

A 'double cloth' fabric is made of two interlinked layers of cloth. The two layers each have their own set of warp yarns and are woven independently yet simultaneously. They then interlace at predetermined points, forming a join between the two layers. Plain weave is the most commonly used structure to weave the two layers, but other structures are applicable too (Fig. 7.11).

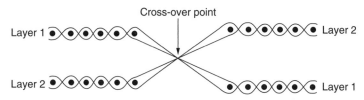

7.11 Double cloth cross-section.

The double cloth structure can create pockets in between the two layers. This can make a quilted construction, particularly if a wadding yarn is 'floated' in these pockets to add thermal bulk or padding.

Dimensional effects can be produced by using yarns of different character for each layer. If one layer is woven with a stretch elastane weft and the other layer with a non-stretch yarn, when the stretch layer retracts, it forces the silk layer to protrude forward in a blister. Matelassé works on a similar principle whereby the two separate warps have highly contrasting tensions. Once off-loom and relaxed, the taut layer retracts further and likewise has a blistering effect on the layer woven at a looser tension.

It is also possible to weave a triple cloth, quadruple cloth or an even higher number of layers too. However, the more layers, the greater the thread-count, which makes it heavier, more expensive and more time-consuming to produce. Also an extremely high ends per cm can make it very difficult for the loom to 'shed' properly and the ends can start to chaff on each other if they are too closely packed. There would need to be a very specific technical need to warrant this construction.

Key characteristics:

- Insulating
- Heavy weight
- Thick
- Each layer could have different yarns, thus different characteristics
- Reversible
- Durable (depending on the yarn)
- High thread count
- More time-consuming to weave
- Stable
- Firm
- Can be three-dimensional.

7.3.8 Pile weaves

Pile weaves can be classed into two categories: loop pile and cut pile. An example of loop pile would be terry cloth, and cut pile fabric examples include velvet and corduroy.

Pile fabric structures can be extremely durable, depending on the fibre used and density of the weave. Pile structures are always tightly woven in order to prevent the cut pile from shedding and the loop pile from snagging. The resulting densely woven fabric serves as a good thermal insulator and is also very absorbent.

Key characteristics:

- Durable
- Firm
- Dense
- Insulating
- Surface texture
- Absorbent
- Abrasion resistant
- Strong
- High yarn content.

7.3.9 Triaxial

Triaxial weaves are 3-way constructions with the threads running at sixty degree angles to each other. This type of construction can create a very open, lightweight yet robust fabric which has four times the tear resistance of a conventional biaxial weave. A triaxial weave also has no bias stretch, making it a very valuable construction for strong and robust fabrics. In addition, the structure resists shearing, bursting and slippage as the yarns are firmly locked into place (Fig. 7.12).

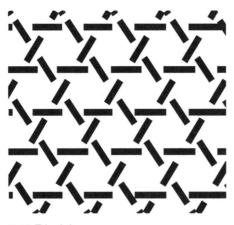

7.12 Triaxial weave.

Key characteristics:

- No bias stretch
- High tear / shear resistance
- Very strong
- Stable
- Lightweight
- Breathable.

7.4   Woven structures – ease of movement

Ease of movement has become a highly desirable fabric property, not only in high performance wear, but also in everyday clothes. In each case, stretch fabrics give comfort and flattering fit, both of which are of great value. High-power elastomeric stretch can also provide a valuable supportive function, applying compression to injured limbs or support to a bust line.

A stretch fabric is defined as having a high degree of elastic stretch and a rapid recovery to its original state. Unless using a stretch yarn or exploiting the bias stretch, a woven fabric generally does not have any inherent stretch abilities like a knitted fabric. However, certain 'counter-leno' woven constructions can offer a degree of inherent stretch. Also, pleating a fabric can give some similar characteristics, such as ease of movement and comfort.

7.4.1 Weaving with stretch yarns

Stretch properties can be integrated into a woven fabric through the character of the yarn. There are three types of stretch yarns:

- elastomeric yarns, where elastic recovery is a fibre property;
- crepe or 'over spun' yarns, where the high degree of twist in the spinning process makes the yarn pull back on itself; and
- texturized yarns, whereby a false twist is manufactured during the production of synthetic yarns.

Stretch yarns can be woven both into the warp and weft of a fabric. However, in the vast majority of cases it is used just in the weft. This is because all warp yarns must be woven at equal tension, otherwise puckering or distortions will occur in the fabric. Whilst modern warping machinery can control and maintain even warp tensions, it remains much easier to maintain an even weft weaving tension. Often it is the case that a weft stretch is sufficient on its own.

Stretch yarns will behave differently when woven in different weave structures. A loosely woven weave structure with longer floats will stretch more than a tightly woven plain weave. This is because a plain weave structure locks and packs the weft yarn into place with its high number of yarn intersections, thus giving a stretch yarn less 'room to move'. However, if the structure has long weft floats, such as a sateen, the weft yarn has space to easily expand and contract to higher degree.

7.4.2 Woven engineered pleats for movement

Folding and applying heat is well known and is the most common pleating process. However, there are also some options for integrating pleats into the actual weave constructions:

Woven pleats

Selectively weaving a weft stretch yarn in small floats in a vertical line on the face or reverse of a non-stretch ground weave fabric can create a pleated fabric without having to fold and press.

Smock weave

A smock weave works on the exact same principle of a stitched smocking process. Extra 'drawstring' threads are woven into a base structure, floating over blocks of threads and inter-weaving in the remaining blocks. Once the fabric is off-loom, depending on this structural formation, the drawstring threads are pulled to pucker and pleat the ground fabric. With the drawstring threads still in place, the fabric can then be heat-set (this would need a minimum of 50% thermoplastic fibre content to be permanently heat-set). The drawstring threads can then be removed leaving a pleated fabric (Fig. 7.13).

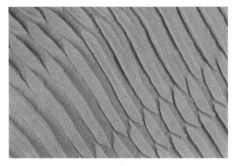

7.13 Heat set smocked pleats (design – Laura Thomas, photograph – David Bryson).

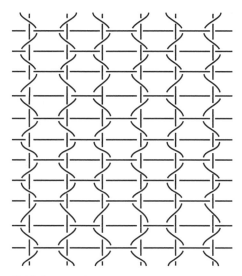

7.14 Counter leno.

7.4.3 Counter leno weave structure

The characteristic twisting action of warp threads in a leno set-up can be manipulated to give a woven fabric a slight stretch. In a counter leno set-up, the twisting action is mirror repeated across the fabric width as opposed to a regular block repeat. This 'mirror twist' forces the weft yarn into a contracted position which can be pulled, thus creating stretch (Fig. 7.14). Akin to other leno structures, a counter leno creates a very open, light-weight fabric that is also very stable. It is commonly used to make gauze bandages, where its stretch properties are usually enhanced by use of an elastomeric weft yarn.

7.5   Woven structures – wicking and thermoregulation

One of the primary functions of apparel is to help regulate the body temperature. This may be to protect from the cold or to help cool the body down when too hot. The temperature of the body is influenced by:

- the climate,
- the activity being undertaken, and
- physiological factors such as illness or injury.

7.5.1 Absorbency, wicking and keeping cool

When in a hot environment or undertaking strenuous activity, the body temperature increases and we start to perspire. There are certain fibres (cotton, silk, rayon) and weave structures (honeycomb, counter leno) that are very absorbent, yet they hold onto this moisture. For a garment to be comfortable to wear, especially during sport, it needs to dry as quickly as possible, so not to cling to the skin, cause irritation, smell or be affected by wind chill. A wicking fabric draws moisture from the skin to the outside of the garment where it can then evaporate. Quick moisture transport away from the skin and then evaporation is the key to keeping the wearer cool.

The success of a wicking fabric is determined by the speed and how widely moisture spreads laterally across the surface of the fabric and then evaporates. Most successful wicking fabrics are made from manipulated hydrophobic fibres which do not absorb moisture but allow moisture to be carried along their surface. Open weave structures which are naturally absorbent help the wicking process, as do pile structures and brushed or raised surfaces. However, these need to be complimented by the most appropriate yarns. The wicking fabric also has to be close to the skin and so will nearly always need an elastomeric content for ease of movement. For an outer layer, which does not need wicking properties, a lightweight, breathable weave structure such as plain weave would be most appropriate. The air pockets in a lightweight double cloth structure could be exploited by either trapping a cooling agent or opened out to allow greater air circulation.

7.5.2 Keeping warm

Historically, the answer to keeping warm was to wear many thick heavy layers of clothing as possible. However, this is a great hindrance as it is bulky and slows us down. This is especially undesirable for cold-climate performance sportswear. The key is to use a lightweight yet thick flexible fabric that is not too bulky. If the wearer is likely to perspire and the garment is to be worn next to the skin, it needs to have wicking properties. Moisture is a rapid conductor of heat so, if a garment is wet in a cold environment then the body will lose heat rapidly. Trapping 'dead air' (air that is not moving) in the textile fibres and structures is key to effective thermal insulation. This can be achieved in a number of ways:

- Densely woven fabrics used to trap air between the body and the garment

- An inherently cellular weave structure such honeycomb and certain leno constructions trap air in their cell-like pockets
- At the spinning stage, e.g. woollen spun yarn – the more loosely spun yarn traps air amongst the fibres
- Pleating is also very effective as dead air is caught between the folds of fabric
- Pile structures such as terry cloth and velvet
- Brushed, raised or textured surfaces trap air in the same manner as pile structures
- Woven layers such as double cloth, piqué and matelassé all have, literally, structural pockets.

Laminated fabric assemblies for cold harsh climates allow the garment to have the most desirable fabric characteristics on the exterior and interior, which may not be possible in just one construction. The process of layering also allows for air to be trapped in-between to act as further insulation. Often, an open-weave wicking inner layer might be bonded to a densely woven outer layer. When an athlete rests, moisture would be drawn from the skin through to the outer layer to evaporate. The inner layer would aid cooling by moisture dissipation and the dense outer layer would insulate the body and prevent wind chill.

7.5.3 Phase Change Materials (PCMs)

PCMs are fabrics that respond to dramatic fluctuations in temperature. The fabric absorbs heat when hot, thus creating a cooling effect. It then releases the heat later, once the ambient temperature is lower, thus warming the wearer. The technology is applied either as a fabric coating, or PCM molecules are encapsulated at the fibre stage. However, the potential for this technology is limited as the heating or cooling effects are short lived. Once the 'phase change' process is complete, it is no longer effective. Its benefits can really be felt only if the wearer repeatedly alternates between a hot and cold environment.

7.6   Woven structures – protection and safety

Protection from the climatic and physical environment is one of the inherent properties that we demand from technical apparel. A protective fabric will always be designed to meet a particular need and thus its individual characteristics will vary accordingly. These protective characteristics may be imbedded at the fibre or spinning stage, or applied as a coating onto the finished fabric. Below are some types of fabric protection where the weave structure is integral to its functionality.

7.6.1 Protection from water

At its most basic, the purpose is to stop water passing through the fabric, so it has to be as tightly woven as possible. There are different levels of protection from water that a textile can provide:

- A *water repellent* finish on a textile is effective in light to medium rainfall for a short period of time. Individual fibres are coated in a water repellent chemical rather than creating a complete layer over the fabric surface. A water repellent fabric makes water droplets 'bead' on its surface before rolling off. There are both durable and non-durable water repellent finishes, chosen on the fabric's intended end use. If the fabric is likely to be laundered, then a durable repellent is required rather than the cheaper non-durable, which is removed in dry-cleaning and washing processes.
- A *waterproof* textile is totally impermeable, so does not allow water or air to pass through it. This can be achieved by coating a fabric with rubber, vinyl or polyurethane.
- A *breathable waterproof* fabric makes a very effective and comfortable garment, offering complete protection from water and wind, whilst still allowing perspiration and body heat to pass through it. This is achieved by a microporous membrane film laminated in between a lining and an outer shell textile (a three-layer laminate). The micropores exploit the dramatic size differential between a water droplet and perspiration (moisture and air) molecules which are a fraction of the size. The micropores allow perspiration to be released but water droplets cannot get through. Gore-Tex™ is a well-known trademarked brand producing fabrics of this type.

7.6.2 Protection from ultraviolet rays (UV)

Wool and polyester inherently absorb UV rays, whilst cotton and nylon allow UV rays through. Dark colours tend to be better at absorbing UV rays than lighter shades. Where UV rays pass through fibres and yarns, the more compact and heavier the fabric, the greater the level of protection. The potential greater weight of a densely woven fabric can be countered by using a microfibre. UV protective coatings both on fibres and fabrics are the most commonly used and effective method to maximise solar protection.

7.6.3 Protection from falls, impact, abrasion and tearing

Falls

Protection from a fall is highly important, particularly in sports such as motorcycling, horse riding and ice hockey. Such pursuits all have their own unique garment requirements, so a fabric would need to be designed to a specification. However, we can consider some generic needs and factors that could be integrated in a woven textile. Densely woven, padded double cloths or quilted rib weaves would be very usefully applied to the vulnerable areas of the body which are most likely to bear the brunt of a fall, such as knees and elbows. Likewise, foam-backed laminated fabrics are also extensively used for their ability to absorb shock. Robust yarns such as aramid, carbon and glass fibre could be used in both warp and weft, to weave fabrics with added strength. Depending on cost and protection needs, they could well be combined with other qualities such as those provided by nylon or polyester. Protective coatings can also add another layer of strength.

Abrasion

Abrasion occurs when there is repetitive friction in a specific area, or in the case of a fall from a motorbike, when the rider skids across the ground at high speed. If composed of an appropriately robust yarn, tightly woven, dense weave structures such as twill and satin/sateen are the most effective against abrasion. A 6-end satin/sateen has a particularly effective level of abrasion resistance. Cut and loop pile fabrics can also withstand high levels of certain types of abrasion.

Uniform, tightly spun yarns woven into uniform flat fabrics are the most resistant to abrasion as they allow for wear to be evenly distributed.

Tearing

Structures with fewer intersections have greater tear and shear resistance, as the weft threads bunch together and are more different to tear together. Plain weave is the weakest weave structure for tear resistance due to its high number of yarn intersections. Weft threads are unable to group together in a plain weave, so each thread has tear pressure applied to it individually, making it much more likely to rip. However, a 'ripstop' fabric is a variation on plain weave and as its name suggests, is very effective at stopping tears. This is due to a grid of thicker warp and weft yarns that inhibit the tearing action. Twills and satin/sateen structures are also very tear resistant, as are triaxial weaves, because they are able to withstand pressure from more angles. Protective coatings and finishes can also enhance a fabric's tear strength.

7.6.4 Safety: light reflective/emitting yarns

Light reflective yarns are highly effective, bouncing back light from up to 200 m away, thus illuminating a cyclist or workman at night. The light-reflective yarn is usually used only in the weft as its effectiveness means that a high concentration is not needed. An intermittent stripe is often sufficient, or a garment trim: high cost is a factor in this case.

Whilst the yarn is very strong, it is advisable for it to be used in a carefully designed weave structure that offers it some protection from abrasion, for example, in the central dip of a honeycomb unit, or alongside another pronounced textured weft yarn.

Fibre optic yarns can also be woven into a fabric as a weft with almost any other fibre and, once connected to a power supply, emit light along their length. They can be programmed to create patterns and display messages. The need for a power supply limits their usefulness in functional apparel but they are undoubtedly going to become more widely used as alternative power sources are developed.

7.7   Three-dimensional fabrics and fully formed seamless woven garments

Woven textiles are generally perceived to be two-dimensional and flat, having to obey the biaxial rules of the warp and weft. As we have already discussed, there are exceptions to this assumptive rule. At its most basic, cellular weave structures such as honeycomb and counter leno can produce a subtly three-dimensional fabric. Selective use of loop and pile structures results in a relief surface and triaxial weave structures disprove the biaxial rule of the vertical warp and horizontal weft, introducing new woven geometries.

7.7.1 Three-dimensional weave structures

Shape3, based in Germany, have developed two different kinds of weaving technologies for the production of truly three-dimensional fabrics: 'Shape Weaving Technology' and tubular weaving technology.

Fabric produced on the 'Shape Weaving' loom is a truly 3D fabric, with a shell-shaped appearance. An adapted shuttleless rapier loom has been fitted with various devices to individually control the weave in every section of the fabric. This includes a warp creel with special yarn brakes to control the tension and 'feed' of every warp thread.

Their tube weaving technology uses traditional shuttle looms, i.e. the weft thread is inserted continuously, and not cut at the selvedge, thus making it possible to weave a double cloth 'seamless tube'. Shape3 have adapted the loom's shed and reeds to allow them to also weave tapered tubes, branched tubes, tubes with well-defined openings, and tubes with bellow-like structures. The variation of the diameter of the tube (for tapered tubes) is controlled by a fan reed. According to the required diameter, the reed (which has changing distances between the bars over its height) is placed at the necessary height.

7.7.2 Vacuum-formed thermoplastic fabric

In much the same way that pleats are permanently pressed into place by heat and pressure, relief surfaces and 3D shapes can also be created through vacuum forming. The fabric needs a high polyester content so that is thermoplastic memory can be 'set' to the shape of the mould. A compact fabric construction such as a twill or satin holds its shape far more successfully than a plain weave or open structure.

7.7.3 Fully fashioned seamless weaving

It is unlikely that fully fashioned woven garments will ever be as flexible and as controllable as 'whole garment' knitting technology. However, Issey Miyake and Dai Fujiwara have been making interesting inroads into this concept following on from their successful cut-out knitted whole garments. This project and brand is entitled A-POC 'A Piece of Cloth'.

All the garment pattern pieces are woven as a jacquard double cloth into one piece of cloth, side-by-side, thus eliminating waste. Once off loom, the fabric has then to be cut following the single cloth woven seams around the pattern pieces. These seams hold the two layers of cloth together in the same way that a stitched seam would. The seams will not fray due to the tight weave structure, the character of the yarns and the finishing process.

A limited amount of joining is necessary with some pattern pieces, but the amount of labour that is involved in making an A-POC garment has been greatly reduced. There is scope for selectively using elastomeric yarns to help to shape the garment for comfort and fit.

7.8   Woven structure: conductive fabrics

Conductive fabrics that dissipate static energy and protect from electromagnetic fields alongside other attributes such as thermal regulation, anti-allergy properties and anti-bacterial properties, have been widely used for nearly two decades. However, it was soon realised that fabrics constructed with conductive fibres such as carbon, gold, stainless steel, silver, or copper could offer great potential by facilitating the integration of 'soft' networks into fabrics, thus making them 'smart'. Unlike most technical textiles, smart textiles are not passive in their function: They can sense and respond to stimuli such as touch, temperature or heartbeat. Intelligent textiles can incorporate antennas, global positioning systems (GPS), mobile phones and flexible display panels, without compromising the inherent characteristics of the fabric: The conductive yarns can look, feel and behave like a traditional fabric.

The fabric itself is often used as a 'switch' in an electronic circuit to perform a function for another external electronic device. In order for a 'switch' to happen, a connection between two conductive fabrics or yarns needs to occur.

Holding back the widespread take-up of smart textiles are the not insignificant anxieties surrounding washability, durability and most importantly, power source. To function, all electronic textiles (e-textiles) need an electric charge. Whilst much development has been done in the miniaturisation of batteries, they are still cumbersome and need to be removed for washing. Development of many smart application therefore depend on the development of improved power sources.

7.9   Future trends

Technology itself is undoubtedly the driving force for most future textile developments and much has been written on this subject. However, caution is necessary, as it has been proven that consumers will not buy technology for technology's own sake. It is imperative that future developments are undertaken on solid foundations where the market need is proven.

Also, it is important not to look only for new concepts driven by contemporary technology, but also to have an appreciation for textile craft history, as there lies great sources of inspiration for textile innovations.

7.9.1 Learning from textile craft

As the performance demands of textile structures increase, there will be a re-evaluation of traditional simple and complex hand-weaving techniques, analysing what properties they might offer. Examples may include leno and triaxial constructions, which are currently under-used and under-explored, despite having highly desirable unique characteristics. At first glance, the gap between technologists who are developing the most innovative new fabrics and art/craft weave designers seems to be vast. However, there are many examples of hand-weavers lending their expertise and collaborating with industry to open up new possibilities for hi-tech problem solving. Just because a particular quality of fabric cannot currently be woven on a power loom, does not negate its potential to be commercially produced. Whilst it is obviously significantly more expensive to hand-weave a fabric, if it is a crucial component, with a customer willing to pay a premium, then all avenues, including hand-weaving, should warrant exploration.

Commercial power looms are improving all the time and becoming increasingly flexible. However, it is important not to be blinkered to textile

7.15 Curved weft hand woven fabric by Mike Friton.

construction possibilities offered outside of current commercial boundaries. Indeed, these could pave the way for new types of machinery, new processes and new fabric qualities. Having an appreciation of the potential offered by a hand loom, complimented by a thorough understanding of modern power looms, is essential. Looking at contemporary woven textile art and design, as well as traditional textile techniques from around the world, can offer practical inspiration and suggest routes for technological problem solving. There are many examples of woven textile artists and designers who are pushing the boundaries of construction, but often for aesthetic purposes or just for the 'challenge'. There is a wealth of information and skills that can be tapped into to support the development of creative technical textiles. In addition, the design input of weavers from such backgrounds is needed in order to make the fabrics destined for consumer markets far more desirable.

Mike Friton, a sports footwear innovation engineer working for Nike, is currently exploring curved weft hand-woven fabrics, an example of which is shown in Fig. 7.15. The fabric does not have the usual bias stretch and is therefore more malleable to form the toe of a shoe. His ultimate aim is to be able to weave a shoe in its entirety, reducing wastage and costly make-up time.

7.9.2 Multifunctionalism and other considerations

The demands that we now have of our attire will continue to grow. Seasonless multifunctional garments will become ever more common and necessary, due to climate change making extreme weather a regular occurrence. Technical garment consumers will become more sophisticated and demand an improved aesthetic/functional relationship. Pattern, colour and other decorative effects will no longer be alien to the technical apparel market. Price and easy-care factors will remain extremely important and will be the deciding factor in the mass take-up of smart textiles. Recycling will become compulsory in the near future so sustainability issues will increase in importance and the 'end of life' for technical garments will need to be considered and planned for.

7.10   Conclusion

This chapter aimed to highlight the importance of giving due consideration to the woven structure of a fabric and how this can impact on the overall textile function and performance. The various individual properties of the basic weave structures have been exemplified, then applied and discussed in relation to performance needs.

For the non-weave specialist, having a basic understanding of weave structures and their individual merits will aid communication with a textile mill or supplier to specify a fabric. For the weave specialist, it is hoped that this chapter has prompted some thoughts regarding the development of future fabrics, particularly by addressing end user needs to drive the innovation.

7.11   Bibliography

K ANDERSON *Smart Textiles Update*, August 2005, http://www.techexchange.com/thelibrary/smarttextiles.html – accessed August 2006.

A BOLTON *The Supermodern Wardrobe*, V&A Publications, London, 2002.

S E BRADDOCK and M O'MAHONY *Sportstech: Revolutionary Fabrics, Fashion and Design*, Thames and Hudson, London, 2002.

S E BRADDOCK and M O'MAHONY *Techno Textiles: Revolutionary Fabrics for Fashion and Design*, Thames and Hudson, London, 1998.

S E BRADDOCK CLARKE and M O'MAHONY *Techno Textiles 2: Revolutionary Fabrics for Fashion and Design*, Thames and Hudson, London, 2005.

P COLLINGWOOD *The Maker's Hand: A Close Look at Textile Structures*, Bellew Publishing, London, 1998.

D GOERNER *Woven Structure and Design: Part 2 – Compound Structures*, British Textile Technology Group, Leeds, 1989.

V HENCKEN ELSASSER *Textiles: Concepts and Principles*, Second Edition, Fairchild Publications, Inc., New York 2005.

R HIBBERT *Textile Innovation: Interactive, Contemporary and Traditional Materials*, Second Edition, Line, London, 2006.

S LEE *Fashioning the Future: Tomorrow's Wardrobe*, Thames and Hudson, London, 2005.

C MCCARTY and M MCQUAID *Structure and Surface: Contemporary Japanese Textiles*, The Museum of Modern Art, New York, 1998.

M MCQUAID *Extreme Textiles: Designing for High Performance*, Thames and Hudson, London, 2005.

H NISBET *Grammar of Textile Design*, Chapter IX Gauze and Net Leno Fabrics, Scott, Greenwood & Son, London 1906.

A PRICE, A C COHEN and I JOHNSON *J.J. Pizzuto's Fabric Science,* Eighth Edition, Fairchild Publications, Inc., New York, 2005.

W S SONDHELM *Handbook of Technical Textiles* Chapter 4 Technical fabric structures – 1. Woven fabrics, Woodhead Publishing, Cambridge, 2000.

A-POC Making: Issey Miyake and Dai Fujiwara, Vitra Design Museum GmbH, Germany, 2001.

Shape3 http://www.shape3.com/

8

Nonwovens in smart clothes and wearable technologies

F. KANE, De Montfort University, UK

Abstract: This chapter provides an overview of the production technologies used to manufacture nonwovens, the characteristics of the resulting fabrics, their application in smart clothes and wearable technologies and 'end-of-life' considerations. Key web-bonding, web-forming, fabric finishing and product make-up processes are summarised and an overview of the raw materials used in production is given. The use of reclaimed and bio-degradable fibres in nonwoven production and the application of various nonwovens as: support materials; alternatives to traditional outer garment fabrics; performance and art fabrics and; medical and protective clothing fabrics are also discussed.

Key words: nonwoven materials, nonwoven production processes, nonwoven applications.

8.1   Introduction

Originally perceived as economical alternatives to traditional textiles for obscure product components such as interlinings and carpet backings, nonwovens have evolved into unique engineered fabrics that are used in medical textiles, architectural and geo-textiles, domestic and personal hygiene products, fashion, art, performance and protective wear, including smart clothes and wearable technologies. Nonwovens are applied within these areas as component parts performing unseen technical functions and as visible surfaces providing aesthetic appeal.

8.1.1 What are nonwovens?

The term 'nonwoven' implies something that is simply 'not woven', but there are a number of precise definitions that give specific details as to what a 'nonwoven' is. The British Standards Institute, European Disposables and Nonwovens Association (EDANA), The International Nonwovens and Disposables Association (INDA) and The Textile Institute all provide specific definitions. The British Standards definition states that a nonwoven is: 'A manufactured sheet, web or batt of directionally or randomly orientated fibres, bonded by friction, and/or cohesion and/or adhesion, excluding paper and products which are woven knitted, tufted, stitch-bonded incorporating

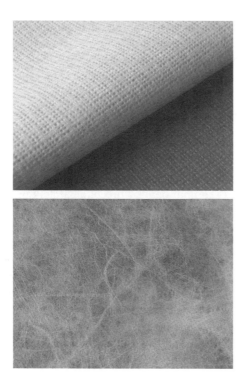

8.1 Top: embossed nonwoven; bottom: spunbonded nonwoven.

binding yarns or filaments or felted by wet-milling whether or not addition-ally needled. The fibres may be of natural or man-made origin. They may be staple or continuous filaments or be formed in situ' (British Standards Institute, 1992).

The processes and materials used to make nonwovens, and the qualities of the end products themselves, link them to the chemicals, plastics, felt and papermaking industries. Although there are similarities with these industries and products, the various elements of nonwoven production create a unique industry, and products with specific and unique properties (see Fig. 8.1 showing two examples of nonwoven fabrics).

8.1.2 Chapter overview

This chapter provides an overview of the production technologies used to manufacture nonwoven and the characteristics of the resulting fabrics. Some 'end-of-life' considerations of nonwovens are then outlined and finally the current and potential use of nonwovens in smart clothes and wearable technologies are discussed. The aim of the chapter is to provide an overview of nonwovens, equipping the reader with enough understanding of subject specific terminology and language to enable appropriate materials selection, subsequently enabling effective design development. It does not provide in-depth technical descriptions of process parameters, machinery or specific fabric.

8.2   Nonwovens technologies and manufacturing processes

Nonwovens are one of the newest and fastest growing branches of textile production. However, they originate in ancient textile production methods that involve the processing of animal hair by mechanical action involving water, heat and chemicals, resulting in felt-like materials (Jirsak and Wadsworth, 1999). Although most definitions of nonwovens technically exclude paper and felt, making links between these ancient processes and industrial nonwoven production methods enables the technologies involved to be more easily envisaged and understood from a design and making perspective. Similarly to felt and paper making processes, the production of nonwovens can be described in three main stages:

(i)   Web formation
(ii)   Web bonding
(iii) Finishing

There are a variety of manufacturing processes and technologies used at each of these stages, the choice of which is dependent upon the end-use of the fabric and the performance properties required. Nonwovens are often classified by production process. As shown in Table 8.1, web forming methods are usually classified within three distinct categories and a further three categories are used to classify web bonding methods.

8.2.1 Materials for nonwovens

The basic raw materials for nonwovens are fibres and binders. Albrecht notes that 'virtually all kinds of fibre can be used to produce nonwovens' (Albrecht, 2003, p 15). However, Wilson highlights that market statistics show man-made fibres accounting for 90% of nonwoven production, over half of which is said to be polypropylene (Wilson, 2007). When choosing a fibre, the required profile of the end fabric needs to be considered along with the cost-effectiveness of the fibre and its suitability for nonwoven processing.

Binders for nonwovens can be subdivided into two main groups – fluids or fibres. Both serve to bond fibre webs. Ehrler (2003) writes that the term binder fluid can be used as an overall concept that covers latex, adhesive, synthetic and polymer dispersions and emulsion polymer. Binder fluids can

Table 8.1 Nonwoven production processes

Nonwoven production

Fibre forming

Fibre forming
Film forming

Fibre preparation

Opening
Blending

Blending
Production of fibre/water suspension

Web forming

Polymer-laid

Spun-laid
Melt blown
Flash spun
Film treatment

Dry-laid

Carded
Airlaid
Combined
Cross, Parallel or Random laid

Wet-laid

Wet laid
Airlaid

Web bonding

Mechanical	**Thermal**	**Mechanical**	**Thermal**	**Chemical**	**Mechanical**	**Thermal**	**Chemical**
Needlepunching	Calendering	Needlepunching	Calendering	Coating	Needlepunching	Calendering	Coating
Hydroentangling/	Through air	Hydroentangling/	Through air	Spraying	Hydroentangling	Through air	Spraying
Spunlacing	(oven)	Spunlacing	(oven)	Printing	Spunlacing	(oven)	Printing
		Stitch bonding	Ultrasonic	Powder	Stitch bonding	Ultrasonic	Powder
				Foam			Foam
				Saturation			Saturation

Finishing

Mechanical
Embossing, moulding, shrinking, creping, cutting

Chemical
Dyeing, printing, coating, laminating

be adapted to provide specific properties for the end fabric and are usually applied at the web bonding stage.

Binder fibres have been described as one of the most convenient ways to bond webs. Spindler defines binder fibres as 'those which are able to provide adhesive bonds with other fibres due to their soluble or fusible character' (Spindler, 2003, p 129). Like binder fluids, binder fibres can be subdivided into two main categories – soluble fibres and hot melt adhesive fibres. Soluble fibres are those which become sticky when influenced by a solvent or alginate. Hot melt fibres are those made from hot melt adhesives or thermoplastic polymers, which have a lower softening point than the main fibres in the web. These polymers can also be applied in powder form to nonwovens. Adhesive fibres are blended with the main fibres being employed in the fabrics at the web formation stage of manufacture.

8.2.2 Web forming

The first stage in nonwoven production is the arrangement of fibres or filaments in a specific orientation to produce a loosely held structure called a web. The structure and composition of the web highly influences the properties of the final fabric. As shown in Table 8.1, web forming methods are categorised into three main areas – dry-laid, wet-laid and polymer-laid. As highlighted by Wilson (2007), these methods relate to textile, paper and plastics manufacturing technologies respectively.

Dry-laid

Dry-laid web formation involves the manipulation of fibres in their dry state. Carding and air-laying are the main methods used to do this. Staple fibres of 15–250 mm in length and 1–300 dtex linear density can be used (Brydon and Pourmohammadi, 2007). Both synthetic and natural fibres can be used.

Carding originates in early felt making methods and involves the 'combing' of fibres into a loose structure, referred to in felt-making terms as a 'batt'. Carding in nonwoven production is similar to that used in the pressed felt industries. As illustrated in Fig. 8.2, staple fibres are passed through a series of rotating drums covered in fine wires that essentially comb them into a parallel or random arrangement, creating a batt. The batt is fed continuously onto a conveyor belt, which travels to a system that layers it in either a parallel or cross-wise manner to form a web of specific weight. The web is then bonded using mechanical, thermal or chemical methods. Dependent on the fibre composition and bonding method applied, the nonwovens produced can be used for a variety of applications including interlinings for clothing and fabric softeners.

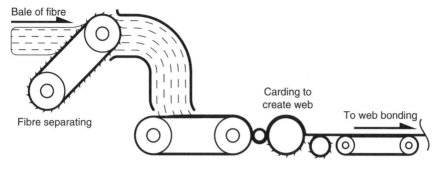

8.2 The carding process.

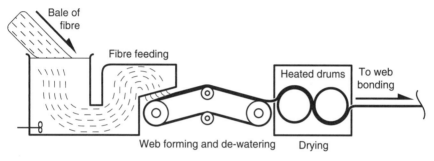

8.3 The wetlaid process.

In the air-laying process, fibres are suspended in the air and collected with the aid of an air stream onto a moving belt or rotating cylinder that separates the fibre from the air, creating a batt of fibres. Again, the web is bonded either mechanically, thermally or chemically to produce fabrics for a range of end-uses including high-loft products (such as wadding) for clothing.

Wet-laid

Both synthetic and natural fibres can be used in this process but they are usually of short length, between 2 and 6 mm (Batra, *et al.* 1999).

Wet-laid web forming methods are essentially a modification of the papermaking process. The fibres are dispersed in water to form a fibre dispersion (this enables different fibre types to be effectively blended). As illustrated in Fig. 8.3, the dispersion is deposited onto a moving screen or perforated drum and the water is drained away by suction to leave a web. The web is further drained and consolidated by pressing between rollers and drying. Web bonding is achieved by use of self-bonding cellulose fibres, as in paper making, or through the addition of chemical binders or thermo-plastic fibres and subsequent heat bonding.

The resulting fabrics often have paper-like qualities and are applied to a range of end uses including teabags, filters, interlinings, fire-retardant protective apparel and disposable medical products. High-performance fibres, such as glass, aramid and ceramic, can be used in wetlaid processes to produce technical fabrics.

Polymer-laid

Polymer-laid (or spun melt) methods of web formation include spunbonding (or spun-laid), melt-blowing and flash-spinning processes. These processes relate to methods of polymer extrusion and involve the formation of webs directly from molten polymers in one continual process. As illustrated in Fig. 8.4, in spunbonding systems thermoplastic fibre-forming granules are melted to form molten polymer that is extruded through spinnerets to form continual filaments. The filaments are rapidly cooled and laid onto a conveyor belt to form a web. The filaments are laid in a continual spiral arrangement resulting in elliptic structures of fibre (Kittlemann and Blechsmidt, 2003). The webs are bonded using mechanical, chemical or thermal methods. The thickness and quantity of the fibres can be controlled to enable a range of fabric densities and weights to be achieved.

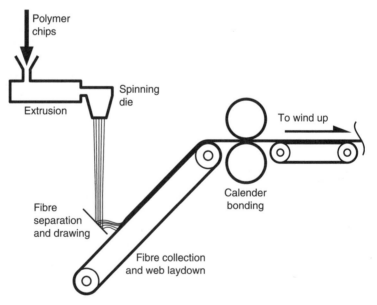

8.4 Spunbonding process.

Other polymer-laid techniques usually involve web-forming and bonding methods that occur simultaneously to enable economical production (Kittlemann and Blechsmidt, 2003). Melt-blowing, like spunbonding, involves the extrusion of thermoplastic fibre-forming polymers. The extruded polymers are blown through an air stream and collected on a screen or belt to form a web. The fibres are laid together and bonded through a combination of entanglement and cohesive sticking (Batra, *et al.* 1999). In the flash-spinning process, polymers are dissolved in a solvent and sprayed into a vessel. The solvent is then evaporated leaving a web of fibres that is further bonded using hot calender rollers.

Spunbond fabrics typically weigh between 10–200 g/m^2 (Bhat and Malkan, 2007) and often have a smooth surface texture. It is possible to produce spunbonds with near isotropic properties but the majority of commercial spunbonds are anisotropic (machine direction). Due to their structural characteristics, spunbonds are utilised in a range of application areas including automotive, geotextiles, hygiene and medical. They are used widely in medical products due to an ability to adjust their breathability, resistance to fluid penetration, lint-free structure, sterilisability and impermeability to bacteria. Within hygiene and medical applications, such as sanitary napkins and disposable operating gowns, spunbonds are used because of their ability to keep the skin dry and comfortable (Bhat and Malkan, 2007).

With regard to meltblown nonwovens, Bhat and Malkan (2007) explain that their fine fibre network and large surface area provide good insulation. They note that the large surface area enables air to be trapped through the drag force on air convection currents moving through the fabric. These properties are exploited in outdoor clothing products.

Polymer-laid nonwovens are often combined to form composites but are also used in their uncombined state.

8.2.3 Web bonding

Most nonwoven webs have little strength in their unbonded form. Bonding processes can be mechanical, chemical or thermal. The method and level of bonding determines to a large extent the properties of the resulting fabric, in particular the strength of the fabric.

Mechanical bonding

Mechanical bonding essentially involves the strengthening of webs through the physical entanglement of fibres. Mechanical methods include needle-punching, hydroentangling and stitchbonding.

Starting in the nineteenth century, needlepunching was developed as a means of processing natural and regenerated fibres that could not be

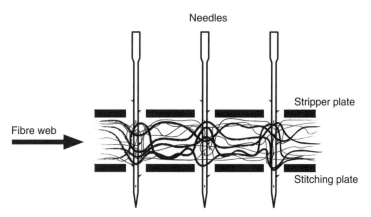

8.5 Needlepunching.

used in the felting process (Kittlemann, *et al.*, 2003). Since the development of synthetic fibres in the 1940s, needlepunching has developed into
an efficient means of processing such fibres. As illustrated in Fig. 8.5, this
process uses barbed needles, which are repeatedly pushed and pulled
through the web, The needles hook tufts of fibre and pull them vertically
through the web, entangling and interlocking the fibres. The web moves
continuously, at a given speed, between two holed surfaces through which
the needles travel. A number of parameters impact on the quality of the
fabric produced, including needle design, needle arrangement, the penetration depth of the needle, and punch density (number of punches in a
given area). Applications areas for needlepunched fabrics are broad,
including geosynthetics, filters, floorcoverings, wadding and padding, automotive fabrics, wipes, roofing, insulation and synthetic leather (Swarbrick,
2007).

In the hydro-entanglement process, fine, high-pressure jets of water are
used to entangle the fibres. The fibre web is placed onto an entangling
surface, for example, a wire mesh, and fine water jets are applied to the web
causing the entanglement of individual fibres. After one or both sides have
been entangled, the fabric is dried (Jirsak and Wadsworth, 1999). The resulting fabric is usually then perforated in a pattern that mirrors that of the
supporting surface. The hydro-entanglement process is also known as spun-
lacing, as the arrangement of jets can give a variety of visual effects that
have lace-like qualities (Edana, 2007). The process results in fabrics that, in
comparison to other nonwovens, are soft and have good drape, handle and
strength (Jirsak, 1999). Dependent upon the fibre used and properties
achieved, hydroentangled nonwovens can be used as high temperature
protective clothing, synthetic leather, surgical fabrics, medical gauze, linings
and outerwear clothing (Anand *et al.*, 2007).

Stitch-bonding is a process by which fibre webs and/or yarns are bonded together by stitching with continuous filament or staple yarn, creating a series of loops that hold the web together. This process is sometimes excluded from definitions of nonwovens, but the products produced are closely linked with the production and consumption of other nonwovens (Jirsak and Wadsworth, 1999). A number of mechanical systems and methods exist in which webs are stitched. The yarn or filament can be inserted in a cross, parallel or multiaxial fashion (Anand *et al.*, 2007).

Chemical bonding

Chemical bonding usually takes place in two stages – the application of a bonding agent and the triggering of the bonding agent using heat (Ehrler and Schilde, 2003). These bonding agents are usually in the form of polymer dispersions (latex) or polymer solutions (Jirsak and Wadsworth, 1999). The solutions are predominantly water-based but some powdered adhesives, foams and organic solvent solutions are also utilised (Edana, 2007). Chemical binders are applied uniformly using impregnation, coating, spraying or printing methods. Chemical bonding can also be used to colour webs by adding pigments to the binder solutions.

Chemical bonding through impregnation essentially involves immersing the web in a binder, as illustrated in Fig. 8.6. The saturated web is then squeezed between rollers to remove excess binder, dried and cured using heat. In coating techniques, binder paste or foam is applied to the nonwoven by knife or scraper systems which are supported by air, blanket or rollers (Chapman, 2007). In spray bonding, the binder is applied to one or both sides of the web using high-pressure spray guns (Jirsak and Wadsworth, 1999). Traditional printing methods have been adapted as a further means

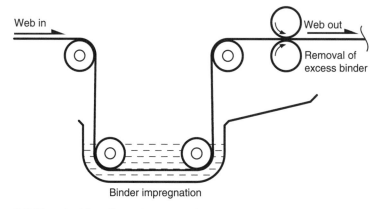

8.6 Chemical bonding – impregnation.

of applying chemical binders to webs. The binder is printed onto the web using either patterned rollers or rotary screens, applied either in specific areas or over the entire surface of the fabric.

Chemically bonded nonwovens are used as wipes and linings and as component parts in footwear, furniture and automotive interiors.

Thermal bonding

Thermal bonding utilises the thermoplastic properties of certain synthetic fibres to form bonds under controlled heating (Edana, 2007). These binder fibres are blended into the main matrix of the web and heated using either hot calenders, 'through-air' methods or ultrasonics. When heated to their glass transition temperature, the bonding fibres liquefy and surround the main fibres. This bonds the web at the fibre intersection points (Spindler, 2003). Bonding fibres are often selected because of their glass transition temperature and subsequent suitability for use with other fibres under heated conditions. The thermal resistance required of the end-product also impacts upon choice of bonding fibre. Fusible adhesives in powder form are also used in thermal bonding.

In calender bonding, heat and pressure are applied through rollers to weld fibre webs that pass between them at high speeds, as shown in Fig. 8.7. Either one or both of the rollers are heated and, as the web passes through, the fibre layer is compressed and heated. The heat liquefies the thermoplastic fibres and, aided by the pressure, they surround the other fibres in the web, locking them into place as they cool (Spindler, 2003). The temperature at which the fibres are bonded is based upon the temperature at which the fibre begins to soften (this is the glass transition temperature). Up to a certain point, the higher the temperature is raised above this, the more the fibre runs and the greater the extent of bonding that occurs. Embossed or engraved rollers can be used to bond webs at specific points – usually at the raised points of their surfaces. This enables strong, lightweight and low-

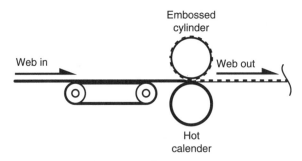

8.7 Thermal calender bonding.

loft fabrics to be produced (Batra, *et al.* 1999). Calendering is also used to create patterned or textured embossed surfaces as a finishing process.

Through-air bonding takes place through the use of a controlled hot-air stream. The web, which contains a thermoplastic component, passes through a chamber in which controlled hot air accumulates. The hot air is either blown over or sucked through the web by a fan, causing the thermoplastic components to fuse. Due to the absence of pressure or consolidation in this process, voluminous or bulky fabrics can be produced.

In ultrasonic bonding, the fibre web is drawn between an ultrasonic 'horn' and a patterned roller. The horn produces high frequency sound waves and the energy produced is transferred from the horn into web. This energy is changed into heat inside the fibres, increasing their temperature and creating fusion between the raised points of the pattern roller (Jirsak, and Wadsworth 1999). The process is used to bond a single nonwoven web, or to laminate several webs together (Batra, 1999). It is a comparatively cool and energy-efficient method of thermal bonding and therefore enables fibres that would be adversely affected by the higher temperatures of other processes to be used.

8.2.4 Finishing processes

The finishing stage of nonwoven production is central to the development of fabrics, being the primary means of achieving technical functionality and specific visual appearance in the fabric. Like nonwoven construction processes, nonwoven finishing processes have their origins in the textile and paper industries as well as links to the leather industry (Ahmed, 2007). Therefore, a variety of processes and technologies are available to create fabrics with specific properties. The selection of a particular process is therefore dependent on the intended application of the fabric.

Nonwoven finishing is characterised as either chemical or mechanical.

Chemical finishing

Chemical processes include, dyeing, printing, impregnation, coating and lamination. These processes are used to add colour and chemical finishes to nonwovens.

Dyeing

Nonwovens can be dyed with conventional textile dyestuffs using continuous or batch systems. In selecting dyestuffs (and auxiliaries), consideration of the fibre composition of the fabrics must be made. It has been noted that nonwovens made of conventional textile fibres tend to dye to a deeper shade than their woven or knitted counterparts due to the greater accessible surface area of fibre available and the absence of yarn twist and yarn intersections in the fabric construction (Ahmed, 2007). Pigments can also be used to colour nonwovens by printing onto the surface with a suitable binder. Ahmed explains that nonwovens made from extruded molten polymer can be dyed using 'dope' dying methods. Dyes or pigments are added to the molten polymer before the extrusion process (Ahmed, 2007).

Printing

Nonwovens can be printed using a range of traditional textile printing procedures, including rotary and flat screen printing, sublimation and transfer printing, engraved roller printing, pigment printing, discharge printing and digital ink-jet printing. As with dyeing, the fibre composition of the nonwoven needs to be considered, as does fabric weight. Pigment printing has the advantage of being suitable for any fabric regardless of fibre type but often produces a firm or stiff handle. Transfer printing is most suitable for polyester fabrics and, due to the high temperatures required by the process, the thermal stability of the fabric needs to be considered.

The increasing visibility of nonwovens in end-use areas such as home furnishings has led to a growth in the demand for printed nonwovens (Stückenbrock, 2003).

Coating

Coating nonwovens is one of the primary means of applying chemical finishes to nonwovens. Methods include the use of rollers through which the fabric is passed and a controlled amount of chemical preparation applied. Preparations can also be applied through the use of rotary screens extrusion techniques, scattering (of powder) and applying dots of paste (Ahmed, 2007).

Chemical finishes used in the nonwovens industry include flameproof and waterproof finishes, lubricants, stiffeners and softeners, anti-static agents, antimicrobial finishes and UV stabilisers. The technical functions achieved through the applications of such finishes can now also be integrated into fibres or filaments prior to construction by means of chemical additives (Ahmed, 2007).

Laminating

Laminating is the joining of two or more sheets of prefabricated materials. Two or more nonwovens can be joined together or nonwovens joined with other substrates including films, scrims and other textile materials. The process can be wet or dry. In the wet process, an additional adhesive is applied in the form of a solvent or dispersion to one of the substrates. In dry laminating, thermoplastic resins including powders or adhesives, or scrims made from thermoplastic filament fibres are applied to the substrates and activated by heat and often pressure. Methods of application and heating include, calender lamination, extrusion laminating, flat bed lamination and hot melt spraying (Ahmed, 2007).

Mechanical finishing

Mechanical finishing processes that are used in nonwoven manufacture include drying, shrinking, compacting and creping, calendering and embossing, moulding and stamping, perforating and slitting.

Drying

During production, nonwovens are often subjected to processes that cause distortion in the fabric structure. This is particularly true of wet processes. Drying is a necessary stage and, when the appropriate drying mechanism is used, fabric distortions can be corrected through relaxation (Stückenbrock, 2003). Drying methods include the use of ovens, through-air drums, hot flue dryers and infra red heaters (Ahmed, 2007).

Shrinking, compacting and creping

Shrinking processes are utilised to achieve fabrics with greater weight per unit area, greater density, more bulk and often more strength. Dry shrinkage methods are usually used on primarily synthetic fibres and can be achieved by heat treatments. Wet shrinkage methods are used on natural fibre webs through the use of hot-water and drying methods that do not involve tensioning of the fabric. Decorative relief surfaces can be achieved by combining shrinkable and non-shrinkable webs in the needling process and then shrinking. Such fabrics have been used as carpets or wall coverings (Stückenbrock, 2003).

Various compacting and creping processes have been developed from traditional mechanical compressing shrinking techniques involving the pressing and drying of wet or steamed fabrics (Ahmed, 2007). These processes are used to achieve nonwovens with better drape and handle. Visible creping can also be achieved (Stückenbrock, 2003).

Calendering and embossing

As outlined above, calendering involves the application of pressure and heat to fabrics through the used of heated rollers. In this process, heat and pressure are uniformly applied to the fabric. The fabric is usually compacted through this process and the surface smoothed or glazed. Patterned rollers can also be used to achieve embossed pattern for decorative or functional purposes.

Moulding

Moulding and shaping processes exploit nonwovens with thermoplastic properties (due to their fibre or chemical composition). The fabrics are moulded to a specific shape through the application of heat and pressure. Lamination is sometimes combined with this process. Products are often used in automotive interiors to create dashboards, rear shelves and head-liners. Waste fibres can be used in conjunctions with appropriate resins (Stückenbrock, 2003).

Perforating and slitting

Traditionally, nonwovens have been considered as having poor handle and drape in relation to clothing applications. Perforating and slitting processes are often employed to improve these qualities. The perforating process involves the use of hot needles on chemically bonded nonwovens. The needles perforate the surface creating small holes which are sealed by the heat. Slitting involves the cutting of slits into the surface of the nonwoven. Optimum cutting lengths and distances between cuts to give maximum softness with minimum decrease in strength have been established in indus-try. Stretching and setting slit nonwovens can enable three-dimensional effects to be obtained. Such fabrics are used in padded bras (Stückenbrock, 2003).

Surface effects

A range of surface effects can be achieved on nonwovens using techniques such as flocking, raising, shearing and singeing, suede finishing and softening as applied in traditional textile industries.

Flocking involves the application of synthetic fibres to specific areas of the fabric surface using adhesives. A three-dimensional pile (protruding fibres) on the surface is often applied to create decorative effects. Raising is also used to achieve pile effects – fabrics are passed between raising wires that produce surface pile. This can be done uniformly across the entire surface of the fabric or by using a stenciling method to create pattern effects.

Suede finishing involves the abrasion of the fabric surface to produce a dense pile and subsequent soft quality and matt appearance. Nonwoven fabrics can be polished to achieve better handle and a lustrous, shiny surface. In polishing, the fabric is subjected to rotating, heated, etched drums.

Mechanical softening processes are used to improve the handle and drape of nonwovens. In such processes, fabrics impact against grid-like structures at high speeds and steam as chemical softening agents are sometimes introduced for further effects. The entire fabric is affected in this process, as apposed to the surface only.

Shearing is used to remove surface fibre from fabric by blades. Singeing performs a similar function to shearing but is more intrusive and utilises gas flames to remove the fibres. These processes are used when smooth, clean surfaces are required; for example, when a fabric needs to be printed (Ahmed, 2007).

Making-up processes

One of the first successful applications of nonwovens was as interlinings for clothing (Assent, 2003). Nonwovens are still widely used for this purpose but are also used as the main material for protective clothing (Haase, 2003) and increasingly as the outer layer in fashion-based and technical garments (University of Leeds, 2007). The making-up of nonwovens is therefore an important consideration. Patterning, cutting and joining are considered very basically here.

Nonwovens for interlinings are processed in such a way as to give them an adhesive surface. The patterns for these nonwovens are designed and made together with the patterns for the upper fabric and garment lining during the pattern design and grading stages. Table 8.2 outlines the functional aspects of nonwovens that impact on their performance as interlinings, and highlights the stages of fabric production that affect these aspects.

The cutting of nonwovens into garment pieces is done using conventional fabric cutting techniques, using rotating-blade machines, band knife and vertical-blade machines, and automatic cutting machines. Laser and ultrasound cutting can also be employed. Laser cutting is useful when very fast, precise cutting is necessary and when special materials or small numbers of pieces are required. Due to the thermal effect of lasers, laser cutting essentially fuses the fabric edges, but can sometimes cause singeing or discoloration. Ultrasound also has a thermal effect but, compared with laser cutting, does not discolour fabric edges. Ultrasound can also be used to reduce energy costs in the cutting process (Rödel, 2003).

Nonwoven pattern pieces can be joined, as conventional textiles are, by sewing, and also by welding and sticking.

With regard to joining nonwovens to other fabrics, Rödel notes that the fabric's 'strength against slip should be sufficient for the nonwoven's

Table 8.2 Functional aspects of nonwoven interlinings and performance properties

Function	Coverage	Resistance to mechanical stress	Suitability for apparel	Processability	Serviceability
Properties	Weight per unit area Thickness	Tensile strength Abrasion resistance	Hand Surface smoothness	Cutting Setting, Sewing, Ironing, Appropriate dimensional change	Resistance to care processes Colourfastness
Functional elements/ Production considerations					
Fibres (all materials)	○	○	○	○	○
Web formation	○	○	○		○
Web bonding		○	○		○
Dyeing and finishing			○		○
Binder				○	○
Application techniques and formulation of binder			○		○

Source: Rödel, H. 'The Making Up of Finished Products' in *Nonwoven Fabrics: Raw Materials, Manufacture, Applications, Characteristics, Testing Processes*, Ed Albrecht, W, Fuchs, H, and Kittelmann, W, Wiley-Vch, Weinham, 2003, p. 527.

structure not to be disturbed in case of loads attaching the structure next to the solid seam' (2003, p 477). As well as performing a function in the final garment, nonwoven interlinings can improve sewing if positioned accurately.

Welding processes can be used to join nonwovens made from thermoplastic fibres or coated with a thermoplastic. Welding involves the joining of two such substrates by applying heat and pressure. Techniques used include high-frequency welding, which utilises the electric field between two electrodes to join the fabric, ultrasonic welding, hot wedge welding in which heat is directed to the seam by a heat wedge immediately before the seam is pressed between two rollers, hot-air welding in which the heat is directed to the seam by hot air before pressing, and gluing. The gluing process is commonly used to join interlinings to the upper/outer garment material. Adhesive tapes, which are usually nonwovens that have been imbued with adhesive properties during manufacture, are activated by heat and pressed to the seam between rollers (Rödel, 2003)

Assent (2003) suggests that, currently, 80% of interlinings are incorporated by bonding and 20% by sewing.

8.3   End-of-life considerations for nonwovens

Growing environmental awareness and the introduction of landfill legislations have prompted the necessity within all areas of industry to consider the whole life cycle of products from their initial stages of design to final manufacture. Environmental sustainability needs to be addressed when developing new products such as smart clothes and wearable technologies. Sustainability needs to be considered in regard to all product components, and nonwovens are not exempt from this. Two ways in which these issues have been addressed within nonwoven production are briefly outlined here – the use of reclaimed fibres and the development of bio-degradable polymers.

8.3.1 Reclaimed fibres

It has been suggested that, up until the 17th century, reclaimed fibres were used exclusively in the paper-making industry. From this point, technology developed to enable the spinning of longer reclaimed fibres into yarn. Gulich (2003) notes that recycling in this way was encouraged by the high cost and limited availability of textile raw materials. However, since the development of synthetic fibres, fibre availability has not been an issue and the reclaiming and re-using of fibres has continued mainly for environmental reasons, but also as an economical option in production. Due to the processing required, reclaimed fibre properties are different from those of primary fibres, often characterised by their being varied in length and quality and available primarily as blends. However, fibre properties can be maximised by using the correct reclaiming technologies. If specific minimal requirements are met, they can be used in a range of nonwoven production processes. Nonwovens made from reclaimed fibres are used in geotextiles and agrotextiles, building and insulation materials, and automotive textiles. Gulich (2003) notes the following as further examples of nonwoven applications in this area: reclaimed fibres from wool used in laminates, reclaimed aramids in protective clothing and reclaimed micro-fibres used in insulation.

8.3.2 Biodegradable polymers

As noted, man-made fibres dominate nonwoven production – in particular, polypropylene and polyester (Wilson, 2007). In spunbonding processes, these fibres are commonly used due to their low cost and relatively good strength. A study carried out at The Saxon Textiles Institute (STFI) noted, however, that the chemical composition of these fibres is problematic with regard to disposal and composting, and that from an ecological perspective, biologically degradable nonwovens would be desirable (Schilde et al., 2006). In response to this, a series of investigations into the development of bio-degradable, hydrodynamically bonded spunbond nonwovens were carried out from which the use of bio-polymers was identified as a means of opening up innovative product development in this area. Suitable production parameters were identified and the properties of spunbond nonwovens made from four polymers – polyester amide (BAK®402/006), aliphatic–aromatic co-polyester (Easter Bio® GP), polylactide (NatureWorks® H2000 D) and recycled polylactide (NatureWorks® 2002 D) – were accessed. The study showed that all four polymers resulted in fabrics that retained a soft handle and good elasticity. The polyester amide and aliphatic–aromatic co-polyester showed the best results in terms of bio-degradability. Research such as this points the way to ecological developments within nonwoven production.

8.4   Smart and perfomance clothing applications

Nonwovens can be considered as an integral element of smart clothes, wearable technologies and performance fabrics. Their uses as key support and composite materials within such products, their growing importance as outer (or shell) garment fabrics, their use within dance and performance art and their central role in medical and protective fabrics are outlined here.

8.4.1 Support materials

Smart clothes rely on the good design of the garment that makes the technology wearable. Nonwovens play a central role in garment and footwear technology as support components. Key areas are their use as interlinings and as uppers and lining materials in footwear.

Interlinings and insulation

Assent (2003) summarises the history and use of nonwovens as linings in *Nonwoven Fabrics*. It is highlighted that the term 'nonwoven interlinings' is used to describe nonwoven materials that are incorporated into articles of clothing to fulfil a range of functions. Assent writes that these functions can be broadly divided into three areas – shaping and supporting, stabilising and stiffening, and providing bulk. Nonwovens are bonded to large areas of garments to provide shaping and support without impacting on the properties of the shell fabric. They are used on smaller garment areas such as collars and cuffs to provide stiffness and stability.

The ability of nonwovens to trap a high percentage of air in comparison with other textile materials makes them ideal insulating components in garments. Bhat and Malkan (2007) note that meltblown nonwovens in particular create good insulation due to the large surface are of the web, which creates a significant drag force on air convection currents passing through the fabric. This trapping of still air provides high insulation which, they note, has been successfully exploited in products such as Thinsulate® by 3M (2008).

The high loft of certain nonwovens is also used to provide a backing for quilting or decorative embroidery.

Different performance properties and characteristics are required from nonwovens, depending on their function within the garment. The method of manufacture, raw material and finishing processes used impact on the fabric properties.

Footwear

Nonwovens are used in both the upper and inner components of footwear. They can produce 'leather-like' polymeric materials that are used as upper components. Stoll and Brodtka (2003) suggest that, although these fabrics do not have the same quality with regard to the performance properties of natural leathers, they provide a necessary substitute. 'Clarino' and 'Parcassio' by Kuraray (JCN Newswire, 2001) are examples of such products. They consist of polyurethane nonwoven substrates combined with layers of microfibres which are covered with a moisture and air permeable polyurethane layer. Finishing methods similar to those used on real leather are then applied.

Nonwovens are also used widely as inner materials in footwear, due to their high stability and non-fray qualities. Products such as Cambrelle® shoe linings, which are made from nonwovens composed of bi-component fibres, provide comfort, abrasion resistance and moisture management within footwear. Cambrelle's (2006) Cambrelle®+ utilises a specialty fibre called AMICOR which has locked-in anti-bacterial and anti-fungal agents that enable increased product performance in athletic shoes, and outdoor and casual footwear.

8.4.2   Outer garment alternatives

Nonwovens are now becoming viable alternatives for the outer shell of garments as well as providing support as interlinings. Developments in spunlaid and bi-component technologies by Freudenberg (2008) have resulted in Evolon® fabrics. These fabrics are similar to conventional textile products in that they are soft, strong, have good drape and are light-weight and washable. Their ability to provide UV and wind protection, thermal insulation and breathability, and their fast-drying qualities, mean that they can be successfully applied in sportswear to prevent post-exercise chill effects.

The Wool Research Group has conducted research into the development of wool and wool blend nonwovens for sports and fashion garments (Campbell, 2006). Fabrics produced using intimate blends of synthetics and wool within the hydroentanglement process were developed alongside wool-based needlepunched fabrics to which shrink-resist treatments were applied to create surface effects. It was reported that as little as 10 or 20 per cent wool fibre in these fabrics gave a 'wool' quality to the fabric. Similar developments by the Canesis research group in New Zealand have exploited shrink-resistant finishes on stitchbonded nonwovens to produce fabrics with built-in surface patterning texture for fashion applications (McFarlane, 2006; Anderson, 2005).

For some time now nonwovens have been used for creative purposes by a number of high profile designers. Hussain Chalayan's use of Tyvek in his 1995/1996 autumn/winter collection to produce the envelope dress, the use of a mouldable nonwoven by 'Boudica' to create a sculpted bodice, and the decorative use of layers of nonwovens by design duo 'ieuniform' are a few examples (Braddock and O'Mahoney, 2005). As Braddock points out, nonwovens are utilised by designers for their technical advantages, but also because their economic advantage as a base material means more money can be spent on expensive finishing processes.

The possibilities of using nonwovens as serious alternatives for conventional textiles in fashion has been brought even nearer by extensive exploration of the use of nonwovens by fashion students at Leeds University. Numerous men's and women's collections that incorporate light and heavy-weight nonwovens, combined with fabric manipulation processes, are traditional and non-traditional tailoring techniques have been developed and shown during a series of catwalk shows from 2005 onwards (University of Leeds, 2007).

8.4.3   Performance and art fabrics

The unique technical and aesthetic properties of certain nonwovens have been exploited by textile artists within both performance and installation contexts since the 1980s. More recently, Marie Blaisse's work (Braddock and O'Mahoney, 2004) has exploited the technical properties of industrial materials including nonwovens, rubbers and foamed polyamides to produce sculptural and flexible forms for dance and performance. Blaisse's explorations with thermoplastics, lamination and vacuum forming have resulted in extreme three-dimensional geometric forms that are wearable (Betsky, 2004). Figure 8.8 shows an example of Blaisse's work with thermoplastic nonwovens.

Nonwovens are also the base material of Janet Emmanuel's conceptual garments that explore the notion of Acoustic Shadows. Emmanuel's work is the result of research involving the welding of textile substrates using ultrasonics and pinsonics. The technical strength and fragile aesthetic of extruded nonwovens are exploited in Emmanuel's pieces which 'demonstrate the animation of man-made materials'. Various synthetic nonwoven structures are used within the work and are bonded with silk fabrics and component objects, such as feathers, that extend from the surface. Emmanuel writes that the tonal depth created makes 'acoustic shadows', echoing the inaudible sound used to bond the materials (Emmanuel, 2003).

8.4.4   Medical and protective clothing and products

Medical and protective fabrics form a substantial part of the nonwoven market. Protective clothing is often single-use and so the economical advantages of nonwoven production alongside their ability to be engineered make them suitable materials in this area.

Medical clothing and products

Within medical and surgical environments, nonwovens are found in various product forms including disposable surgical drapes, gowns, masks, head and footwear and sterilisable packaging. The ability to engineer the fibre matrix, combine various layers of fibre, and apply a range of finishes through the

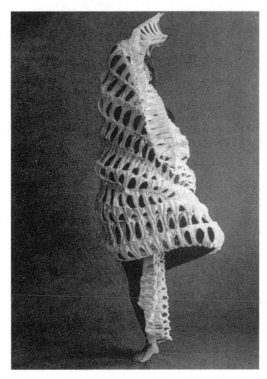

8.8 Nonwoven costume, Maria Blaisse in collaboration with students from the University of Kassel (Braddock and O'Mahoney, 1994).

various production technologies available, enables fabric functions essential to such products to be achieved. Bhat and Malkan (2007) note that spunbond technology enables fabric breathability, resistance to fluid penetration, sterilizability, impermeability and a lint-free fabric structure to be obtained. Hydroentanglement technology is also employed in this area to achieve soft and comfortable fabrics (Anand, *et al.* 2007). The insertion of continuous films into the fibre structure and the application of chemical coatings are also employed as a means of creating barrier properties. Hoborn notes that the production of nonwovens suitable for use within medicine relies on an understanding between the raw material suppliers, and fabric and device manufacturers, of the totality of the work and the need for an unbroken chain of safety measures (Hoborn, 2003)

Protective clothing

Protective and safety textile forms are considered as technical textiles. Nonwovens can be developed to fulfil a range of protective functions including protection against chemicals, micro-organisms, radioactivity, moisture and cold, heat, fire and flame, and physical attack. Various technologies are employed to give nonwovens such capabilities. Aspects of chemical, heat, moisture and flame resistance technologies are outlined below.

Hasse (2003) writes that the most important aspects of chemical protection are the barrier protection properties of the fabric relating to both penetration and permeation of chemical substances. When seeking chemical and dust protection, spunbond and meltblown technologies are often combined to produce a barrier effect (Haase, 2003). Additional protection properties can be achieved using specialist finishes. Hasse notes that products within the Tyvek® range utilise antistatic and moisture-absorbing finishes for products used in contexts where sparks caused by minute amounts of static electricity triggered by unfinished synthetic fibre surfaces are enough to cause an explosion.

In moisture- and cold-protection, insulation technology in the form of linings described in the section on support materials, is employed. Phase-change technology based on NASA research in the 1980s is now employed in heat-insulating or cooling materials for use in extreme temperature conditions. Paraffins are enclosed in micro-capsules that are incorporated into fibres, nonwovens, coatings or foams. Hasse explains that these materials possess the ability to change their state of aggregation within an adjustable temperature range and to store or release energy during the phase transition according to the direction of the temperatures (Haase, 2003).

Hydroentangled aramids are well established as moisture barrier substrates in fire fighting garments (Anand, et al. 2007). The fabrics provide high temperature resistance as well as softness and drape. These nonwovens are often used as linings within protective, woven, shell fabrics.

8.5   Summary

Nonwovens are a growing sector within industry, providing economical and often environmentally advantageous alternatives to conventional textile materials for application in a range of product areas. They currently form essential component parts within specialist and non-specialist apparel, including the sports, performance and technical clothing and footwear. Their engineering ability, adaptability and economical and environmental viability make them important fabrics to consider in the development of smart clothes and wearable technologies.

8.6   Sources of further information and advice

The reference material cited provides more depth on the information and ideas presented in this chapter and the following organisations specialise in nonwovens:

The Nonwovens Network, www.nonwovensnetwork.com
European Disposables and Nonwovens Association (EDANA), www.edana.org
Association of the Nonwovens Fabric Industry (INDA), http://www.inda.org/

8.7   References

3M (2008) 3M *Thinsulate™ Insulation – Products – Thinsulate™ Insulation (US)* [URI   http://solutions.3m.com/wps/portal/3M/en_US/ThinsulateInsulation/Insulation/Thinsulate-Products/Thinsulate-Insulation, accessed May 21st 2008]

AHMED, A. I. (2007) 'Nonwoven Fabric Finishing.' In: Russell, S.J. (ed) *Handbook of Nonwovens*. Cambridge: Woodhead Publishing Limited.

ALBRECHT, W. (2003) 'Fibrous Material'. In: Albrecht, W, Fuchs, H, and Kittelmann, W. (eds) *Nonwoven Fabrics: Raw Materials, Manufacture, Applications, Characteristics, Testing Processes*. Weinheim:Wiley-Vch.

ANAND, S., BRUNNSCHWEILER, D., SWARBRICK, G. and RUSSELL, S. J. (2007) 'Mechanical Bonding.' In: Russell, S.J. (ed) *Handbook of Nonwovens*. Cambridge: Woodhead Publishing Limited.

ANDERSON, K. (2005) *Nonwoven Fabrics in Fashion Apparel*, [TC]² Bi-weekly Technology Communication Newsletter. [URI http://www.techexchange.com/thelibrary/nonwovenfabrics.html, accessed May 21st 2008].

ASSENT, H-C. (2003) 'Nonwovens for Apparel' In: Albrecht, W, Fuchs, H, and Kittelmann, W. (eds) *Nonwoven Fabrics: Raw Materials, Manufacture, Applications, Characteristics, Testing Processes*. Weinheim:Wiley-Vch.

BATRA, S. K., *et al.* (eds). (1999) *Nonwoven Fabrics Handbook*. Raleigh: INDA Association of the Nonwoven Fabrics Industry.

BETSKY, A. (2004) *A Modern Movement – The Work of Maria Blaisse*. Presented by the Central Museum, Utrecht in association with PICA.

BHAT, G. S. and MALKAN, S. R. (2007) 'Polymer-laid Web Formation.' In: Russell, S.J. (ed) *Handbook of Nonwovens*. Cambridge: Woodhead Publishing Limited.

BLAISSE, M. (2004) 'Tube Design'. Keynote Presentation at *The Space Between Conference*, 15th–17th April, Perth, Australia. [URI http://www.thespacebetween.org.au/dialogue.html, accessed May 21st 2008].

BRADDOCK, S. and O'MAHONEY. (1994) 'Maria Blaisse' in *Textiles and New Technology 2010*, published to accompany the Crafts Council Exhibition '2010' by Artemis, London, pp 54–55.

BRADDOCK, S. and O'MAHONEY, M. (2005) *Techno Textiles 2 – Revolutionary Fabrics for Fashion and Design*. London: Thames and Hudson.

BRITISH STANDARDS INSTITUTE. (1992) *Textiles – Nonwovens – Definitions, BS EN 29092: 1992, ISO 9092:1988*, London: British Standards Institute.

BRYDON, A. G. and POURMOHAMMADI, A. (2007) 'Dry-laid Web Formation'. In: Russell, S.J. (ed) *Handbook of Nonwovens*. Cambridge: Woodhead Publishing Limited.

CAMBRELLE (2006) *Cambrelle⁺ with Amicor*. [URI www.cambrelle.com/innovations/cambrelle.asp, accessed May 21st 2008].

CAMPBELL, M. (2006) 'Spinning the Web or Weaving the Magic. Expanding the Boundaries', *Nonwovens Network 1st International Conference*, 25th–26th May, Leeds, UK.

CHAPMAN, R. (2007) 'Chemical Bonding.' In: Russell, S.J. (ed) *Handbook of Nonwovens*. Cambridge: Woodhead Publishing Limited.

EDANA (2007) *Making Non-wovens* [URI http://www.edana.org/story. cfm?section=|edana_nonwovens&story=making.xml#webbonding, accessed May 21st 2008].

EHRLER, P. (2003) 'Binders.' In: Albrecht, W, Fuchs, H, and Kittelmann, W. (eds) *Nonwoven Fabrics: Raw Materials, Manufacture, Applications, Characteristics, Testing Processes*. Weinheim: Wiley-Vch.

EHRLER, P. and SCHILDE, W. (2003) 'Chemical methods.' In: Albrecht, W, Fuchs, H, and Kittelmann, W. (eds) *Nonwoven Fabrics: Raw Materials, Manufacture, Applications, Characteristics, Testing Processes*. Weinheim: Wiley-Vch.

EMMANUEL, J. (2003) 'Art and Industry an Innovative Approach: New Textile Bonds using Ultrasonic Technology.' In: *Book of Proceedings for Fibrous Assemblies at the Design and Engineering Interface, International Textile Design and Engineering Conference INTEDEC* September 22–24th Heriot-Watt University, Edinburgh.

FREUDENBERG & CO. (2008) *Clothing – An Unexpected Performant Fabric for Sportswear, Leisurewear and Sportswear.* [URI www.evolon.com/microfilament-clothing-fabric.html, accessed May 21st 2008].

GULICH, B. (2003) 'Reclaimed Fibres.' In: Albrecht, W, Fuchs, H, and Kittelmann, W. (eds) *Nonwoven Fabrics: Raw Materials, Manufacture, Applications, Characteristics, Testing Processes*. Weinheim: Wiley-Vch.

HAASE, J. (2003) 'Nonwovens for Protective.' In: Albrecht, W, Fuchs, H, and Kittelmann, W. (eds) *Nonwoven Fabrics: Raw Materials, Manufacture, Applications, Characteristics, Testing Processes*. Weinheim: Wiley-Vch.

HOBORN, J. (2003) 'The Use of Nonwovens in Medicine – Safety Aspects.' In: Albrecht, W., Fuchs, H., and Kittelmann, W. (eds) *Nonwoven Fabrics: Raw Materials, Manufacture, Applications, Characteristics, Testing Processes*. Weinheim:Wiley-Vch.

JCN NEWSWIRE (2001) 'Kuraray Announces Renaissance in Man-made Leather Field Development and Marketing of "Parcassio".' [URI http://www.japancorp.net/article.Asp?Art_ID=785, accessed May 21st 2008].

JIRSAK, O. and WADSWORTH, L. C. (1999) *Nonwoven Textiles*. Durham, North Carolina: Academic Press.

KITTLEMANN, W and BLECHSMIDT, D. (2003) 'Extrusion Nonwovens.' In: Albrecht, W., Fuchs, H., and Kittelmann, W. (eds) *Nonwoven Fabrics: Raw Materials, Manufacture, Applications, Characteristics, Testing Processes*. Weinheim: Wiley-Vch.

KITTLEMANN, W, DILO, J P, GUPTA, V P and KUNATH, P. (2003) 'Needling Process'. In: Albrecht, W, Fuchs, H, and Kittelmann, W. (eds) *Nonwoven Fabrics: Raw Materials, Manufacture, Applications, Characteristics, Testing Processes*. Weinheim: Wiley-Vch.

MCFARLANE, I. (2006) 'Nonwovens in Apparel Expanding the Boundaries,' *Nonwovens Network 1st International Conference*, 25th–26th May, Leeds, UK.

RÖDEL, H. (2003) 'The Making Up of Finished Products.' In: Albrecht, W., Fuchs, H., and Kittelmann, W. (eds) *Nonwoven Fabrics: Raw Materials, Manufacture, Applications, Characteristics, Testing Processes*. Weinheim: Wiley-Vch.

SCHILDE, W., BLECHSCHMIDT, D., ERTH H. and GEBHARDT, R. (2006) 'Latest Results in Processing Biologically Degradable Polymers by Spunbonding Technology.' *Autex Conference*, 11th–14th June, NC State University, College of Textiles.

SPINDLER, J. (2003) 'Adhesive fibres.' In: Albrecht, W., Fuchs, H., and Kittelmann, W. (eds) *Nonwoven Fabrics: Raw Materials, Manufacture, Applications, Characteristics, Testing Processes*. Weinheim: Wiley-Vch.

STOLL, M. and BRODTKA, M. (2003) 'Nonwoven Support Materials for Footwear.' In: Albrecht, W., Fuchs, H., and Kittelmann, W. (eds) *Nonwoven Fabrics: Raw Materials, Manufacture, Applications, Characteristics, Testing Processes*. Weinheim: Wiley-Vch.

STÜCKENBROCK, K H. (2003) 'Chemical Finishing.' In: Albrecht, W., Fuchs, H., and Kittelmann, W. (eds) *Nonwoven Fabrics: Raw Materials, Manufacture, Applications, Characteristics, Testing Processes*. Weinheim: Wiley-Vch.

SWARBRICK, G. (2007) 'Mechanical Bonding.' In: Russell, S.J. (ed) *Handbook of Nonwovens*. Cambridge: Woodhead Publishing Limited.

UNIVERSITY OF LEEDS (2007) 'Fashion with Nonwovens – Galleries.' Leeds: University of Leeds. [URI http://www.nonwovens.leeds.ac.uk/fashionwithnonwovens/galleries.html, accessed May 12th 2008].

UNIVERSITY OF LEEDS, SCHOOL OF DESIGN, 'Fashion with Nonwovens – Galleries,' http://www.nonwovens.leeds.ac.uk/fashionwithnonwovens/galleries.html, accessed Sept 8th, 2007.

WILSON, A. (2007) 'Development of the Nonwovens Industry.' In: Russell, S.J. (ed) *Handbook of Nonwovens*. Cambridge: Woodhead Publishing Limited.

9

Sensors and computing systems in smart clothing

A. J. MARTIN, University of Wales Newport, UK

Abstract: 'Wearable electronics' embodies the convergence of mobile, personal and wireless communications technologies, with emerging textile, fibre and polymer based components. This chapter provides a summary of sensors and output devices used within wearable applications along with examples of their usage. It also describes the underlying principles of contemporary computer architecture and some of the challenges faced by wearable technology designers. The chapter is aimed primarily at those new to the field of wearable electronics, especially clothing and interaction designers.

Key words: wearable electronics, wearable sensors, wearable user interface, wearable location systems, wearable computing.

9.1   Introduction

There are many hundreds of different electronic sensors that have been developed for use in a wide variety of industrial, medical, scientific, military, consumer, communication and other fields since the discovery of electricity and development of electronic devices during the late 1800s. This chapter focuses primarily upon sensors and computing systems used within wearable applications. The field of wearable electronics embodies the convergence of mobile, personal and communications technology with emerging textile, fibre and polymer based components. As part of this transition, engineers from both textile and electronics fields are finding ways of moving from the relative stability of a desktop or handheld device to the harsh environment of flexible, washable and ultimately practical clothing. The following chapter provides a summary of sensors and output devices used within wearable applications, along with examples of their usage. It is aimed primarily at those new to the field of wearable electronics, especially clothing and interaction designers.

9.2   Transduction

A sensor converts some form of physical property such as a chemical process, light intensity, sound waves or mechanical change into a form that can be interpreted by an electrical circuit. This process of converting one form of energy into another is called transduction. Similarly, an electronic signal can be converted into another form of energy such as motion or sound. In electronics, a transducer that converts electricity into motion is called an actuator.

There are many different types of transducer that convert one form of energy to another within the field of wearable electronics. These include:

- Electroacoustic – conversion between electrical signal and sound waves.
- Electrochemical – conversion between chemicals and electricity.
- Electromagnetic – conversion between magnetic phenomena and electricity.
- Electromechanical – conversion between a mechanical process and electricity.
- Thermoelectric – conversion between temperature and electricity.

The number of discrete transducers within an electronic system may range from only one or two for simple applications, to many tens or hundreds in more complex systems. A wearable system could include several of each of the above transducers within it: electroacoustic for sound input and output; electrochemical for battery power; wireless communications using an electromagnetic radio; electromechanical motion sensors and pressure sensitive buttons; thermoelectric components for temperature management and monitoring. Indeed, laptop computers and some mobile phones already incorporate similar systems. Table 9.1 is by no means exhaustive, but it does highlight a wide range of wearable-related input sensors and output devices. Given that there are over 40 sensor types listed, it is impossible to describe each one in any depth within this chapter. What follows is a sample of applications that have been developed using some of these sensors.

9.2.1 The application of sensors

Sensors are the eyes, ears and sense of touch within an electronic circuit; however, it is important to realise that a sensor in isolation tells you very little. It is the role of the accompanying circuitry, and often specially written software, to interpret and act upon readings from any given sensor. A light sensor circuit, for example, provides only information regarding the current light level, or to be more accurate the electrical resistance across the light dependent resistor (LDR) component. If the LDR were to be used as a trigger to turn-on a lamp when it gets dark, either an analogue or digital circuit could be designed to perform this action. Many outdoor security lamps use a simple analogue circuit to achieve this functionality. In the case

Table 9.1 Range of wearable-related input sensors and output devices

Type	Input	Output
Optical	Light dependent resistor (LDR) Photodiode CMOS (e.g. camera phone) CCD (e.g. video camera) Laser range-finder Infrared camera Passive Infrared (PIR) Image recognition/gestures	Light emitting diode (LED) Organic LED (OLED) Polymer LED (PLED) Liquid crystal display (LCD) E-ink display Projected laser display Infrared LED (non-visible)
Acoustic/ atmospheric pressure	Microphone Speech recognition Audio recording Ultrasonic detector Barometer/altimeter	Loudspeaker Piezospeaker Headphones In-earphones (wired or wireless) Pre-recorded sound Speech synthesis Ultrasonic transmitter
Movement/ vibration	Tilt-switch Vibration sensor Accelerometer (2D and 3D) Potentiometer Electronic induction Piezoresistive fabrics Pedometer	'Tactors' (phone vibrator) Electronic motors Solenoids Shape memory alloys (SMA) Electroactive polymer Actuators Pneumatics
Buttons/touch input	Mechanical switches Textile switch (i.e. Elektex) Fabric keyboard Polymer switch (i.e. softswitch) Laser keyboard Capacitive touch screen	
Temperature	Thermistor Resistance temperature detectors (RTD)	Heat (via electrical resistance) Thermochromic Inks
Biometric	Electrocardiogram (ECG) Heart-rate (e.g. polar sports) Galvanic skin response (GSR) Electroencephalography (EEG)	Electricity (i.e. pacemaker) Mechanical (exo-suit)
Location	Global positioning system (GPS) Ultra wideband (UWB) radio WIFI, Bluetooth and Cell ID Ultrasonic Infrared beacons Visual markers (barcodes) Radio frequency identification (RFID)	Location can be transmitted via a communications system, such as a 3G or WIFI network. Sensors within the environment could also detect wearable transmitters, e.g. Bluetooth, RFID, Image recognition, infrared beacon, visual markers

of an analogue circuit, a user-adjustable dial called a potentiometer is moved to calibrate the device. Alternatively, for a digital circuit, a value commonly between '0' and '255' would be 'programmed in' to trigger the lamp to switch on when it gets dark enough. In a digital system, the value of '128' would switch on the light when the sensor reported light levels dropping below 50% of its range, whereas an analogue circuit would rely on a physical dial being turned half way.

Although this example is simple, it illustrates that sensor information is only a particular value captured at a specific moment in time. It is the role of hardware developers, software designers or even end-users to determine how a system responds to changes from a given input – for example, a wearable camera automatically taking a photograph when sudden changes in light-levels occur.

9.2.2 Digital and analogue electronics

In digital electronics, the electrical signals produced by sensors are converted into numeric form; this is called analogue-to-digital conversion. Once the sensor's information is in a digital format, the signal can be analysed mathematically using specifically written software.

A digital signal will commonly be delivered as a numerical value ranging from 0 to 255. As in the previous example, the number '0' may represent complete darkness and '255' denote bright sunlight. This is known as bit depth or resolution. Table 9.2 shows the number of possible range of values available within a given bit depth. In digital photography, 8 bits are used for each point or 'picture element' (pixel) within the image, with ranges of 0–255 for red (R), green (G) and blue (B), giving 24 bits in total for an RGB image.

Most sensors provide information, in analogue form, as a range of changing voltages or resistive properties. In order to use these sensors within a digital system a process called analogue-to-digital conversion (ADC) is

Table 9.2 Comparison of bit depth and resolution possible

Bit depth	Resolution range possible
1	0–7
2	0–4
4	0–16
8	0–256
16	0–65 536 (thousands)
24	0–16 777 216 (millions)
32	0–4 294 967 295 (billions)
48	0–281 trillion

undertaken. This process involves 'sampling' the analogue signal – essentially measuring the electrical characteristics of a sensor component many tens, hundreds, thousands or millions of times a second. The speed at which a component is sampled will largely depend upon the application; this is called 'sample-rate' and is usually denoted as Hertz (Hz). For simple applications this may only be a few times per second – 15 Hz would equal a sample rate of 15 times per second and would be more than adequate for the outdoor light sensor circuit described previously . If an analogue signal is required to be sampled more accurately, the sample-rate is increased. CD-quality audio, for example, is sampled at a rate of 44 100 Hz. If the sample-rate is lowered, the quality decreases. A telephone operates at a sampling frequency of around 8000 Hz. This introduces artefacts and distortions to the sound, especially noticeable with words containing 's' sounds.

The reverse process of analogue-to-digital conversion is digital-to-analogue, DAC. One of the most common applications is to convert digital audio data into electrical currents that are used to drive the magnetic coils inside a loud speaker, therefore moving a diaphragm to create sound. Other applications may include controlling variable-speed motors or generating radio frequencies with wireless transmission systems.

In analogue electronics, the functionality of a system is largely determined by the design of the internal circuitry. Usually the operator or 'end user' of the device has only a limited set of pre-defined controls. These controls would directly affect specific properties of the circuitry, such as sliding a potentiometer to adjust the volume level of a radio or pressing a button to turn on a light. Similarly, sensors attached to an analogue device may effect specific functions, such as automatically turning on a security light when it gets dark. In order to fine-tune these types of device, a small amount of manual adjustment is usually required.

As applications increase in complexity, so too does the circuitry in an analogue system. If the product requirements change, then the design of the circuitry has to be altered. Such changes have to be done by the manufacturer – perhaps new electronics would be designed or components adjusted or replaced. In contrast, a digital circuit would contain generic re-programmable 'chips' into which variations of software could be loaded.

With digital systems, minor alterations are often trivial. Hardware devices can be re-programmed in the factory, so one hardware 'platform' can form the basis of many separate products. If an adjustment is required, embedded software within the device can be updated before or even after the product leaves the factory. The embedded software may enable some features in one product and disable certain features in another, perhaps the only physical difference being the model number printed on the case. The term 'firmware' is commonly used to denote this kind of embedded software, newer models having more recent 'firmware versions' installed.

With more sophisticated digital electronic devices, capabilities such as remote updating can be added. The widespread use of the Internet means that device manufacturers can now update software remotely and it is now commonplace to receive updated firmware from device manufactures, sometimes free and sometimes at a cost. This is the model Apple is currently pursuing with their latest iPod Touch and iPhone product lines.

With the wide availability of digital components, most, if not all, wearable electronic devices are digital rather than analogue. As discussed, there may well be some analogue sensors included, but information gathered from these components would be converted into digital signals for further processing or transmission over wireless networks.

9.3   Information processing

Deciding when and how to act upon a sensor's input is often called the 'processing' stage within a system. This concept is fundamental to how electronic systems, and in particular computing systems, work.

A computer is, in essence, a highly sophisticated processor. At the very heart of a computer is the central processing unit (CPU). A modern CPU can perform many millions of calculations per second. Increasingly, high-performance CPUs are being incorporated into mobile devices such as mobile phones and portable entertainment devices, and these devices now rival the computing power of desktop machines of a few years ago.

The CPU's role is to take a piece of 'inputted' information or 'data'; perform a mathematical operation upon it, termed the 'process', and send that new information back out into other parts of the system, known as 'output'. The input could be derived from any number of devices: light level information, key press events, mouse movements or information sourced from a network. The output could comprise a multitude of purposes, from signals to generate sounds, images to be displayed on a screen, commands to activate motors or data to be sent over the Internet.

9.3.1 Microprocessors

The terms microprocessor usually refers to a central processor which is attached to external chipsets for communication, memory and other operational features. A PC contains a microprocessor or central processing unit CPU, upon which the operating system runs. A microcontroller commonly denotes a processor used within embedded applications and includes various input/output connections. This distinction between microprocessor and microcontroller is becoming increasingly blurred as computer equipment becomes smaller and more embedded into objects; and embedded processors become more sophisticated, gaining internet connectivity and networking with other devices.

Microprocessors are now an integral part of our everyday lives as they are embedded within many of the electronic devices we use today. A typical household contains tens if not hundreds of microprocessor devices, which can be found within central heating systems, wrist watches, washing machines, car engines, televisions, fixed-line telephones, radios, cookers, dishwashers, clocks, computers, set-top boxes, games consoles, keyboards, mice, mobile phones, burglar alarms, children's toys, greetings cards and credit cards. Such is the proliferation of this technology that almost any electronic device will contain some form of embedded microprocessor chip.

The venerable Zilog Z80 microprocessor, created in 1976, has been used in a wide variety of applications and configurations for different markets. Many of the early home computer systems and games consoles were based upon variants of the Z80 chip. Zilog continues to sell variations of the Z80, promoting their use within 'industrial control, communication devices, automation, security, consumer electronics, medical equipment, entertainment devices, vending machines, and other embedded Internet applications.' (Zilog Inc, 2008).

Other popular processor families include the 'ARM' architectures from ARM Ltd, used in many mobile phones, portable games consoles, PDAs, sat-nav systems and portable media players. Other commonly used processors are available from Intel, Freescale Semiconductor, Advanced Micro Devices (AMD), Texas Instruments, VIA Technologies, and many other suppliers.

9.3.2 Computer hardware

As the personal computer industry evolved, microprocessor manufacturers strived to provide the most appealing and powerful CPU chip sets. Common slogans such as 'Intel Inside' were used to differentiate products and vendors. As the hardware became more capable, so software developers invented more resource-intensive applications and operating systems. This constant evolution of hardware was predicted by Gordon Moore (1965) – his law states that 'the number of transistors on a chip will double about every two years' (Intel Corporation, 2008). This has remained true for the last 40 years. Moore's law can sometimes be interpreted as the processing speed increasing year on year; however, it is more interesting from the perspective of wearable applications to interpret the law as a continuous 'shrinking' of chip sizes, i.e. going from 10 mm to 5 mm, to 2.5 mm, to 1.25 mm and so on. With this interpretation, albeit somewhat simplistically, one can see how chips have gone from 'coin size' chip to 'fibre size' equivalent in little over a decade. Many other factors are involved though, particularly in terms of power supply to the device, heat dissipation from such a small surface area, and moving from a personal computer centric 'hard case' to a flexible and washable form factor necessary for wearables.

Currently, much microchip development is focused not on shrinking the actual form of the component, but on squeezing more processors into a single 'multi-core' chip, therefore delivering improved computing performance for personal computers. This focus would have to change before any major advances in chip size could be realised for wearable applications. Even mobile-specific hardware components are still too big for embedding into actual textile fibres. So, for the time being, wearable hardware developers will have to miniaturise as much as possible, encasing hard components into waterproof clip-on components.

9.3.3 Computer software

The personal computer (PC) software industry is founded upon the basis that a generic hardware platform provides a consistent set of technologies upon which software applications can be built. This hardware platform would include things such as a screen, keyboard, mouse, data storage, central processor and system memory. Once a hardware standard has been agreed, then an 'operating system' can be written for these generic 'boxes' and, in theory, should work on any generic 'box' supplied by different hardware vendors.

A computer's operating system, often-abbreviated to 'OS', abstracts the hardware layer from applications that run on the system. For example, if a word processor wants to read a file from the hard disc, the software author would have only to write a simple line of code such as: 'fileOpen('shoppinglist. txt')'. The operating system and firmware would then perform all the actions required to do that, including spinning the hard disc up to speed, moving the read-heads to the correct location, counting the magnetic ones and zeros as they spin past the head, moving the data into memory, verifying the data and converting the binary data into text.

Similarly, the operating system would provide functions to show text on the screen, save altered data back to the disc, communicate via the Internet, activate fans to prevent overheating, and many thousands of other functions and features. Figure 9.1 shows a block diagram of a typical computing platform. User interface with software applications through a graphical user interface (GUI), which then in-turn sends information to increasingly lower levels of the system. The user's action eventually reaches physical hardware where the electronic signals are present. Information is then sent back up the chain to provide further information to the user. In the case of saving a file to a hard drive, the user clicks on the 'save' icon which triggers a chain

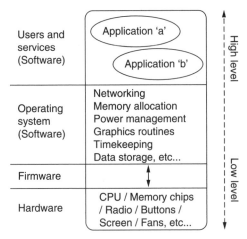

9.1 Typical model of a computer's software, firmware and hardware architecture.

of events through the various layers of the system. Eventually this action results in an electromagnet pulsing over a rapidly spinning disc platter. The user is informed of the activity via a graphical 'progress bar' shown on screen. The progress bar is being updated by operating system which in turn is communicating with firmware within the hard drive. The firmware is controlling the motors and magnets of the hard drive, converting the magnetic signals from the disc head into digital signals that the computer's operating system understands.

Commonly used PC operating systems include Microsoft's 'Windows', Apple's 'Mac OS', and 'Linux', developed by the open source community. Most personal computer (PC) hardware comes pre-installed with an operating system and a small range of applications. Users will then install their own separately purchased software applications, such as word processors, spreadsheets, image editors, video editors and games. It is also possible for end users to upgrade or install alternative operating systems onto the hardware at a later date.

Often software applications are created by companies or individuals separate from the hardware and operating system vendors. For example, one may use an application from company 'A', running within operating system by company 'B', upon hardware from company 'C'. This multi-vendor approach usually works reasonably well, but there can be compatibility issues between different versions of hardware, operating system and application. Sometimes these incompatibilities are accidental, sometimes they happen because the hardware or operating system is missing a specific new feature, and sometimes software vendors deliberately engineer them to protect their market position.

9.3.4 Mobile operating systems

Within the mobile phone market, the hardware platform and operating system model is increasingly being adopted. These operating systems include Microsoft's 'Windows Mobile' (Microsoft Corporation, 2008), Symbian's 'Symbian OS' (Symbian, 2008) used by Nokia, and Apple's 'iPhone OS' (Apple Inc., 2008). These so-called 'smart phones' incorporate a hardware platform, an operating system and a suite of applications, in much the same way as personal computers. There is also an emerging range of open source mobile operating systems that is set to change the way that mobile phone manufacturers develop their devices in future.

Unlike the PC industry, end users cannot easily install alternative operating systems onto their mobile handsets (although some users do 'hack' or modify their equipment to run different operating systems, but this is fairly uncommon and not usually encouraged by handset makers). It seems likely that wearable computers would adopt a similar operating system and application model. Of the few wearable computing platforms available, most require a PC type operating system, such as Linux or Windows, or are simply based upon existing mobile phone or PDA computing platforms. There is still work to be done before a common platform for wearable computing can be established. Just as in the early days of the personal computer, the field of wearable computers is still the domain of the experts, enthusiasts and researchers, rather than mass-manufacturers, retailers and consumers.

9.4 Application examples

The following wearable projects illustrate how sensors have been used in real-world applications and research scenarios. Most of the examples given highlight how information is gathered from the body and relayed to a separate PC computer for real-time or post-event analysis. It is important to realise that the initial capture and transmission of sensor data is only the first stage within the system. The subsequent viewing and analysis of the data is equally important and is the interface with which the end users interact with information captured from the wearer.

9.4.1 Light sensors

One of the simplest forms of sensor is the light sensor, which comes in two forms – a Light Dependent Resistor (LDR) or a photodiode. These inexpensive devices alter their electrical characteristics depending upon the amount of light falling upon them. Because they detect only ambient light levels, their usage is limited to simple tasks, but they can be used to provide or trigger events, such as turning on lighting as it gets dark. Within wearable applications, they have been used to detect a person's transition between rooms and outdoor spaces as light levels change. Cheok and Li (2007) have described a method of transmitting indoor location data using modified fluorescent lamps and a photodiode sensor circuit linked to a wearable computer. Randall *et al.* propose a method of indoor tracking using a wearable solar cell and light irradiated from existing fluorescent lighting (Randall *et al.*, 2004).

Modern digital camera technologies employ an integrated circuit that contains an array of millions of colour-sensitive points known as pixels. These individual pixels act like a unique photodiode, so that when a light is focused through a lens, each pixel detects a different segment of the image. There are several technologies that can be used to implement as a digital camera – charge-coupled devices (CCD) are often used within camcorders and digital-still cameras, and complementary metal-oxide–semiconductor (CMOS) sensors are commonly used within products such as 'camera phones' or PC web-cams. Due to ongoing improvements with both technologies, these application boundaries are increasingly blurred.

9.4.2 Image capture

SenseCam (Microsoft Research, 2007), developed by Microsoft Research, is a wearable camera that captures images continuously throughout the day. It incorporates a range of electronic sensors, including a light-intensity and colour sensor, passive infrared 'body heat' detector, temperature sensor, and a motion sensing accelerometer. Information from these sensors are continuously monitored by the device and images are recorded onto the SenseCam's internal memory card. The camera can be configured to take images automatically at pre-set intervals, such as every 30 seconds. However, by utilising the on-board sensors, the camera can respond and capture events such as a person walking past or sudden changes in light levels. To allow for this functionality, the SenseCam contains a microprocessor that constantly monitors the sensors. If, for example, the light-intensity changes by a certain amount within a specific time period, such as when walking outdoors, the processor would trigger the camera to capture that event. The captured image is then saved to memory and also logged with a time stamp and any other relevant sensor information. The SenseCam's internal memory can store over 30 000 images, which is enough for a week's worth of pictures. As the SenseCam does not include a screen, it works in conjunction with the viewer software to allow users to view and interpret the captured images and information on a PC. In practice, the users of SenseCam transfer images daily onto a PC where they can be viewed using software such as MyLifeBits (Bell *et al.*, 2008).

Applications for the SenseCam include memory aids to assist those with Alzheimer's Disease and amnesia. Microsoft have worked with the Memory Clinic and Memory Aids Clinic at Addenbrooke's Hospital, Cambridge, UK. and are supporting a number of collaboration projects to investigate how these technologies can be used to improve memory recall in patients. (Berry *et al.*, 2007).

9.4.3 Acoustic sensors

The microphone is a familiar acoustic sensor found within many devices today. Sound is essentially variations in air pressure that our ears detect and interpret as music, speech and other noises. An electronic microphone contains a small diaphragm that vibrates in response to variations in air pressure, moving a magnetic coil. The electrical signals produced when the magnetic coil moves can be converted into a digital format and used for applications such as noise level sensors, audio recording, or with sophisticated software, speech recognition.

Other forms of acoustic sensor include ultrasonic range finders which use short pulses of sound outside the range of human hearing to determine distance from solid objects. As the ultrasonic pulse is generated, it travels through the air until it hits an object, then 'bounces' back toward the sender where the receiving unit detects the reflected pulse. The time taken between sending a pulse and receiving the reflected ultrasonic sound gives an accurate indication of the distance to the object. Work by Chandra *et al.* (2004) demonstrated how it is possible to use body-worn ultrasonic range finders as an indoor location aware system.

9.4.4 Motion sensors

Basic motion can be detected using vibration sensors or simple tilt switches. However, they are of limited use within wearable applications. More commonly, a sophisticated electronic device known as an accelerometer is used to detect motion in two or three directions.

The term g-force is used to describe the amount of accelerative or decelerative force being applied to the device. A value of 1g is the equivalent to the earth's gravitational pull: 0 g would be weightless and indicate that the accelerometer is falling. Accelerometers are used to measure position, motion, tilt, shock, and vibration, and are available as one, two or three-dimensional sensing versions.

Traditionally, accelerometers contain a small weight suspended by springs; as the accelerometer is moved, the weight is 'pushed' either towards or away from the springs by the force being applied. The movement of the suspended weight within the device can be measured electronically and has been used for a wide variety of consumer and industrial applications. Modern 'chip-sized' versions of the accelerometer are based upon Micro Electro Mechanical System (MEMS) technology. These MEMS components contain a series of microscopic structures which detect the accelerative forces applied to the device. This has recently enabled a variety of 'motion aware' mobile devices to be produced. MEMS accelerometers are becoming increasingly inexpensive; at the time of writing (2008) a basic three-axis accelerometer costs around five dollars.

Analog Devices (2008), a key supplier of MEMS devices, describe how MEMS accelerometers have been used in a wide range of consumer, industrial and automotive applications. These include low-g applications (± 1 g to ± 20 g) such as mobile phone handsets, hard-drive protection systems and safety-critical automotive electronic stability control systems. High-g accelerometers (± 20 g to ± 250 g) are most commonly found in car airbag control systems.

As with any other input device, an accelerometer provides information regarding only the current accelerative forces acting upon it. It tells you very little about the context in which those forces are happening. The information provided is a constant flow of data representing the forces acting upon the x, y and z co-ordinates. It is the role of software and hardware designers to interpret what those forces represent and configure the rest of the system to act accordingly. This will depend largely upon the application and can range from moving an animated character around a game's screen, inflating an airbag at up to 160 mph depending upon the speed of impact, or alerting a doctor that a patient has fallen over. One of the most recent high-profile consumer applications incorporating an accelerometer is Nintendo's Wii games console. An embedded accelerometer within the controller has allowed games developers to create new and innovative forms of physical game interaction, including golf, tennis and boxing games. Figure 9.2 shows a photographic composition of a Wii controller in use and Fig. 9.3 shows a chart of the corresponding accelerometer data captured as a result of that motion. This data is interpreted by the Wii console software to control game character movements.

Increasingly, mobile device manufacturers are incorporating accelerometers into their hardware as an additional input device. Apple's iPod Touch and iPhone include accelerometers, which are used to detect screen orientation when the device is rotated from portrait to landscape mode.

In addition to accelerometers, gyroscopes can be used to provide additional orientation and angle information. A traditional gyroscope comprises of a spinning disk mounted on gimbals. As the gyroscope is moved and rotated, the internal spinning disc remains relatively static to its original orientation. The discrepancy between the disc and the outer structure can

9.2 Composition of a Wii controller being used in an upward swing motion.

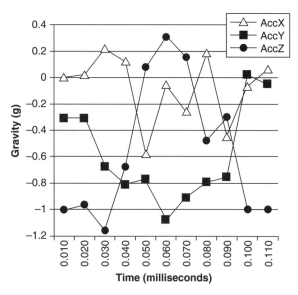

9.3 Accelerometer data captured during an upward swing motion from a Wii controller.

be measured electronically and provides angular movement data. MEMS-based gyroscopes work on a similar principle but contain vibrating microscopic surfaces instead of a spinning disc.

9.4.5 Activity recognition

Within the wearable research community, MEMS-based accelerometers and gyroscopes have successfully been used to detect not only motion, but also to accurately identify the actions a person is undertaking. This type of research is especially relevant for wearable applications because it does not require any prior 'training' of the software by its designers, or pre-set patterns of movement to be 'programmed in'.

Minnen *et al.* (2006) demonstrated an automatic activity discovery technique that uses accelerometer and gyroscope data to identify unique patterns called 'motifs' from physical activity. The algorithm described in that paper automatically detects and classifies the six different activities and recognises them again when they reoccur. Although further work is required, their approach was able to distinguish between six different dumbbell exercises, performed 846 times, over a period of 27 minutes. The algorithm described by Millen *et al.* achieved an overall accuracy of 86.7%, recalling 832 of the 864 repetitions, but falsely identified 51 and missed 32 repetitions.

Huynh and Schiele (2006) have also investigated techniques of automatic activity recognition, using combined generative 'automatic' recognition and discriminatory 'human trained' mathematical algorithms. Their research suggests that a combined approach can yield improved results for real-world activity recognition systems, especially when training data is minimal or impractical to gather. In the future, these techniques could be used in wearable systems to assist athletes in training, injury rehabilitation, the monitoring of vulnerable adults or automatic sign-language translation.

9.4.6 Location systems

Global positioning system

The Global Positioning System (GPS) consists of a constellation of at least 24 satellites, stationed in low earth orbit. Each satellite transmits timing signals, which are used by GPS receivers on earth to determine location and altitude. It is considered a dual-use technology for both military and civilian purposes and is provided worldwide free of charge by the United States government (PNT, 2008). GPS receivers such as hand-held units, vehicle sat-nav systems, or mobile phones, detect the timing signals sent by the satellites and apply a mathematical algorithm to calculate its location,

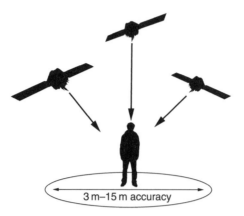

9.4 Simple illustration of GPS signals being received and level of accuracy expected.

speed, direction, and time. No information is sent back to the GPS system so individuals can't be tracked using GPS alone. GPS *trackers* can operate only by relaying the co-ordinates calculated by the receiver back to a monitoring service. This is often done via the cellular phone network.

For a GPS receiver to operate effectively, it must be able to simultaneously receive the signals from at least three satellites, and ideally four or more. Figure 9.4 shows a simple visualisation of how the timing signals from the GPS satellites are simultaneously received to calculate a user's location. When indoors, or if surrounded by high structures such as buildings, steep hillsides or trees, it becomes difficult for the GPS receiver to detect the satellite signals; therefore, its location cannot be determined as accurately. Tall buildings and narrow streets create so called 'urban canyons' which block the timing signals, this preventing GPS receivers in these locations from accurately calculating their location. Other degradations in the signal can occur because of weather conditions or solar radiation. A typical level of accuracy for a consumer GPS receiver would be between 3 and 15 metres but could decrease to 25 metres under poor conditions. High sensitivity GPS receivers are also becoming increasingly available, improving accuracy when indoors (QinetiQ, 2008). Vehicle sat-nav systems are able to improve their perceived accuracy by assuming vehicles will always be on a road, going in the same direction and at predictable speeds. By calculating a trajectory, sat-nav developers can predicted where a vehicle will be several seconds in advance. People on the other hand, are not constrained by pre-mapped roads. They can abruptly stop and change direction, making it difficult to achieve accurate predictions should the GPS signal deteriorate. To improve the accuracy of GPS, several augmentation systems have been developed:

- Differential GPS (DGPS), which uses a series of ground-based reference stations to improve maritime navigation. In the UK this is maintained by the General Lighthouse Authority (Trinity House, 2008).
- Assisted GPS, or A-GPS. This augmentation is found within many GPS capable mobile phones; it uses the cellular network to assist in determining the device's location (Nokia, 2008).
- Wide Area Augmentation System (WAAS). This is a GPS augmentation system designed for aviation navigation and has an accuracy of around one meter. EGNOS is the European version of this system and is operated by the European Space Agency (ESA, 2008).

The GPS system is operated by the US military, who until May 1st 2000 applied incorrect timing signals to reduce the accuracy of civilian GPS receivers to around 100 metres. This 'Selective Availability' could be increased during military operations to prevent enemy forces using GPS technology to target weapons against American forces. Since the removal of selective availability, a non-degraded signal has been available for public use, with the US able to use selective availability on a regional basis leaving the remaining global signal intact (FGCS, 2000).

Defence Advanced GPS Receivers (DAGR) provide enhanced accuracy for military personnel of around one metre. This is achieved by utilising additional encrypted signals that the GPS satellites transmit. These signals are not available to civilian GPS receivers.

Wearable GPS

GPS has been a popular choice for location-aware applications; aside from conventional uses of GPS technology such as navigation, other uses of the technology have been explored.

The wearable computer for field archaeology (Baber *et al.*, 2001), a system designed to aid archaeological surveys, uses a wearable GPS system to record site surveys, along with imaging and annotation functions.

Riot! 1831 (Reid *et al.*, 2005), developed as part of Mobile Bristol, used GPS technology along with the 'Media Spaces' content authoring tool, to create a virtual historic re-enactment situated within Queen's Square. Visitors to the event were given GPS-equipped wearable computers that would play different segments of audio soundtrack, depending on where they were walking.

The 'Know Where Jacket' (Interactive Wear AG, 2006), developed as a demonstration of the company's technology, incorporates specially modified GPS components that enable the garment to determine its location, without the need for bulky add-ons. The creators, Interactive Wear AG, state that their jacket can be used for a range of applications including luxury garments, entertainment, leisure, safety, medical, traffic and logistics.

Indoor location systems

Indoor positioning is seen as more difficult to accomplish because GPS does not work well indoors; this is because walls and steel structures can block the radio signals. To achieve accurate indoor location tracking, infrastructure has to be incorporated into the building. One of the earliest attempts at this was Olivetti's 'Active Badge' location system (Want, 1992), which used infrared enabled badges to track employees around the building. Aside from initial privacy concerns from employees, the trial found that the system became widely accepted by staff members because of the efficiency benefits it brought to their work. The active badge system allowed employees to 'find' others within, to find out who people were 'with', to 'look' at who is at a particular location, to 'notify' others when a particular person returned to the office, and it provided a 'history' which contained one hour's worth of location data presented in a report. The most important discovery made by Want and his team during the research at Olivetti was not 'Can we build a location system?', but 'Do we want to be part of a location system?' To protect the employees' privacy, the system was intentionally designed not to permanently record any location data.

Privacy is a significant issue, raising significant debate within contemporary computing research. Although privacy falls outside the scope of this chapter, it should be noted that technology is amoral; it is ultimately people who decide how a technology is to be used.

9.4.7 Biophysical monitoring

One of the most well-established brands of sports monitoring equipment is Polar Electro (http://www.polar.fi), which offers a range of heart-rate monitoring sport apparel. Many of these devices use a combination of a chest strap transmitter, linked via a short-range wireless signal, to a watch. The watch calculates and stores information such as average and peak heart rate, and calories used. This information can then be uploaded onto a PC for further analysis. A small battery provides power to the chest-straps, which is claimed have a lifespan of about 2500 hours usage. Many of Polar's products share a common core technology, with value added by using higher quality materials, aesthetic appearance, communication protocols and analysis software. The chest-straps are commonly made from a rubber-like material and have been criticised for being uncomfortable. In response, Polar also offers a textile-based chest strap in addition to the plastic models.

'adiStar Fusion' is a product range developed collaboratively between Polar and Adidas (http://www.adidas-polar.com) aimed at high-end consumers who run. The garment consists of a sports bra for women or close fitting vest for men, with bonded textile based electrodes. The runner's heart rate is captured and relayed to a sports watch via a Wearlink transmitter. Wearlink is a small electronic device that snaps onto the front of the garment, making an electrical contact with the garments electrodes. An additional accelerometer-based stride sensor is also available, integrated into training shoes: it measures stride, speed and distance. The stride sensor wirelessly relays this information back to the sports watch. The watch allows runners to monitor their heart-rate and to calculate optimal training regimes for themselves.

Wearable health monitoring is also being used for a variety of safety applications. VivoMetrics, a manufacturer of wearable heath monitoring systems, identifies workers using protective clothing being particularly at risk from over-exertion and physical stress. These include fire-fighters, clean-up workers, biohazard workers and hazmat response teams. In response, VivoMetrics (2008) have developed the VivoResponder, a system which, it is claimed, assists safety officers and incident commanders in determining when key personnel are in danger. The VivoResponder chest strap incorporates sensors that monitor respiration, heart rate, activity, posture and skin temperature. The chest strap sends this information in real-time to a command centre PC where the VivoCommand software is used to display the life-sign data. The software incorporates an easy-to-understand interface, designed to show 'granular information' or to enable quick access to key sensor information using colour-coded displays.

The VivoResponder system demonstrates how the person wearing the device is not necessarily the primary user of the system. In this example we see how data is captured from the wearer of the chest strap device, but it is the incident commanders or safety officers who read and act upon that information. The development of this PC-based analysis software is just as important as the wearability and useability of the garment itself. The technology used within VivoResponder is derived from VivoMetrics' LifeShirt, for which other applications include medical research and sports training.

Wearable Health Care System (WEALTHY, 2005) was coordinated by Italian e-textiles development company Smartex. It was funded by the European Information Society Technologies (IST) programme during the period 2002 to 2005, and developed prototype garments that could continuously monitor patients' vital signs. The prototype garment included simple integrated woven sensors, which allowed for information such as respiration, skin temperature, patient orientation and movement to be monitored. Using the mobile phone network, data could be sent directly to remote monitoring centres, triggering messages about the wearer's health status. It was proposed that future versions of the garment could transmit the exact location of the wearer in an emergency, using location-based technology. Many potential users were identified as beneficiaries of the WEALTHY health monitoring garment; these included soldiers, athletes, high-risk personnel such as fire fighters, older people, newborn babies, sleep apnoeas (people who stop breathing in their sleep) or long-distance drivers. The subsequent MyHeart (2008) research programme examined the use of this technology for patients with cardio-vascular diseases.

9.5   Summary

This chapter has introduced the concept of transducers and how one form of energy can be converted into another. Analogue and digital electronics were examined and the influence of the personal computer architecture upon mobile and wearable platforms discussed. A variety of wearable applications have been explored and it has been shown how the use of software algorithms converts raw sensor data into meaningful contextual information. The chapter also highlighted how wearable computing could be incorporated into our daily lives with projects such as SenseCam and LifeShirt.

9.6   Future trends

Although much work is continuing towards the practical integration of wearable electronics into washable garments, we are still limited to early 21st century manufacture techniques and semiconductor technologies. Silicon chip technology was originally intended for use in computers, mobile phones and other electronic devices housed in hard plastic boxes. The move to flexible and washable garments will not be easy to accomplish. Perhaps for wearable technology to fully realise its ambition of being an always-on, unobtrusive part of our lives, a new generation of electronic hardware is required. As the cost and size of electronic components continues to fall, we may reach a point when printing a computer is as cheap as printing with ink. This would revolutionise the way in which designers perceive the notion of a computer. Just as today where logos, text and images can be printed onto paper and fabrics, future computers will be printed just as liberally onto the surfaces of our daily lives. With this technology, flexible electronics would then not simply be possible, but would be abundantly available. Then, part of the role of a clothing designer would be to develop new uses and applications for this technology.

9.7   References

ANALOG DEVICES, 2008, MEMS Sensors iMEMS® Accelerometers iMEMS® Gyroscopes Analog Temperature Sensors (Temp sensors) Digital Temperature Sensors [Homepage of Analog Devices], [Online]. Available: http://www.analog.com/en/cat/0,2878,764,00.html [accessed, 6/23/2008].

APPLE INC., 2008, March 6, 2008-last update, Apple Announces iPhone 2.0 Software Beta, [Online]. Available: http://www.apple.com/pr/library/2008/03/06iphone.html [Accessed, 8/22/2008].

BABER, C., CROSS, J., WOOLLEY, S.I. and GAFFNEY, V.L., 2001. Wearable Computing for Field Archaeology, 2001, *IEEE Computer Society* p. 169.

BELL, G., GEMMELL, J. and LUEDER, R., 2008-last update, MyLifeBits project – Microsoft BARC media presence group. Available: http://www.mylifebits.com [accessed, 8/22/2008].

BERRY, E., KAPUR, N., WILLIAMS, L., HODGES, S., WATSON, P., SMYTH, G., SRINIVASAN, J., SMITH, R., WILSON, B. and WOOD, K., 2007. The use of a wearable camera, SenseCam, as a pictorial diary to improve autobiographical memory in a patient with limbic encephalitis: A preliminary report. *Neuropsychological Rehabilitation*, 17(4), p. 582.

CHANDRA, M., JONES, M.T. and MARTIN, T.L., 2004. E-Textiles for Autonomous Location Awareness, *ISWC '04: Proceedings of the Eighth International Symposium on Wearable Computers*, 2004, IEEE Computer Society pp. 48–55.

CHEOK, A. and LI, Y., 2008. Ubiquitous interaction with positioning and navigation using a novel light sensor-based information transmission system. *Personal and Ubiquitous Computing*, Vol. 12, No. 6, pp. 445–458.

ESA, 2008 – last update, ESA – navigation – the present – EGNOS [Homepage of European Space Agency], [Online]. Available: http://www.esa.int/esaNA/egnos. html [accessed, 4/13/2008].

FGCS, May 02, 2000, 2000 – last update, GPS & selective availability Q&A [Homepage of Federal Geodetic Control Subcommittee], [Online]. Available: http://www. ngs.noaa.gov/FGCS/info/sans_SA/docs/GPS_SA_Event_QAs.pdf [accessed, 4/13/2008].

HUYNH, T. and SCHIELE, B., 2006. Towards Less Supervision in Activity Recognition from Wearable Sensors, *ISWC*, 2006, pp. 3–10, doi: 10.1109/SW.2006.286336.

INTEL CORPORATION, 2008, Moore's Law: Made real by Intel® innovation [Homepage of Intel Corporation], [Online]. Available: http://www.intel.com/technology/mooreslaw/ [accessed, 6/12/2008].

INTERACTIVE WEAR, 2006-last update, Know Where jacket [Homepage of Interactive Wear AG], [Online]. Available: http://interactive-wear.de/cms/front_content. php?idcat=58 [accessed, 4/13/2008].

MICROSOFT CORPORATION, 2008-last update, windows mobile: smartphone and PDA software. Available: http://www.microsoft.com/windowsmobile/default.mspx [accessed, 4/13/2008].

MICROSOFT RESEARCH, 2007-last update, SenseCam [Homepage of Sensors and Devices Group], [Online]. Available: http://research.microsoft.com/sendev/ projects/sensecam [3/25/2008, 2008].

MINNEN, D., STARNER, T., ESSA, I.A. and J.R., C.L.I., 2006. Discovering Characteristic Actions from On-Body Sensor Data, *ISWC*, 2006, pp. 11–18, doi: 10.1109/SW.2006.286337.

MOORE, G.E., 1965, 'Cramming More Components onto Integrated Circuits', *Electronics Magazine*, [Online], Vol. 38, No. 8, Available from: http://download.intel.com/ research/silicon/moorespaper.pdf. [accessed, 12/6/2008].

MYHEART, 2008, 28/2/2008-last update, IST-2002-507816- MYHEART, [Online]. Available: http://www.hitech-projects.com/euprojects/myheart/objectives.html [accessed, 6/23/2008].

NOKIA, 2008, Assisted GPS – Nokia Maps 1.0 [Homepage of Nokia], [Online]. Available: http://europe.nokia.com/A4668007 [accessed, 6/24/2008].

PNT, 2008-last update, global positioning system [Homepage of National Executive Committee for Space-Based Positioning, Navigation, and Timing (PNT)], [Online]. Available: http://www.gps.gov/ [accessed, 4/13/2008].

QVINETIQ, 2008-last update, GPS – high sensitivity indoor GPS. Available: http://www. qinetiq.com/home/capabilities/gps/GPS_Products/high_sensitivity_gps.html [accessed, 4/13/2008].

RANDALL, J., BHARATULA, N.B., PERERA, N., VON BÜREN, T., OSSEVOORT, S. and TRÖSTER, G., 2004. Indoor Tracking using Solar Cell Powered System: Interpolation of Irradiance *Ubicomp 2004 Adjunct Proceedings* – Posters, September 7–10 2004.

REID, J., HULL, R., CATER, K. and FLEURIOT, C., 2005. Magic moments in situated mediascapes, ACE '05: Proceedings of the 2005 *ACM SIGCHI International Conference on Advances in Computer Entertainment Technology*, 2005, ACM pp. 290–293.

SYMBIAN, 2008-last update, symbian OS [Homepage of Symbian Limited], [Online]. Available: http://www.symbian.com/symbianos/index.html [accessed, 4/13/2008].

TRINITY HOUSE, 2008-last update, trinity house | aids to navigation | satellite navigation [Homepage of The Corporation of Trinity House of Deptford Strond], [Online]. Available: http://www.trinityhouse.co.uk/aids_to_navigation/the_task/satellite_navigation.html [accessed, 4/13/2008].

VIVOMETRICS, 2008, View Our Products [Homepage of VivoMetrics], [Online]. Available: http://www.vivometrics.com/hazmat/view_our_products/ [accessed, 6/23/2008].

WANT, R., HOPPER, A., FALCÃO, V. and GIBBONS, J., 1992. The active badge location system. *ACM Trans. Inf. Syst.*, 10(1), pp. 91–102.

WEALTHY, 2005, 28/2/2005-last update, IST-2001-37778- W E A L T H Y. Available: http://www.wealthy-ist.com/ [accessed, 6/23/2008].

ZILOG INC, 2008, Product Overview [Homepage of ZiLOG, Inc], [Online]. Available: http://www.zilog.com/company/products.asp [accessed, 6/12/2008].

10

The application of communication technologies in smart clothing

P. LAM, Wireless Edge Communications Limited, UK

Abstract: The use of communications in textile and fashion is a relatively new discipline and is an exciting development in its rapid evolution and the pervasive nature of its incorporation into everyday life. Many of the developments in communications have been in the local and wide area networks, to make them faster and more feature-rich. The development of piconets – micro communication networks around the human body – sees communications venturing into and straddling areas from which it has traditionally been excluded.

Key words: piconet, local area network, wide area network, personal connector, Wifi, WiMax, Bluetooth, RFID, network security, wireless, EDGE, micro communications, GPRS, 3G, real-time.

10.1 Introduction

Communication technologies and communication applications have been rapidly changing and developing, especially in recent years. In the area of wireless technologies and applications, these have found their way into many areas that previously have been considered as science fiction. The communicator device in *Star Trek* is now a functional device used in hospitals and schools, which is activated by pressing on the device and, coupled with voice recognition software and call processing software, calls are made to various doctors or teachers on duty. A further development of these technologies in communications is to combine textile technologies and fashion with wireless communications, to provide piconets – micro communications networks around the human body. This chapter looks at the end-to-end communications flow of how we retrieve the information provided by the various sensors, buffer and store this information, and send it across a wide area communications network to a data centre for processing, analysis and reporting. The final step in this end-to-end process is to inform end users automatically, as well as via a call centre, when the results of the information processed are outside of the normal envelope for a particular person.

There are three areas of analysis:

(i) Personal Communications Networks (PCN) – how to receive the information provided by the various sensors in the smart clothing, and to store and prepare this to be transferred for analysis.

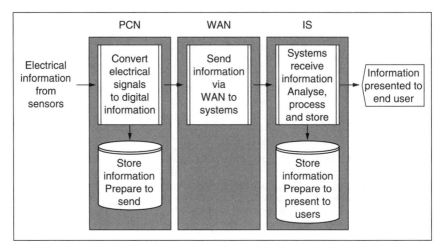

10.1 End-to-end process flow.

(ii) Wide Area Networks (WAN) – how to take the information stored by the PCN and transfer it to another physical location; this could be in another part of the country or anywhere in the world.

(iii) Information Systems (IS) – how to analyse the data received, to monitor the subjects under observation and provide reporting information to end users.

These steps can be visualised in Fig. 10.1.

The final section of the chapter will look at some of the potential opportunities and the technological and commercial challenges for the proposed solutions.

10.2   Personal communications networks

There are a number of different common technologies that are available, including Wireless Fidelity (WiFi), Worldwide Interoperability for Microwave Access (WiMax), Radio Frequency Identification (RFID), and Infrared, but the one that is the most appropriate for this application is Bluetooth, as this is readily available, low in cost and the most energy efficient. Power consumption (from 1 mW to 100 mW) is a major consideration as the device needs to be able to function over several days before it needs to be recharged. Cost of components also needs to be low as the garment needs to be affordable for everyday use.

10.2.1 Bluetooth

Bluetooth wireless technology is a short-range communications technology to connect portable and fixed devices while maintaining high levels of security. In this case, we are using this to connect our PCN to the WAN via a WAN mobile device. The key features of Bluetooth technology are robustness, low power, and low cost. The Bluetooth specification defines how a wide range of devices can connect and communicate with each other.

Bluetooth technology has achieved global acceptance such that any Bluetooth-enabled device, almost everywhere in the world, can connect to other Bluetooth-enabled devices in proximity. Bluetooth-enabled electronic devices connect and communicate wirelessly through short-range, micro networks known as piconets. Each device can simultaneously communicate with up to seven other devices within a single piconet. Each device can also belong to several piconets simultaneously. Piconets are established dynamically and automatically as Bluetooth-enabled devices enter and leave radio proximity.

A fundamental Bluetooth wireless technology strength is the ability to simultaneously handle both data and voice transmissions. This enables users to enjoy a variety of solutions such as a hands-free headset for voice calls and synchronisation with Personal Digital Assistant (PDA), Apple iPhone, laptop, and mobile phone applications.

Spectrum

Bluetooth technology operates in the unlicensed industrial, scientific and medical (ISM) band at 2.4 to 2.485 GHz, using a spread spectrum, frequency hopping, full-duplex signal at a nominal rate of 1600 hops/sec. The 2.4 GHz ISM band is available and unlicensed in most countries.

Interference

Bluetooth technology's adaptive frequency hopping (AFH) capability was designed to reduce interference between wireless technologies sharing the 2.4 GHz spectrum. AFH works within the spectrum to take advantage of the available frequency. This is done by detecting other devices in the spectrum and avoiding the frequencies they are using. This adaptive hopping allows for more efficient transmission within the spectrum, providing users with greater performance even if using other technologies along with Bluetooth technology. The signal hops among 79 frequencies at 1 MHz intervals to give a high degree of interference immunity.

Range

The operating range depends on the device class:

- Class 1 radios (100 mW, 20 dBm), which are used mainly in industrial applications and have a range of approximately 100 metres

- Class 2 radios (2.5 mW, 4 dBm), which are most commonly found in mobile devices and have a range of approximately 10 metres
- Class 3 radios (1 mW, 0 dBm) which have a range of up to 1 metre

Power

The most commonly used radios are Class 2 Bluetooth radios, which use 2.5 mW of power. Bluetooth technology is designed to have low power consumption, with the radio units able to be powered down when inactive.

Data rate

- 1 Mbps for Version 1.2;
- Up to 3 Mbps supported for Version 2.0 + Enhanced Data Rate (EDR)
- 53 Mbps to 480 Mbps, WiMedia Alliance (Proposed)

10.2.2 Personal Connector

The Personal Connector is the device on the garment connecting the various sensors. Also, the Personal Connector wirelessly interfaces to a WAN mobile device (e.g. iPhone, PDA, mobile phone). Class 2 Bluetooth is used to communicate between the Personal Connector and the WAN mobile device as this associated device is normally on the person or on a desk or table close by, which is normally less than 10 metres away. In terms of data rates, only a small amount of data is required to be sent on a regular period and so a 1 Mbps data rate is sufficient for this application.

Figure 10.2 shows the functional blocks for the Personal Connector, which provides the interface design to a garment and the transfer of information to far-end systems for monitoring and analysis.

A network of sensors in a garment is connected back to the Personal Connector. This information is buffered and processed by its on-board processor. All information is stored in its Information Store, which is a non-volatile random access memory store. The Radio Device, which in this case is using Bluetooth, checks for connectivity to the far-end servers via a mobile phone using General Packet Radio Service (GPRS), Enhanced Data rates for GSM Evolution (EDGE), or 3rd Generation Mobile Service (3G). When connectivity is available to the far-end servers, data stored in the Information Store is sent. Cyclic Redundancy Checks (CRC) are computed on both ends to ensure the validity of the sent data, which also remains on the Information Store until confirmation is received from the far end that the data has been received correctly.

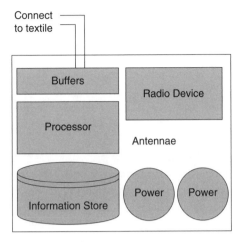

10.2 Personal connector.

Power is provided by rechargeable batteries, which should have enough power for a few days of continuous use. Advancements in battery technologies and further reductions in power consumption, especially in wireless technologies, will lead to extended periods between the charging of batteries.

10.3   Wide area networks

For the Wide Area Network, it is envisaged that the existing GPRS, EDGE or 3G mobile network is used, as this is commonly available with coverage available in most areas of the UK. The importance here is that the information on the Personal Connector needs to be transferred on a regular basis to the far-end servers for analysis. A software application is installed on a mobile device, which communicates with the Personal Connector as well as the server application: this is the service integration software and is the key to provide the communications service for analysis of information on Smart Clothing.

There will be occasions when there is no signal on the mobile network and so the information cannot be transferred to the far-end servers. However, in these cases, the information is polled from the Personal Connector and stored on the mobile device – this information is still held on the Personal Connector until the end servers can be reached and the information transferred successfully.

All information is time-stamped and a message sequence number attached. This two-stage process ensures that once the wide area mobile network is available, information can be sent to the far-end servers without having to wait for the successful transferring of information from the Personal Connector as well – this increases the reliability of the overall service.

In a normal scenario,

(i)   this application polls the Personal Connector,
(ii)   retrieves the information,
(iii)   checks the time stamps and message sequence numbers of this information,
(iv)   sets up a connection to the far-end servers,
(v)   sends the information to the far-end servers,
(vi)   receives an acknowledgement from the far-end servers with the message sequence number of the last information received,
(vii)   completes Cyclic Redundancy Checks (CRC) successfully, and
(viii)   sends a 'Success' message back to the Personal Connector with the message sequence number and receives an acknowledgement, sent by the Personal Connector.
(ix)   All pointers and information are now updated.

In failure scenarios, where acknowledgements are not received or CRC fails, a timeout occurs and the information is re-sent.

This software application will provide the interface to the chosen mobile network, whether it is GPRS or 3G. In the future, the software application can be upgraded as the wide area network changes, which could be whatever the chosen technology for 4G mobile, or it may be WiMax. Of importance is the definition of a reliable protocol for the end-to-end process of data communications from the personal network to the far-end servers, via a wireless wide area network.

10.4   Monitoring systems and service

Once the information reaches the far-end servers, it needs to be checked and analysed. A process needs to be defined as to what to do if the analysed information highlights an issue with the data – whether the information is corrupt or the sensors have picked up an anomaly with the individual's information. See Fig. 10.3.

To complete this end-to-end process, a small communications software application resides on a server, which communicates with the mobile devices to enable the transfer of sensor information to the server farm. Once the user is checked for authenticity and the information for validity, a positive acknowledgement is sent back to the mobile device: a mobile device receives this positive acknowledgement, updates its information and advances its pointers and message sequence numbers. A positive acknowledgement is propagated to the Personal Communicator, which in turn will update its information.

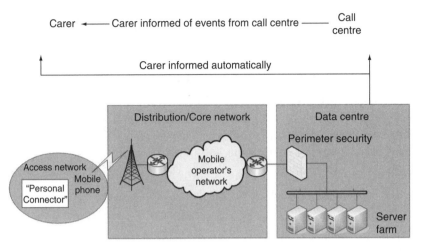

10.3 Monitoring.

Information picked up by the sensors is stored and sent via a wireless network to a secure data centre using a secure Virtual Private Network (VPN) connection between the mobile device and the perimeter security appliance. Although the information sent is very small (but frequent), it is important that a VPN tunnel is built and the data is encrypted using 256-bit AES encryption to ensure maximum protection of the sent data. Even though the meaning and format of the data is proprietary, it is good practice to send private information in a secure way. Once the encrypted data is unencrypted by the perimeter security, this is sent to an information store, ready for processing. For added security, an extra layer of encryption could be added, with the final un-encryption being performed by one of the servers in the server farm. A back-end server will retrieve this information and being the processing and analysis of the data. In normal circumstances, the data received regarding the wellbeing of an individual will show that they are inside their personal 'health' envelope. However, if the results show that the individual is outside or approaching the outside of their normal envelope, a trigger is sent to another server for processing. The result of this could be that a text message is sent to a carer on a regular basis until it is acknowledged or a call centre representative could call a carer to check on the user.

A number of checks and balances could be added to this process to ensure nothing is amiss. This whole process is not designed for instant reaction but rather it is a continuous monitor to check an individual's health over time. As an individual approaches the edge of their envelope, this system could act as an early warning system of a drift towards a potential issue with the individual's health.

10.5   Opportunities and challenges

For all electronics, consistent and reliable power is required to ensure operational effectiveness. The issues for the personal connector are its power requirements, how long it can continue to work, how the unit is recharged, and what happens if the unit fails.

10.5.1 Power consumption

Ideally, the unit is self-charging: once attached to a garment, the unit uses different methods to recharge its batteries. This is going to be a challenge and in the near term this could be just a simple set of batteries charged via an external charger. The whole personal connector needs to be waterproof and withstand being put in a washing machine.

10.5.2 Unit failure

When a personal connector is perceived to have failed, there will be a failure in the periodic information sent to the far-end servers. This could be interpreted by the far-end algorithms as:

(i)   The individual is simply out of range with no mobile coverage,
(ii)   a unit failure, or
(iii)   a serious problem with the individual.

The latter obviously is the most serious problem.

Further work is required to define the different failures and what action is required for each failure category.

10.5.3 Integration testing

Further analysis is required for the integration testing of the whole end-to-end process. A protocol will need to be defined with failure scenarios. The use of a mobile network for data transmission in a real-time application is not new but to do this reliably and efficiently is a challenge.

10.6   Applications

There are a number of areas in leisure and sports where piconets could be applied. For example, during football training, it is important to check on the performance of individuals and how quickly they recover. Also, it would be interesting to analyse positioning during a game and see how these can be changed and improved to give more effective results. The less obtrusive the technology in a garment, the less hindrance during a game and the more pervasive the technology becomes. However, adding a two-way communications element into the device means that it is possible not only to analyse positioning but sports coaches could also provide instructions directly to an individual, based on real-time computer simulations. This not only would enable coaches to analyse more effectively the whole game but also provide individuals with specific enhancements to their game.

10.7   Future trends

A two-stage process from the piconet and personal connector interfacing to the WAN device has been described. This could be merged into one, where the personal connector communicates directly over the WAN – this has a number of challenges including a bigger power requirement. However, there is research into the use of the human body as a source of power to power such devices.

Another area of development is the personal connector being assimilated into the human body, which then picks up the information directly from the human body and sends it to the WAN for analysis. The ability for two-way communications for audio, video and data means that we can choose to be constantly in touch and have access to information while we are out and about.

The idea of 'bionic' parts in the human body is not new but more and more this is becoming a reality, where the use of textiles, coupled closely with fashion and the embedding of communications into the whole human body, is an exciting area that will move much of science fiction to science fact. As a major manufacturer has coined the phrase, 'Welcome to the Human Network'.

10.8   References

BLUETOOTH SIG INC. (26th July 2007). Bluetooth Specification, Version 2.1 + EDR. Bluetooth SIG Inc.
ERICSSON. (February 2007). The Evolution of EDGE. White Paper.
STALLINGS, WILLIAM. (2005). *Wireless Communications & Networks*. Upper Saddle River, NJ: Pearson Prentice Hall.

11

Power supply sources for smart textiles

G. MIN, Cardiff University, UK

Abstract: This chapter begins with a brief review of a number of portable power sources such as polymer batteries, and photovoltaic, thermoelectric and piezoelectric devices. It then discusses their suitability for smart textile applications and the prospects of developing them into a fully wearable power sources. These power sources are capable of self-recharging through energy harvesting from body heat, movement and surrounding light.

Key words: wearable power, human power, energy harvesting, thermoelectric, piezoelectric, photovoltaic, flexible batteries.

11.1 Introduction

Microelectronic technology continues to shrink the world of electronics. Sensors, actuators and circuits can now be incorporated into textile structures to realise functional clothing that sense and react to the wearers or their environments. Success in this endeavour will substantially enhance the functionality of textiles and find immediate application in numerous fields, particularly in the areas of medical and health care. The smart textiles that can monitor hydration, body temperature, electrocardiogram (ECG) and heart rate will provide non-invasive biomedical sensing for remote monitoring of patients. Gas sensors can be embedded or woven into textiles to monitor and warn of imminent risks in hazardous environments. Even consumer products such as music players or fitness pulse rate monitors can be incorporated into textiles, belts, shoes or headgear for the convenience of wearers. Currently, a number of companies and research institutions are exploring the possibilities in these areas. A smart shirt has been developed by the Georgia Institute of Technology, USA, for medical purposes. The smart shirt can monitor both heart and breathing rates using optical and electrical conductive fibres that are woven into the textile of the shirt (Park and Jayaraman, 2004). In Europe, Levi is testing the market for a musical jacket developed by the Massachusetts Institute of Technology Media Lab (Machover, 2007). The jacket consists of a miniature Musical Instrument Digital Interface (MIDI) synthesiser and a fabric capacitive keyboard.

The three key components of smart textiles are sensors, electronic circuits and actuators. Sensors are required to detect signals, electronic circuits to process the signals and then make an intelligent decision for actuators to react on it. The operation of these devices requires energy and they cannot function without electrical power. Batteries are an obvious choice, but they are bulky and require frequent replacement or recharging. Ideally, an electrical power source that is suitable for smart textile applications should be flexible, light weight and self-renewing. Energy harvesting devices are promising solutions to meet these requirements. Electrical power can be generated from sources of energy that are available to a garment. For instance, body heat can be converted to electrical power using thermoelectric devices (Rowe, 2005), mechanical motion using piezoelectric converters (Antonio, 2004; Yang, 2006) and light using solar cells (Luque and Hegedus, 2003). In this chapter, after a brief review on power consumptions of typical portable electronic devices, the suitability of available power sources for smart textile applications will be examined and the prospects for future development of these energy harvesting devices discussed.

It is to be noted that all devices discussed in this chapter are sophisticated energy converters. Accurate descriptions and understanding of these devices require complex theoretical models, which are outside the scope of this chapter. Our intention is to introduce basic concepts and relevant facts that are associated with smart textile applications. For a comprehensive understanding of these devices, readers can consult the references provided.

11.2   Power requirements of portable devices

Electronic devices have become a necessity of modern life. Almost everybody carries something that is electronic by nature: watches, mobile phones, music players, cameras and laptop computers. More importantly, some special electronic devices are essential for health care applications such as cardiac pacemakers, hearing-aids, and temperature or pulse rate monitors. All of these devices require a power supply to function. Because of the portable nature of these devices, batteries have been employed in almost all of them. Table 11.1 shows the typical power levels required by some of the commonly used portable electronic devices, together with corresponding battery type and operating period. It can be seen that power consumption by portable consumer electronics covers a wide range of power levels, from a few micro-watts to tens of watts.

11.2.1 Low end

At the low end, the power required by watches, pacemakers and hearing-aids is only about tens of micro-watts. Typical button-type batteries (such as silver oxide) have a nominal capacity of around 10–120

Table 11.1 Power levels required by commonly used portable electronic systems

Electronic systems	Power	Battery type	Operating period
Watches	3–10 µW	Silver oxide button	1–2 years
Pacemakers	25–80 µW	Lithium button	7–10 years
Hearing-aids	N/A	Zinc-mercury oxide	25–30 days
Digital clock	13 mW	Silver oxide button	6–10 months
Red LED	25–100 mW	Silver oxide button	6–12 months*
Pedometer	250 mW	Silver oxide button	1–2 years*
Portable radio	500 mW	AAA	3–6 months*
RF circuits	300–800 mW	AA	<10 hours
iPod	N/A	Li-ion rechargeable	16–32 hours
Mobile phones	4–10 W	Li-ion rechargeable	7–10 days
Laptop computer	50–80 W	Li-ion rechargeable	3–10 hours

Note: The data in the table are typical values, which can vary depending on manufacturer and specific model, etc.; * The devices operate in non-continuous mode and the period depends strongly on the frequency of the usage.

micro-ampere-hour (µAh) at 1.5 V. This can provide an operating period over 1–2 years for these devices. Button batteries are small and light weight. They can readily be embedded into smart textiles without much trouble or causing inconvenience to wearers.

11.2.2 Mid range

For wireless communications, current radio frequency (RF) circuits require an average power of around 500 mW. They are usually powered using two AA batteries, which provide a continuous operating period of around 10 hours. Embedding these batteries into smart textiles for RF circuits will become awkward due to their size and weight.

11.2.3 High end

The power required by an average laptop computer is about 50 W, which is currently provided by a Lithium-ion (Li-ion) rechargeable battery. This is bulky, weighing typically around 500 g and lasts for only 2–3 hours between recharging. The size, weight and operating period of the battery have been 'bottle-neck' problems in the development of laptop computers. This will become even more challenging for wearable computing. Solutions may require research efforts in two directions:

(i)   Reducing power consumption of electronic circuits
(ii) Developing new power sources.

11.3   Established portable power sources

The battery, which was invented by Alessandra Volta in 1800, is a well-established electrical power source (Buchmann, 2001). It is not surprising to see that all devices shown in Table 11.1 are powered by batteries. In fact, the battery is the only electrical power source available for portable applications at the moment. Currently, conventional batteries cannot be fully integrated into textile structures because of their rigidity. However, they may be embedded into clothing without causing significant inconvenience to the wearers if the power level required is low. For example, button batteries have been considered, apart from providing power, as buttons or decorations for smart jackets.

A battery is an electrochemical device that converts chemical energy into electricity. Figure 11.1 shows a typical structure of the most commonly used 'dry cell' batteries. Electrochemical processes involving electrolyte paste produce negative charges on the zinc anode and positive charges on carbon cathode. The electrical potential established between two electrodes drives electrons (negative charges) flowing from the anode to the cathode when a load (e.g. a light bulb) is connected across the two electrodes. Modern batteries use a variety of chemicals to fuel their reaction. Naturally, batteries are classified based on the chemistry of their electrochemical reactions. The most commonly used batteries on today's markets are lead acid, alkaline, Nickel–Cadmium (NiCd), Nickel–Metal-Hydride (NiMH), Lithium ion (Li-ion) and Lithium polymer (Li-polymer). See Table 11.2 for the main characteristics of the above six most commonly used batteries. The science, technology and applications of batteries have been described in a number

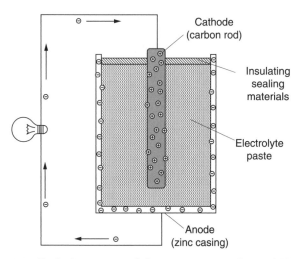

11.1 Typical structure of the most commonly used 'dry cell' batteries.

Table 11.2 Characteristics of most commonly used rechargeable batteries

Battery type	Energy density (W h/kg)	Cell voltage (V)	Charge cycle	Internal resistance (mΩ)
Lead acid	30–50	2.0	200–300	<20
Alkaline	~80	1.5	50	40–400
NiCd	45–80	1.25	1 500	20–40
NiMH	60–120	1.25	300–500	40–60
Li-ion	110–160	3.6	500–1 000	75–125
Li-polymer	100–130	3.6	300–500	100–150

of books (Colin *et al.*, 1984; Hahn and Reichl, 1999; Buchmann, 2001). In this section, a brief discussion is given to the aspects that are relevant to smart textile applications. These include the energy density, service life, costs and compatibility.

For high-power wearable applications, batteries become inadequate due to their limited energy density. The energy density is the amount of electrical energy stored in a battery of a given weight, usually measured in watt hour per kilogram (W h/kg). The operating period of a battery is determined by the power requirement of the electronics and the energy density of batteries employed. For example, if an electronic circuit consumes one watt and the batteries employed weigh 20 g and have an energy density of 150 W h/kg, the operating period will be 3 hours. However, if the same battery is used to power an electronic circuit that requires only 1 mW, the operating period will be 3000 hours (125 days). Even for such a small power requirement, the operating period is still significantly shorter than the life expectancy of a garment. Therefore, in order to provide adequate service life for smart textile applications, rechargeable batteries have to be employed.

11.3.1 Rechargeable batteries

The lead acid battery was the first rechargeable battery for commercial use and was invented in 1859. It is inexpensive, easy to manufacture and capable of high discharge rates. However, its relatively low energy density limits its use to stationary and wheeled applications, where weight is of little concern. The lead acid battery is the preferred choice for automobiles, hospital equipment, powered wheelchairs, and emergency lighting. The alkaline battery has a higher energy density compared to the lead acid battery. However, it has a relatively shorter charge cycle, which makes it unsuitable for smart textile applications.

The Nickel-Cadmium battery has the largest charge cycle of up to 1500 times. It can provide long service life with periodical recharging. In addition, it has a high discharge rate and low manufacturing cost. Most of the present day AA or AAA type rechargeable batteries are based on Nickel-Cadmium chemistry. They are widely used for portable radio receivers, biomedical equipment, video cameras and power tools. However, the NiCd battery has a relatively low energy density and contains toxic metals, making it undesirable for smart textile applications. The European Union has recently recommended that the use of cadmium and several other toxic materials should be restricted, because of the risks they pose to health and environment (http://eur-lex.europa.eu/LexUriServ/LexUriServ.do?uri=OJ:L:2003:037: 0019:0023:EN:PDF).

Nickel–Metal Hydride batteries were developed as a non-toxic alternative for NiCd. In addition, they have higher energy density and have been gradually replacing the NiCd in some markets, particularly in wireless communication and mobile phone sectors. However, the NiMH has a relatively low charge cycle compared with NiCd.

The Lithium Ion battery provides the highest energy density with a large charge cycle, making it the fastest growing and most promising battery for numerous portable applications. A unique advantage of the Li-ion battery is that it has no memory effect* and the recharging can be done whenever it is convenient. Currently, the Li-ion battery is more expensive and the technology is not fully mature. Potentially higher energy densities may be achievable.

Lithium Ion Polymer is a potentially lower cost version of the Li-ion. The chemistry is similar to that of the Li-ion battery in terms of energy density. However, the Lithium Ion Polymer battery uses a dry polymer electrolyte to replace the traditional porous separator. This enables very slim geometry and simplified packaging, and the battery can be potentially flexible.

11.3.2 Flexible batteries

Currently available batteries are bulky and inflexible. They cannot be fully integrated with smart textiles. Developing flexible batteries is widely seen as an indispensable enabling technology for smart textiles. Over the past five years, significant progress has been made in the development of flexible batteries. The breakthrough came from the successful fabrication of polymer

---

*Some batteries remember their discharge point and require a charge when the batteries reach that point. For example, if one starts to recharge a battery when it is discharged to only 50% of its capacity, this battery will be unable to operate below that 50% mark again. For those batteries, they need to be fully discharged before recharging to avoid the loss of capacity.

electrolytes using screen-printing or inkjet techniques. The cells are made of proprietary inks that can be printed or pasted onto virtually any substrate, to create a battery that is thin and flexible.

It has been reported that the Slimcell, developed at MIT, has an energy density of 200 W h/kg (Jha, 2006). The battery is a sandwich of lithium electrodes and a polymer electrolyte, which is not only light and flexible, but also much safer. Because there is no liquid, the battery cannot leak. If it is somehow punctured in one part, the rest of the battery can continue working regardlessly. Researchers at NEC Corporation have developed similar flexible batteries. Instead of using metal electrodes, they employ an organic polymer as the cathode. NEC claims that the battery has a very thin structure (300 μm), and is very flexible, capable of fast recharging and environmentally friendly (http://www.scenta.co.uk/Engineering/437895/flexible-battery-developed.htm). On the other hand, researchers at QinetiQ have developed a flexible primary battery based on lithium/carbon monofluoride chemistry, which has a very large energy density of 650 W h/kg (http://www.qinetiq.com/home/commercial/space/space_technology/batteries_and_fuel.html).

Serious attempts in developing polymer batteries began in 1990s. The research has been driven by the need to improve the energy density and obtain flexible structures. Although the technology is far from mature, there are some trial products already on the market. Table 11.3 shows the power characteristics of flexible batteries from Power Paper (Power Paper, 2007). The power papers have typically a thickness of 0.6 mm and bending radius of 25 mm (http://www.powerpaper.com). This type of battery is much easier to incorporate into textiles than conventional batteries. The ultimate goal is to develop a fully compatible battery that is totally comfortable to wear and can be laundered. Towards this goal, researchers at Stanford Research Institute (SRI), USA, are developing a technology to fabricate anode and cathode materials onto thin carbon fibres that can be packaged around thin strands of electrolytes. These fibre batteries can then be woven into textiles, turning a garment into a mobile power source.

11.3.3 Fuel cell

High performance computing requires a relatively large power supply, usually in the range of 50 W. Rechargeable batteries that are currently

Table 11.3 Power characteristics of flexible batteries

Manufacturer	Nominal voltage	Nominal capacity	Thickness	Outline dimension	Bending radius
Power Paper	1.5 V	5.0 mA h/cm^2	0.6 mm	30 × 30 mm^2	25 mm

available are heavy (typically 300 g), bulky (approximately $6 \times 2 \times 20$ cm^3) and inflexible. They need to be recharged frequently to provide an operation period of only 2–4 hours. It is unlikely that any of these current batteries will be an elegant solution to the power problem of wearable computing. Fuel cells, on the other hand, have a substantially larger theoretical energy density of over 50 000 W h/kg and are seen as a possible technology to meet the power requirement of wearable computing at the high performance end (Vielstich *et al.*, 2003).

A fuel cell is an electrochemical device that converts hydrogen and oxygen into water and, in the process, produces electricity and heat (Vielstich *et al.*, 2003). In many ways, a fuel cell resembles a battery. However, it needs a continuous supply of oxygen and hydrogen from the outside rather than periodic recharging through power adapters. Usually, oxygen is drawn from the air and hydrogen is carried as a fuel in a pressurised container. Apart from hydrogen, fuels such as methanol, propane, butane and natural gas can be used as alternatives. Because the energy density of these liquid fuels is extremely high, the fuel cell has potentially a much higher energy density than any other established power source.

Fuel cells are classified based on the type of electrolyte they use. The proton exchange membrane fuel cell (PEMFC) is the most promising form of the technology (Barbir, 2005). Commercial products based on this technology have already been available on the market. However, in terms of high energy density, small dimension and flexibility, significant effort toward research is still needed to realise its full potential as a solution to the power problems of wearable computing.

11.4 Energy harvesting devices

It is envisaged that wearable computing will revolutionise our everyday life by providing effective access to computation and communication. Although the size and weight of modern computational hardware has been reduced considerably to accommodate this vision, power systems are still bulky and inconvenient. The energy density of the present-day batteries is typically about 100 W h/kg (or 100 mW h/g). At this level, a battery which weighs about 10 g could have an operating period of only a few hours to a few hundred hours, depending on the power consumption of the electronic circuits. To develop a sustained power supply which covers the lifetime of smart textiles is an extremely difficult challenge. A medium-term solution is to combine power harvesting with energy storage. Power harvesting devices continuously convert various forms of ambient energy into electricity, while the storage devices accumulate the generated electrical energy for the intermittent consumption usually required by smart textiles.

The history of energy harvesting dates back to windmills and water-wheels, which convert gravitational energy into kinetic energy to ease or replace manual labour. The first wearable power harvesting device was a 'self-winding' watch that winds-up automatically from the wearer's arm movements. Modern energy harvesting devices that generate electricity include photovoltaic cells from light, piezoelectric devices from human motion and thermoelectric modules from body heat. All these devices have the potential of being fully integrated with smart textiles. The challenge is to efficiently convert the limited energy available from/around the human body into electrical power.

11.4.1 Energy from the human body

The human body is a tremendous storehouse of energy with around 4×10^8 Joules stored in the average person (Starner, 1996). The body also consumes energy at a surprising rate to produce heat (thermal energy) and motion (mechanical energy). An average person continuously produces 80 W of heat during sleep and up to 1000 W during long-distance running. Body motion generates mechanical energy, even in seemingly motionless situations such as breath and blood transport. For example, it is estimated that from the mechanical power produced during exhalation, approximately 1 W could be used to drive a miniature turbine generator without affecting the body's performance. On the other hand, walking can, on average, produce more than 300 W of mechanical power with approximately 60 W available for electrical power conversion using piezoelectric devices. Upper limb motion during housekeeping or finger motion for typing can also produce a significant amount of mechanical power available for electrical power generation. In addition, a human body has a typical surface area of about 1.7 m^2, part of which can be covered with solar cells to convert light into electricity.

A human body carries a sufficient amount of energy in various forms. Even if only a very small fraction of this energy can be converted into electricity, the power required by many of the present-day low power devices can be met. The possibility of harnessing human body energy provides an attractive solution to the power problems of smart textiles. Success will depend upon the efficiency and compatibility of these conversion technologies.

11.4.2 Thermoelectric generator

A thermoelectric generator is a semiconductor device which converts heat into electricity based on the Seebeck effect (Nolas, 2001). Figure 11.2 shows the schematics of a conventional thermoelectric generator that consists of n- and p-type semiconductors connected in series and sandwiched between

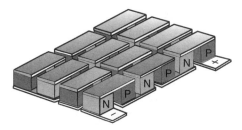

11.2 Thermoelectric modules consisting of n-type and p-type semi-conductors connected in series, and then sandwiched between two electrically insulating, thermally conducting ceramic plates (ceramic plates are not shown in the figure).

two insulating ceramic plates. Electrical power will be generated when a temperature difference is established across the two ceramic plates. Unlike conventional heat engines (such as steam engines and turbine generators), a thermoelectric generator is a solid-state converter employing electrons (holes) as its 'working fluid'. It has no moving parts and is totally scalable.

The efficiency and power level of a thermoelectric generator depend on the temperature difference and a material parameter called the thermoelectric figure-of-merit (or ZT) (Nolas *et al.*, 2001). Currently, the best thermoelectric materials have a ZT value of about 1.0. In a normal environment, the temperature difference between a human body and ambient is around 5–10°C. For an average human body with typical heat generation at a rate of about 100 W, the power output per unit area is estimated at 20–50 μW/cm^2. This indicates that a total possible power of about 300–850 mW may be obtained using thermoelectric harvesting based on an average body surface area of 1.7 m^2. However, it is unrealistic to cover the whole human body with thermoelectric generators with good thermal contact. A more realistic estimate is based on a surface area of 0.1 m^2 (equivalent to an area of 40 cm × 25 cm) that enables heat extraction from a body without causing significant inconvenience. The corresponding power output is about 2–5 mW. Such a power level is sufficient for some low-power applications; in order to increase the power level to meet a larger power requirement, it is necessary to improve the thermoelectric figure-of-merit, ZT. In fact, research efforts over the past 50 years have mainly focused on developing materials with improved ZT. Recently, materials with ZT approximately 2 have been discovered (Harman *et al.*, 2002).

Commercially available thermoelectric generators are bulky and rigid. They are not suitable for wearable applications. The development of miniature thermoelectric devices in recent years opens up a possibility for wearable applications. Seiko have reported a miniature thermoelectric generators (http://www.roachman.com/thermic/) that generates 40 μW electrical power at 3 volts from body heat, which has been employed to power electronic watches (Kotanagi *et al.*, 1999). It has been suggested that such miniature devices can be embedded into smart textiles for power harvesting. However, a more adventurous attempt is to develop flexible thermoelectric thin films or polymers that can be woven into fabrics. Prototypes of flexible thermoelectric generators have also been reported by several research groups (Qu *et al.*, 2001; Goncalves *et al.*, 2007). The success in this attempt will provide a truly compatible power source for smart textiles.

The unique advantage of a thermoelectric generator is that the same device can operate in a reversed mode based on the Peltier effect for heating or cooling (Goldsmid, 1986). When an electrical current is applied to the device, heat will generate on one side and cooling on the other. By changing the direction of electrical current, the heat pumping direction will be reversed. Thermoelectric generators have been incorporated into the pillows of surgical beds for patient comfort (Buist, 1989). Attempts have been made by the USA air force to develop helmets for astronauts and fight-jet pilots, which incorporate thermoelectric devices to keep them cool and alert in a hot environment, or by just flicking the switch, keep them warm in a cold environment.

11.4.3 Piezoelectric devices

A mechanical strain on a piezoelectric material produces an electric field (Yang, 2006). This effect (namely the piezoelectric effect) can be employed to convert mechanical energy into electrical energy. It has become one of the most popular energy harvesting devices which attracts an increasing research effort because of the simplicity of its underlying principle and the device's structure. Electrical energy can be generated simply by compression, bending or slapping of piezoelectric materials. The conversion efficiency of this process is measured by the coupling constant (in a dimensionless unit: coulomb volt/newton meter, C V/N m).

The two common piezoelectric materials are lead zirconate titanate (PZT, P in this abbreviation being derived from chemical symbol for lead Pb), and polyvinylidene fluoride (PVDF). PZT exhibits larger coupling constants (0.69 by compression and 0.35 by bending) than those of PVDF (0.11 by bending). Bending is usually easier to achieve and therefore employed in many applications. However, PZT does not have a wide range of motion in bending direction because of brittleness, which is not ideal for smart textile applications, where flexibility is an essential requirement. On the other hand, PVDF is very flexible, and easy to handle and shape. It has been the material of choice for many applications where flexibility is required, despite its lower coupling constant. Recently, attempts have been made to develop composite PZT materials, which are aimed at achieving improved flexibility while retaining a higher coupling constant (Liu *et al.*, 2005).

The motion of the human body induces strains that can be harvested using piezoelectric devices. Sources of motion include walking, arm swing, typing, blood pressure and shoe impact. While little is known about totally compatible piezoelectric smart textiles, many efforts have been reported on human body energy harvesting from walking. Piezoelectric elements are being imbedded in walkways to recover the human energy of footsteps (Starner, 1996). They can also be embedded in shoes to recover the 'walking energy'. Starner considered the use of PVDF materials as a part of shoe stiffeners for recovering some of the power in the process of walking. He estimated that 5 W of electric power can be generated by a 50 kg user at a brisk walking pace (Starner, 1996). Such a power level is sufficient for many demanding wearable applications.

The development of piezoelectric devices for energy harvesting from body motions such as the swing of the arms or typing is more challenging in both compatibility and power level. It was estimated that the power generated from these motions is limited to the order of milliwatts (Starner, 1996). For instance, typing by a moderately skilled person is estimated to produce about 20 mW of mechanical energy. Employing PVDF materials with about 11% of conversion efficiency, this gives 2 mW that can be generated. It has been suggested that wearable, wireless keyboards may be fabricated by incorporating PVDF materials into the fabrics. Each keystroke may generate enough electrical power to 'announce' its character to a nearby receiver through a wireless circuit.

Currently, piezoelectric-based harvesting can generate only low levels of power, in the order of microwatts. Significant effort towards research is needed to improve the power level without causing inconvenience or requiring significant body movement. A disadvantage of piezoelectric devices is that they are unsuitable for people with disabilities or limited mobility.

11.4.4 Photovoltaic cells

Electromagnetic radiation in the form of light is another source of energy around the human body. Unlike heat and motion, light is not generated by us but is abundant everywhere in daylight. Solar energy (energy in sunlight) can be converted into electrical energy using photovoltaic cells (or solar cells) (Antonio, 2004). A solar cell is a semiconductor device in which solar energy of certain wavelengths can be absorbed to generate free electrons (negative charges) on one side and holes (positive charges) on another.

The conversion efficiency of a solar cell is defined as the ratio of the electric power generated by the solar cell to the light irradiance collected on the surface of the solar cell. The irradiance varies with time, location and weather conditions. For example, the sun's average energy above the atmosphere is 1400 W/m^2, while it is only 1000 W/m^2 in the Sahara Desert, about 350 W/m^2 in Texas and 150 W/m^2 in the UK or some Northern European countries. With the best commercially available solar panels which have an efficiency of approximately 15%, the maximum electrical power that can be obtained using solar cells of 1.7 m^2 (equivalent to the human body surface area) will be about 40 W in a UK environment. However, in reality, such a power level is unlikely to be obtainable because the shadows and angles of solar irradiance are inevitable on parts of human body. A realistic estimate gives about 1 W for a solar panel area of 600 cm^2 (equivalent to an area of 30 cm × 20 cm).

Solar panels based on silicon, cadmium telluride, and other crystalline semiconductors are not flexible. Recently, attempts have been made to develop flexible solar cells for wearable applications. Konarka USA has reported a plastic solar cell that contains a layer of dye-activated titanium dioxide (TiO_2) particles on a clear plastic film (see Press Release on http:// www.konarkatech.com/, 2008). The operation principle is the same as that of silicon solar cells, but it has a flexibility resembling photographic film and can be mass produced on a roll-to-roll process at a significantly low cost. The conversion efficiency at manufacturing level is about 8%. This value is lower than commercial silicon-based solar cells, but is considered to be well over the initial threshold for the market's needs. An exciting aspect of this technology is that the same coating process may be applied to fabricate the TiO_2 solar cells on a fibre. By weaving such 'solar fibres' into textiles, a totally compatiable energy harvesting device can be achieved for smart textiles applications. Similar synthetic fibres have also been reported by researchers in Germany (Schubert and Werner, 2006). They believe that the fibres could be woven into machine-washable clothes to make wearable solar cells. The only downside to solar cells is that the power source does not work in the dark.

11.4.5 Wireless recharger

RF radiation is another potential source of energy, which comes from radio, television, mobile phone and wireless network transmitters (Griffiths, 1998). However, the energy level is very low and to harvest useful amounts of energy from this source requires either a large collection area or close proximity to the radiating source. One idea, which is not strictly energy harvesting, is to deliberately transmit a RF signal to power the devices wirelessly, as shown schematically in Figure 11.3. The operating principle is very similar to passive radio-frequency-identification systems (RFID) tags (Finkenzeller, 2003). However, a problem associated with this concept is the safety regulation that limits power levels. To avoid this problem, researchers at Massachusetts Institute of Technology are developing a method to transfer electrical energy wirelessly by 'non-radiative' induction (Schuman and

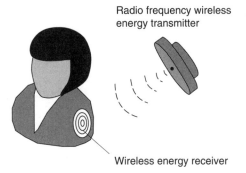

11.3 Recharging wearable power source through a RF wireless transmitter and receiver.

Miller, 2007). A power transmitter generates electromagnetic energy which is limited to a small space around the transmitter. The energy cannot ripple away as an electromagnetic wave and can be picked up only by receiving devices specially designed to resonate with the field. Most of the energy not picked up by a receiver would be reabsorbed by the transmitter. It is hoped that, by employing this technology, laptop computers or mobile phones can be recharged wirelessly when one is waiting in airports or train stations. If flexible and light-weight receivers of this kind can be developed, this technique could be employed to recharge smart textiles.

11.5   Energy storage and power management

Among the body-driven energy harvesting devices discussed above, piezoelectric and photovoltaic devices cannot generate power continuously. They require power storage devices to store some energy during power generation cycles in order to maintain continuous power output during the converter's 'sleeping' periods. Although thermoelectric power generation is continuous by nature, it is also likely to require a storage device so that a large power delivery in a short burst can be achieved. This is particularly useful in the case of wireless communications where a large power delivery is needed only intermittently.

To date, no single energy harvesting method is able to meet the requirement of some power-hungry applications, such as wearable computing. It is likely that a realistic near-future solution will involve a combination of several energy harvesting devices and an energy storage device. Figure 11.4 shows a schematic diagram of a proposed power system for wearable applications. Available electric energy storage devices include rechargeable batteries or supercapacitors. Rechargeable batteries have a larger energy storage density, but lifetime is limited by their recharging cycles, see Table

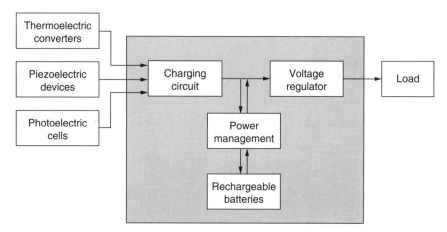

11.4 Schematic diagram of the energy harvesting system, consisting of power generation, storage, and management.

11.2. Supercapacitors, on the other hand, provide a virtually unlimited cycle life. However, their energy storage density is only about one-tenth of NiMH batteries. Recently, progress has been made in the development of carbon nanotube ultracapacitors which have an energy density of up to 75 W h/kg. Since the energy harvested from these sources is random by nature, a power management circuit will be needed to facilitate efficient charging under irregular conditions and to accommodate the varied charging characteristics of individual energy harvesting devices. For example, because the voltage generated by a piezoelectric device is usually very high, a step-down circuit will be needed, while for thermoelectric devices that generate a small voltage, a step-up circuit is required. A regulator circuit would also be needed to ensure efficient power delivery from the storage device to a load. Clearly, intelligent power management circuits will be an essential part of power systems for wearable applications.

11.6   Challenges and opportunities

Compared with the typical power requirements of today's electronic systems, the power level that can be obtained by existing energy harvesting devices is still relatively low. The efficiency of these devices usually decreases further when they are implemented into flexible forms that are required for most wearable applications. A major challenge is to develop flexible energy converters which can be fully integrated into textiles with high efficiency and at low cost.

Existing energy harvesting devices in their present forms can provide certain flexibility, but they are not completely wearable. A truly wearable device should provide not only total flexibility for comfort, but also be able to withstand tough wear and washing. The most exciting idea mentioned above is the development of energy harvesting fibres that can be woven into textiles or form a part of textiles. Recent advances in nanotechnology, molecular electronics and conducting polymers provide a key technological foundation that is urgently needed in this direction of research.

An analysis of the energy inventory of a human body indicates that sufficient energy input is available for harvesting. It is estimated that a total electrical power of 5–10 W is obtainable if these devices can operate close to their efficiency limits. In order to realise such a potential, a key challenge is to increase the conversion efficiency of these energy devices. Over the past a few decades, batteries and existing energy harvesting devices have made only moderate improvements in terms of improving energy density and size reduction. In contrast, microelectronic technology continues to advance rapidly. The size of the electronic circuits and the energy needed for operation have been reduced drastically. Wireless communication systems using only 50 μW are realistically possible. Progress on this front will help to ease the problems encountered by energy harvesting devices required to provide a relatively large power level. It is anticipated that the solutions to the wearable power supply problem may come from the improved energy density of advanced energy harvesting devices, as well as reduced power consumption by future electronic circuits (Fig. 11.5).

Apart from the piezoelectric, photovoltaic and thermoelectric devices, other energy harvesting devices exist, including turbine engines, electrostatic energy generators and magnetic induction generators. Traditionally, these devices are associated with large and bulky machinery. With the recent development in micro-electro-mechanical-systems (MEMS) technology, these devices can be implemented on a microscale and consequently incorporated in smart clothing and wearable technology. Harnessing the

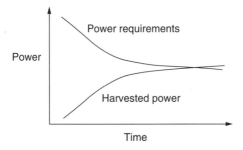

11.5 Key factors in meeting the challenge of wearable power sources: (i) improve energy density of energy harvesting devices; (ii) reduce power consumption of electronic circuits and systems (courtesy David Bryson).

energy from breathing has been considered using a micro-turbine genera-tor. An average person has an approximate air intake rate of 30 litres per minute with a breath pressure of about 2 per cent above atmospheric pres-sure. If only exhalation is employed for the power generation, the available electrical power is estimated to be about 0.4 watts.

Fuel cells, with the largest theoretical energy density, are considered to be one of most promising technology solutions to future power problems. Although they are not energy harvesting devices, their very large energy density makes them the most attractive power source that is being currently investigated. In order to develop fuel cells that are suitable for wearable applications, additional requirements such as small size, light weight and total flexibility will need to be considered.

11.7 References

ANTONIO A V, *Piezoelectric Transducers and Applications*, Springer-Verlag Berlin and Heidelberg, 2004.

BARBIR F, *PEM Fuel Cells: Theory and Practice* (Sustainable World Series), Academic Press, 2005.

BUCHMANN I, *Batteries in a Portable World: A Handbook on Rechargeable Batteries for Non-engineers*, Cadex Electronics Inc; 2nd edition, 2001.

BUIST R J, 'Thermoelectric pillow and blanket', United States Patent: 4859250, 1989.

COLIN A V, SCROSATI B, LUZZARI M and BONINO F, *Modern Batteries: An Introduction to Electrochamical Power Sources*, Edward Arnold Publishing, London, 1984.

FINKENZELLER K, *RFID-Handbook: Fundamentals and Applications in Contactless Smart Cards and Identification*, Wiley & Sons Ltd, 2003.

GOLDSMID J H, *Electronic Refrigeration*, Pion Limited, London, 1986.

GONCALVES L M, ROCHA J G, COUTO C, ALPUIM P, MIN G, ROWE D M and CORREIA J H, 'Fabrication of flexible thermoelectric microcoolers using planar thin-film tech-nologies', *J. Micromech. Microeng.* 2007, **17**, S168–S173.

GRIFFITHS D J, *Introduction to Electrodynamics (3rd ed.)*, Prentice Hall, 1998.

HAHN R and REICHL H, 'Batteries and power suplies for wearable and ubiquitous computing', in *Third International Symposium on Wearable Computers*, Oct 18–19, 1999, San Francisco, The IEEE Computer Society.

HARMAN T C, TAYLOR P J, WALSH M P, and LAFORGE B E, 'Quantum Dot Superlattice Thermoelectric Materials and Devices' *Science*, 2002, **297**, 2229.

JHA A, 'The Slimcell', *Current Events, Electric Auto Association Publication*, 2006, **38**(1&2), 26–27.

KOTANAGI S, MATOGE A, YOSHIDA Y, UTSUMOMIYA F and KISHI M, 'Watch provided with thermoelectric generation unit' Patent No. WO/1999/019775.

LIU W C, LI A D , TAN J, WU D, YE H and MING N B, 'Preparation and characterization of poled nanocrystal and polymer composite PZT/PC films', *Applied Physics A: Materials Science & Processing*, 2005, **81**(3), 543–547.

LUQUE A and HEGEDUS S, *Handbook of Photovoltaic Science and Engineering*, John Wiley & Sons, 2003.

MACHOVER T, Musical jacket project, MIT Media Lab, 2007, http://www.media.mit. edu/hyperins/projects/jacket.html.

NOLAS G, SHARP J and GOLDSMID J, *Thermoelectrics: Basic Principles and New Materials Developments* (Springer Series in Materials Science), Springer-Verlag, Berlin and Heidelberg , 2001.

PARK, S and JAYARAMAN, S., 'e-Health and Quality of Life: The Role of the Wearable Motherboard', in Lymberis A and DeRossi D, *Wearable eHealth Systems for Personalised Health Management*, IOS Press, Amsterdam, 2004, 239–252.

POWER PAPER PRODUCT DATA SHEET, Power Paper Ltd., Yegia Kapayim Street, Kiryat Ayre, Petah Tikva, Israel, 2007.

PRESS RELEASE, Konarka Technologies Inc., 116 John, Street, Suite12, Lowell, MA, USA, http://www.konarkatech.com/.

QU W, PLÖTNER M and FISCHER W J, 'Microfabrication of thermoelectric generators on flexible foil substrates as a power source for autonomous microsystems' *J. Micromech. Microeng.*, 2001, **11,** 146–152.

ROWE D M, *CRC Handbook of Thermoelectrics: Micro to Nano*, CRC Press, London, 2005.

SCHUBERT M B and WERNER J H, 'Flexible solar cells for clothing', *Materials Today*, 2006, **9**(9), 42–50.

SCHUMAN M and MILLER H, 'Wireless power another way', *Engineering & Technology (IET)*, **2**(6), 2007, 32–33.

STARNER T, 'Human-powered wearable computing', *IBM Systems Journal*, 1996, **35** (3&4), 618–629.

VIELSTICH W, LAMM A and GASTEIGER H, *Handbook of Fuel Cells: Fundamentals, Technology, Applications*, John Wiley & Sons, 2003.

YANG J, *Analysis of Piezoelectric Devices*, World Scientific, 2006, http://www. worldscientific.com

Part III
Production technologies for smart clothing

12

Garment construction: cutting and placing of materials

J. McCANN, University of Wales Newport, UK,
S. MORSKY, HAMK Poly, Finland and X. DONG,
China Women's University, PR China

Abstract: The cutting and placing of an appropriate mix of technical
textile structures and assemblies around the body with the adoption of
novel garment construction methods to ergonomically position technical
textiles selected in relation to the particular physiological requirements
of different areas of the body has the potential to address the functional
demands of the modern global 24-hour society, mixing work, relaxation
and everyday activities with each layer of the garment system providing
a potential location for the embedding of wearable technologies.

Key words: functional clothing, layering-system, physiological
requirements, textile structures, novel garment construction, ergonomic
clothing design, embedded wearable technologies.

12.1 Introduction

This chapter discusses a technical design development process for functional clothing that involves a direct relationship with the demands of the human body. It addresses the ergonomics of movement, protection, moisture management and thermal regulation not normally considered in traditional fashion design. The selection and positioning of technical textiles within a well-designed clothing 'layering system' has the potential to address the functional demands of the modern global 24-hour society, mixing work, relaxation and everyday activities. Relevant market sectors include performance sportswear, corporate wear, travel wear and inclusive design with particular relevance to promoting health and wellness and the demands of an ageing community.

The functional 'layering system', traditionally tried and tested in military combat wear, and now adopted in performance sportswear, embraces an inter-dependent base-layer, mid insulation layer and outer protective layer. The effective functioning of the layering system provides a focus for design innovation in the cutting and placing of an appropriate mix of technical textile structures and assemblies around the body. Novel garment construction methods may be adopted to position technical textiles that include light-weight, protective, thermally regulating, moisture wicking and stretch attributes, in relation to the particular anatomical, physiological and ergonomic requirements of different areas of the body. Emerging textile joining and finishing techniques, such as moulding, bonding and laser finishing provide a new aesthetic for the construction of garments in materials made primarily from synthetic fibres.

Each layer of the garment system may now provide a potential location for the embedding of wearable technologies such as vital-signs monitoring in intimate apparel and base layers, thermal regulation in mid-layer garments, as well as impact protection, and textile-based electronic user-interfaces in the outer layer. Smart clothing and related textile products with embedded technologies, provide an important interface between wearers and their environment. To bring emerging technologies to near market, and promote compliance, the aesthetic styling and comfort of functional clothing and 3D textile products must be acceptable, and the technology interface simple and intuitive for an inclusive audience.

12.2   A user-led design approach

When designing garments that are fit for purpose there are many factors to consider in setting an effective design brief. Functional clothing should both work and look good, and be in tune with the aesthetic, technical and cultural lifestyle demands of the target customer (McCann, 2000). In approaching clothing design from a product design, user-needs driven perspective, an analysis is needed of the garment characteristics required, to inform the setting of a clear design brief prior to embarking on the technical design development process. The designer must become familiar with the particular demands of the wearer(s) through observation, semi-structured interviews and appropriate market research (Kettunen, 2001). Requirements-capture of the physiological and style needs of the identified end-user, and an understanding of the end-use, or range of activities, should inform the selection and positioning of specific textile assemblies and trims within an ergonomically designed 'system'. Collaborative design with the user will provide on-going essential feedback throughout the design development process (Risikko and Marttila-Vesalainen, 2006).

The blend of technical and aesthetic design requirements should be finely balanced in relation to issues identified. The designer should consider the specific needs for each garment that may function either independently and/or as part of an adaptable layering system for a particular, or for multifunctional, end-use. The selection of fibres and textile structures will have direct implications in relation to the season, location, weather and environmental conditions, as well as personal protection from impact or a hostile environment, and the maintenance of the clothing system micro-climate in terms of thermal regulation and moisture management in relation to the workload. The functioning and protection of the senses will impinge on the selection of materials surfaces and textures. Antibacterial fibres and finishes and the micro-encapsulation of vitamins will assist in protecting the health of the user. The attributes of stretch fibres in flexible textile structures may address the ergonomics of movement, muscle support and enhanced garment comfort and fit (McCann, 2005). Questions should address the projected life span of the product(s), from initial materials selection and garment construction methods to garment aftercare and disposal at 'end-of-life'.

The performance sportswear sector, an early adopter of technical textile innovations, has now embraced additional functionality through the application of wearable technology for end-uses that include snow sports, mountain biking, running and team sports for the monitoring of fitness, performance and location, as well as for communications and infotainment. Novel textile and garment manufacturing techniques are being adopted to aid both the permanent embedding of textile-based wearable electronics in garments and the incorporation of specific enclosures for removable devices. The embedding and/or incorporation of wearable technology cannot be an add-on to the system but this 'technology layer' must be considered simultaneously throughout the design development process. The introduction of electronic functionality '. . . should be achieved in a way that is imperceptible to the user' (Roepert, 2006). Designers must evaluate the real need of wearable devices in terms of the overall comfort, positioning, functionality and usability of the style, and the clarity of the interface between the technology and the user.

12.3 Considering the benefits of the military/sports layering system

12.3.1 Layering principle

There is a long-established historical process whereby new forms of clothing very often come in through the medium of sports; such as riding coats, hacking jackets and anoraks (Twigg, 2007). Military preparedness has also always driven the development of textile-led functional clothing and most significantly, from the 1950s, dress uniform has been replaced in combat with the functional 'layering system'. The principle of the sports 'layering system' has evolved from military combat dress to enable the selection of appropriate combinations of technical textiles in clothing to protect the body for different activities and/or occupations in contrasting environmental and climatic conditions. The system is normally made up of 'base layer' ('second skin'), 'mid layer' (or insulation layer) and 'outer layer' (or protective 'shell garment'). A new category has come into usage promoted as 'Soft Shell', which is a hybrid mix of outer layer protection with the comfort of the mid layer. In addition, intimate apparel demands innovative technical textile assemblies that offer support and move with the body without chaffing. Personal protective garments, or inserts, for contact sports or for extreme hazardous environments, are also incorporated into the layering system.

12.3.2 Intimate apparel

Intimate apparel has been an early adopter of technical textiles, in combination with novel garment manufacturing processes, since the introduction of Spandex for corsets in the 1960s (Heathcoat, 2008). To achieve comfort, good fit and support, lingerie and corsetry uses a breadth of woven, warp and weft knit, lace and non-woven materials in a range of fibre and yarn combinations. Weft knits offer more stretch around the body while warp knits are more stable in providing support. A variety of garment construction methods rely on a high percentage of synthetic fibres, in combination with elastomers, for the moulding of components and for their suitability in relation to the replacement of traditional garment sewing with new joining and finishing techniques. These processes include bonding, ultrasonic and laser welding, digital embroidery, engineered knits and seam-free knit technology. Integrated textile sensors are now emerging in sports bras for the monitoring of health and performance (Numetrex, 2008).

12.3.3 Moisture-wicking base layer

The term 'second skin' is often used for base-layer garments worn close to the body. Light-weight warp and weft knitted structures are normally selected for ease of movement, providing stretch in their stitch construction to fit like a 'second skin', enhanced through the incorporation of elastomeric fibres. Most sports practitioners are now aware of the attributes of wicking fabrics and the dangers associated with wearing absorbent fibres in cold conditions. Base layer garments, especially for active use, are constituted from fibre/yarn combinations such as polypropylene and polyester, that wick perspiration away from the skin to prevent chilling, especially when stationary and in extremely cold conditions. A close body fit allows for faster moisture absorption to speed the wicking process. A dry inner surface feels more comfortable and reduces chaffing. 'Added value' may be provided through a range of fibres and finishes that offer anti-microbial, antiseptic and anti-UV protection, as well as by the micro-encapsulation of vitamins and natural healing materials (Hibbert, 2004). With particular relevance to wearable technologies, naturally derived antibacterial function is found in silver fibre that also has conductive properties (Noble Biomaterials, 2008). An extreme end-use for close-fitting, stretch base-layers is in custom-made garments for the treatment of burn victims (Gottfried Medical, 2007).

Base-layer garments are made both by traditional 'cut and sew' construction methods and by innovative circular knit construction such seam-free and whole-garment knitting. Strategic 'zone construction' may be used in the placement of fabric structures and properties in relation to the needs of different parts of the body, to provide different functionality (see Fig. 12.1b). Mesh inserts increase air flow for increased moisture wicking in high sweat areas, with mixed fabric weights selected to enhance mobility, give protection and add durability in areas of high abrasion. The moulding of garment components is commonplace in the intimate apparel market, see Fig. 12.1a. Flat-seaming techniques should be adopted for comfort and to avoid friction. Textile sensors are now being incorporated in sports-oriented base layers for the monitoring of vital signs (Anon., 2005). The shape and fit of the body for a breadth of figure types is an obvious consideration for the design development of base-layer garments with embedded wearable technologies.

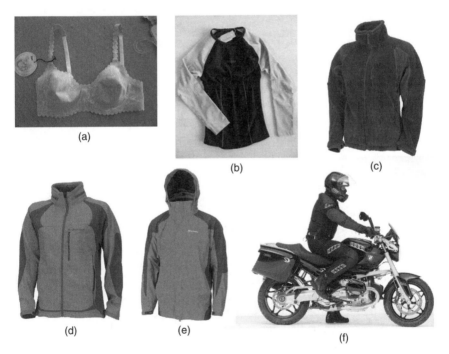

(a)

(b)

(c)

(d)

(e)

(f)

12.1 (a) Moulded bonded bra, Dim. (b) Base layer, Falke Santoni seamless knit. (c) Mid layer pile fleece with non-pile inserts, Sprayway. (d) Soft shell, Sprayway. (e) Outer shell, Sprayway Commanche III S. (f) Protection, Rukka SRO Anatomic motorcycling outfit (Rukka 2007).

12.3.4 Insulating mid layer

The 'insulation layer' varies in thickness and in its ability to trap 'still air' (or 'dead air'). It can be made up of more than one garment (e.g. a fleece jacket and gilet), depending on the range of likely conditions. The insulation layer should have compatible cut and styling with the base and outer layers to maintain an appropriate microclimate and avoid impeding movement. Fleece (brushed pile fabric, see Fig. 12.1c) is now worn more in daily life than in the mountains, often in corporate work wear, with outdoor clothing brands now acceptable in offices and cities. Fibre and fabric mixes for the insulation layer require bulk for the trapping of still air, and include down, synthetic waddings, fleeces, fibrepile and other three-dimensional spacer knit, woven and non-woven assemblies. Fibres may be shaped and oblique or hollow and are often made from polyester to prevent absorption of moisture. Knitted textile developments can be engineered to embrace different 'zones' of insulation in relation to the ergonomic design of the product and the requirement for protection (Patagonia, 2008), thermal regulation and moisture management relevant to different areas of the body. A recent innovation has been promoted as an inflatable fabric construction offering variable insulation (Gore, 2008a). Smart phase-change materials (PCMs), such as those introduced by Outlast Technologies, respond to changes in body temperature in absorbing excess heat and releasing it back to the body when required. Their advanced phase-change formulation, 'Metrix Infusion Coating', may now be printed onto 'virtually any fabric', including polyester (Outlast, 2008). Wearable electronics may be embedded into the mid layer products to provide enhanced thermal regulation, for example the Berghaus Heat Cell jacket and the North Face (Polartec, 2008) heated fleece.

12.3.5 Protective outer layer

The usual requirement for an outer shell garment is to be super lightweight with minimum bulk for easy storage. For end-uses other than sport, the weight will vary according to the demands of the wearer. The 'shell', or protective layer, is selected to provide the most appropriate balance of windproof and 'waterproofness' versus 'breathability' for the specified range of activities. The wind chill factor can lower the ambient temperature dramatically and threaten the clothing microclimate. The outer shell, traditionally also known as the 'hard shell', is intended to protect and maintain the function of the whole 'layering system' (Fig. 12.1e).

'Hard shell' outer fabrics are normally of lightweight nylon or polyester but may have protective areas of more abrasion-resistant fabrics, or particular properties such as fire retardance. Branded fibres can give special properties, e.g. high tenacity nylons offer reduced weight with enhanced strength, nylon Cordura® provides abrasion resistance for knee and elbow areas, and Nomex® provides fire retardance. Outer layer garments are normally made of woven constructions and may be referred to as 'Hard shells' with coatings or laminates incorporated in two or three layer assemblies. 'Two-layer' fabrics have an exposed coating or laminate on the inside, normally protected by a loose mesh lining, and 'three-layer' fabrics have a 'sandwich construction', with a fine single jersey backing or mesh to protect the laminate. The outer shell design must incorporate appropriate ventilation as few textile assemblies cope with the moisture produced from extreme workload. For sport such as outdoor pursuits, details include an adequate map pocket, zip guards, an ergonomically cut slim fit, to avoid billowing, with jacket front hem to be hollowed out and back cut lowered, velcro cuff adjustment, neck adjustment, optional fold-away hood and pit zips for ventilation.

'Softshell' is a new generation of outer layer garment that has emerged as an 'all-rounder' for winter and summer (see Fig. 12.1d). Branded products include W.L. Gore's Gore-Tex® Softshell, a 3-layer Gore-Tex® jacket on the inside of which is a fleece lining. Some 'soft shells' have additional anti-abrasion fabrics laminated onto areas that need greater protection and reinforcement. Designs are becoming increasingly sophisticated in terms of 'Comfort Mapping', the terminology adopted by W.L. Gore (Gore, 2008b), in selecting a mixture of material compositions and constructions to be placed in appropriate positions in relation to the demands of the body in order to achieve maximum comfort and performance. These garments often embrace stretch material constructions for added comfort.

12.3.6 Personal protection

The Italian company, Dainese, originally concerned with protection for motorbike riding, is a leader of innovation in safeguarding the body without restricting movement, for a range of extreme sports. The early 1980s saw the beginning of composite protection gear. Biomimicry provided the inspiration for the 'Aragosta' (lobster), and subsequent back shields such as the anatomically profiled 'man-armadillo', that display 'references connected with the observation of the animal world' (Briatore, 2004, p.30). The role of the prostheses is to absorb impact, distributing the energy over a larger surface than that of the striking point. Two types of protectors are used: 'relatively rigid composite reinforcements positioned at vulnerable points (shoulder, elbow, back, knee-tibia), and soft padding for particularly delicate zones (collarbone, chest, humerus, ilium, femur)' (Briatore, 2004, Fig. 4.1f). In the 1990s materials such as Cordura®, Gore-Tex® and Kevlar® were introduced, together with protective metal components and, since 2000, carbon and titanium knee shields have been 'moulded' into the suits. The Dainese Safety Program was established during 1990–93, later to become the D.Tec (Dainese Technology Centre) to combine the community's continuous crossing of information and sharing expertise among doctors, engineers and technicians (Briatore, 2004).

Dainese garments 'are conceived as structural frameworks in which to insert the protections necessary for different types of challenges' (Briatore, 2004, p.68). The 'Safety Jacket' was developed as a protection system that easily adapts to different types of sports and environments. Protectors are made of a complex mix of 'smart-tech' and 'bio-tech' materials, with plastics or metals which may be rigid or soft, elastic or solid for functions such as anti-intrusion, anti-impact, anti-cut, anti-overheating or anti-abrasion, and may be attached, inserted or embedded to offer shelter and support. 'And walking out the door to confront the weather, the street, work or travel already implies the choice of a wide range of protections. By now it is common knowledge that danger isn't lurking where you think it will be' (Briatore, 2004, p.35). The founder, Lino Dainese, believes in protection for all; 'Elderly people fall down, they slip, almost always in their homes . . . and they break bones. Children engage in lots of sports, often in hazardous urban environments: bicycles, skateboards, roller skates . . . or in natural but still risky settings; skiing, horse back riding, canoeing, etc.' (Briatore, 2004, p.34). 'Design is an important element of active participation in the fate of people and the planet, and it can only become more important as time goes on.' (Briatore, 2004, p.33)

Once again, performance sport provides another example of the early adoption of wearable technology. The Dainese D.Tec division launched the 'Procom' project in 1998, for detection of biometric data about motor bike riders; 'a system of measurement and control of parameters (like temperature, blood pressure, perspiration, adrenaline, etc.) that determine the physical state and comfort of a rider' (Briatore, 2004, p.32) and to gauge levels of stress, fear, competetitive drive and fatigue (Briatore, 2004, p.112). Their Procom system hardware and software were developed in collaboration with research institutes that include MIT, Boston, Technopolis in Bari and CNR in Milan. The system is composed of an 'electronic undersuit' equipped with 32 humidity sensors, 32 temperature sensors and a heart rate monitor capable of taking readings every five seconds, as well as a radio transmission system and GPS plotter. The data are recorded on a small storage system installed in an aerodynamic segment of the suit. They can then be transferred to a computer for processing and interpretation (Briatore, 2004). This suit has been through several phases of evolution. The data has been gathered 'under different conditions of racing, speed, context, climate, duration, with the aim of obtaining a large integrated, representative data sample' (Briatore, 2004, p.114). This type of extreme clothing requirement demands close cross-disciplinary collaboration to integrate the knowledge and skills of a team that embraces garment engineers, doctors, technicians and end-users.

12.4   The traditional approach to pattern development

12.4.1 Design working drawings

The technical design process begins with a drawing or appropriate two-dimensional representation of the style to be developed, see Fig. 12.2. The designer must provide clear guidance from the initial 'drawing' to the final 'toile', or three-dimensional prototype. This should be supported by detailed fabric/materials and trim specifications, with all garment pattern pieces clearly marked with a suggested sequence for their manufacture. Fashion sketches, shown on the Fig. 12.2a, may be highly evocative and illustrative without giving accurate proportions or detailed information. It is normal to provide an additional flat 'working drawing', either for technically oriented designers to follow themselves, or to pass on to the pattern maker. This may still be relatively sketchy, depending on the scale and the location of production envisaged and on the level of communication within the development team. The designer may work closely with a sample maker on the drawing, and the technical specification may be transmitted electronically off-site and/or off-shore, with language differences, and without face-to-face discussion. Technical drawings for functional garments, such as sports, intimate apparel, children's, leisurewear and corporate work wear normally demand clear graphic presentation. Flat diagrammatic drawings, with enlarged details or embellishment are invariably presented in a flat format with fairly accurate proportion. Computer layouts enable reasonably easy alterations of variations on colour and detail. Some buyers are prepared to make range selections from 2D visual representations of designs.

12.4.2 Traditional flat pattern development

Patternmaking is one of the initial steps in the garment design development process. It is 'a craft that has evolved over the centuries into a skilled technical process' (Anderson, 2005). Ethnic and artisan clothing was originally constructed from hand-woven fabric with rectangular-shaped pieces of

12.2 Examples of garment illustrations. (a) Fashion illustrations. (b) Sketch showing functional features. (c) Flat computer drawings.

fabric, incorporating the selvedge, or edge of the weave, where possible, in order to leave the fabric intact and to minimize waste. Garments made up of geometric shapes include gathered skirts and saris, with decorative borders, the kimono, the Indian salwar (loose pyjama-like trousers), the Scottish kilt and workwear smocks (Tilke, 1982). In the fifteenth century, patternmaking evolved as carefully engineered pieces were cut to follow the contour of the body. 'For evermore, fabric would take a back seat to fashion' (As quoted by Anderson, 2005). Prior to the Industrial Revolution it was normal for the elite to have tailors working with their personal measurements to customize patterns. During the Industrial Revolution, standardized patterns were developed, initially resulting in poorly fitting garments. 'After lengthy experimentation and standardized sizing, patternmaking made a triumphant transformation from customization to standardization' (As quoted by Anderson, 2005). Mass-produced ready-to-wear fashion has increasingly been moved from factories in Europe to off-shore locations, although a 'home coming' and the customization of 'value added' products is now being considered for products with wearable technology.

Two-dimensional flat pattern drafting is widely used with patterns developed from size charts or personalized from an individual's body measurements. Standard basic blocks are developed, based on extensive research into the characteristics and measurements of the human bodies of various races in different countries, in order to address their various demands with particular regard to comfort and performance. Blocks are perfected constantly to meet a breadth of everyday design requirements, with appropriate amendments for fitted, semi-fitting or easy fit garments. The function of a standard block serves as a basis for garment designers to make adaptations in flat pattern cutting for the development of their garment designs in a relatively scientific, accurate and speedy process. Blocks are made for particular target customers and garment sectors from intimate apparel to outerwear. A basic block might consist of back and front body and sleeve pattern, with no seam allowances or style lines, providing the basis from which a variety of garment styles may be adapted. The designer/patternmaker develops a new design by adding style lines and garment details. The basic block is symmetric but adaptations can be asymmetric, with side seams and shoulder seams manipulated as basic shapes are converted to individual styles. Details such as facings, linings, pockets, protective panels and seam allowances may be added as demanded by the design and fabric selection.

12.4.3 Reverse engineering

Another flat pattern making method is known as 'reverse engineering'. A garment is taken apart, flattened, the garment pieces traced off, checked and pattern pieces made with the addition of appropriate grain lines and seam allowances. Measurements may also be taken directly from an existing garment to inform the drafting of a flat pattern onto paper. The resulting pattern pieces must be assembled into a garment toile for verification or amendment. This is a method commonly adopted in the development of functional clothing when design teams appropriate and adapt successful pattern developments from competitors or work from and amend tried and tested past styles. It is now normal for resulting flat patterns, however developed, to be plotted onto and stored on a computer with the option for further style manipulation to be developed directly on the screen. Garment sizing adaptations are also developed on computer through the introduction of grading rules in relation to fabric characteristics and the desired width, length, fit and proportion for the specific style and target customer.

12.5   Introducing a flat pattern cutting system for functional clothing

12.5.1 Cutting for movement

There is little text available to explain cutting for movement. While a choice of textbooks exists for the development and adaptation of standard flat pattern blocks for men's, women's and children's wear, with slight variations in their approach to traditional garment cutting, little guidance is available to inform the cutting of clothing for functional end-use. Traditional pattern cutting textbooks cater predominantly for upright posture, for erect figure types, with a limited degree of movement. While the traditional fashion oriented garment market continues as a serious business, there has been rapid growth in the sport, leisure and well-being related markets, with an emphasis on function. In the occupational workwear field, special jobs demand functional wear. For ordinary every day activities, the movement of the human body is relatively limited. For example, when walking, the body is normally erect and both arm and leg movements are not extreme. Traditional standard blocks cater for garments designed for relatively upright posture and restrained movement. These blocks have limitations for more extreme movement, see Fig. 12.3.

Current standardized blocks are not normally appropriate for functional clothing in terms of fit, movement, articulation and predominant posture. Functional clothing development has been driven by technical textile innovation. As fabrics have become more breathable and elements of stretch are added for comfort, leading-edge performance clothing for end-uses such as active sport and corporate work wear is becoming more tailored and streamlined. A close fit that follows the silhouette of the body will promote

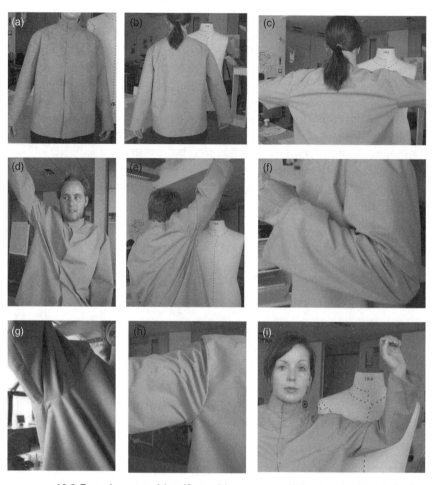

12.3 Experiment to identify problem areas within standard basic jacket block in relation to movement. (a) Original basic block front. (b) Back. (c) Strain across back. (d) Strain across front. (e) Strain with arm lift viewed from back. (f) Excess fabric in elbow area. (g) Strain on arm lift. (h) Excess fabric in sleeve head. (i) Restriction of movement in shoulder area.

agility, while larger, looser shapes may impede movement. Articulated cutting is required for the elbow and knee areas, the underarm, in hoods and sometimes within the overall silhouette of the garment to address predominant posture for a specific activity. Functional wear has quite different requirements from everyday clothing, with prime performance features including support, movement and protection, as well as overall comfort. This is a relatively new design area that is little understood by designers trained in traditional pattern cutting. Current producers of technical clothing have developed their own technical reference materials through trial and error. Consequently, it is necessary to devise a methodology for the drafting of basic blocks that promotes a more efficient and scientific technical design process for the development of performance clothing.

12.5.2 Standard blocks for functional wear

A new approach is proposed in the development of basic blocks that introduces and addresses the ergonomics of movement. Traditional garment construction is primarily done in two-dimensions on a flat surface, with basic blocks that normally consist of elements such as front, back and sleeve (Armstrong, 1987). Risikko and Marttila-Vesalainen in Finland (2006) have concentrated in their book *Vaatteet ja Haasteet* (*Clothes and Challenges*) in general on questions about factors affecting human body temperature, clothing physiology, requirements set by body measurements, body movement and dimensions, as well as how to measure the ease required by movements. In the seventh chapter, they have introduced static and dynamic measurements using wooden mannequins to show differences between these measurements, e.g. bending back and/or knee, lifting arm. They also show briefly some technical construction details of garments.

As we have not been able to identify too many sources – books or otherwise – for technical clothing, we wish to introduce construction methods for the development of functional garment basic blocks. These blocks have been designed to follow the body's three-dimensional contours in providing ergonomic shaping with reduced friction, and enhanced movement. Simple rules have been developed and tested to guide the drafting of the blocks. The resulting blocks address aspects of balance and the introduction of shape that enables unrestricted movement, while removing seams from traditional positions where style lines may be avoided. A similar approach has been taken for the development of mid insulation-layer and base-layer blocks.

12.5.3 Devising rules for the outer layer

Rules have been devised for the development of basic blocks for outerwear garments that provide a starting point for cutting for movement and predominant posture for jackets and trousers for men and women. The blocks developed are suitable for stable woven fabric structures. The development process began with standard outerwear blocks being made up and observed on wearers who were asked to adopt postures common to functional activities. A digital record was made of the testing of the shape and fit of the standard blocks and comprehensive notes and measurements taken as the garments were subjected to movement. Problems identified to do with

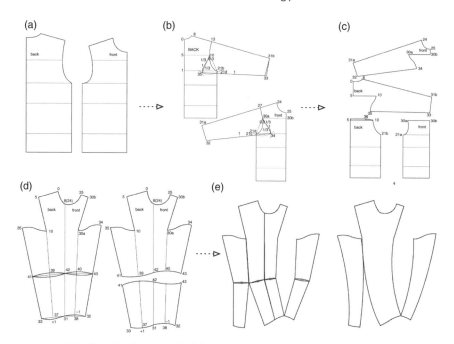

12.4 Developing a basic block for an articulated outerwear jacket.
(a) Basic block back and front. (b) Grown on sleeve development with
drafting of underarm lift. (c) Splitting pattern and lengthening centre
back. (d) Joining back and front sleeve and introducing elbow
articulation. (e) Adaptation showing vertical sleeve articulation.

restriction in movement were rectified and new functional blocks created
with rules tested and refined. Style adaptations devised from these blocks
may have comfort and movement enhanced through the selection of woven
structure fabrics, or stable knit constructions, with a percentage of stretch,
see Fig. 12.4.

The blocks have been developed to address common factors associated
with a range of outdoor activities (see Fig. 12.5). For example, overarm and
shoulder seams have been merged with yolk lines to provide a basic block
that has no joining over the shoulder. This follows the logic that outerwear
garment design normally avoids seams in areas where moisture is likely to
penetrate. Ergonomic movement is introduced by means of articulated
cutting for arm-lift and elbow and knee shaping. This articulation may be
further enhanced with the advantages of stretch fabrics and/or in the
cutting of certain pattern pieces on the bias of the weave as opposed to
making use of the rigidity of the straight grain. Standard functional blocks
may be further developed and refined for more extreme movement in relation to the demands of a particular end-use. The standard block seams may
be manipulated in relation to the desired style. The design of garments with

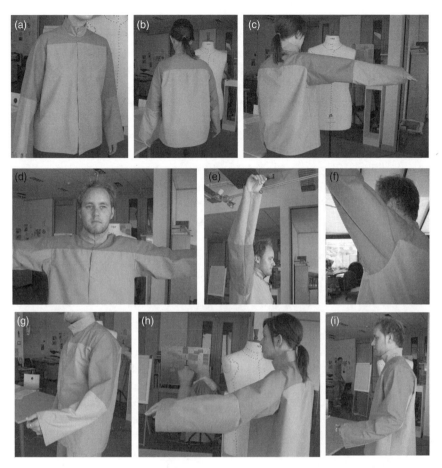

12.5 Testing of new standard articulated block. (a) Front. (b) Back.
(c) Lack of strain across back. (d) Lack of strain across front. (e) Arm
lift side view. (f) Arm lift back view. (g) and (h) Horizontal articulation
at elbow. (i) Vertical elbow articulation through sleeve.

embedded wearable technology should be carried out in on-going liaison
with technology developers to best accommodate the selected devices. The
outer layer is normally a key location for the positioning of wearable technology such as textile switches, soft displays, avalanche detectors, security
pockets and design details such as enclosures for ski passes.

12.5.4 Devising rules for the mid layer

Mid-layer garments must work in harmony with the outer shell or protective layer in terms of styling, shape, fit and articulation. The main purpose
of the mid layer is to provide insulation for thermal regulation. Traditional

mid layers, such as the knitted woollen jumper, are now replaced by, primarily, garments such as lofty warp and weft knit fleeces and fibre pile constructions, and lightweight padded jackets or gilets with woven outer fabrics and natural down or synthetic fillings. Patterns for these garments should be initiated from blocks that are similar in shape and silhouette to the outer layer so that the garments can move together without restriction. In addition, fabrics should be selected to avoid friction between the layers. Predominant posture for bending forward, sitting or climbing may be addressed by hollowing out a front hem line and adding length to the rear of a garment. Rules have been devised to scale down the outer layer functional blocks as a starting point for woven mid-layer garment development. For more flexible knitted fabric constructions, mid-layer garment blocks may be developed from adding tolerance to the base-layer blocks, dependent on the weight and structure of the chosen material.

12.5.5 Devising rules for base layer

The base layer or 'second skin' is the garment within the clothing system that should be most in tune with the body in terms of shape, fit, movement and moisture management. There is often a lack of consideration given to the cut of base-layer garments by comparison with the sophistication of intimate apparel. In addition, this layer has a direct relevance to the embedding of technologies for vital-signs monitoring, where comfort and fit is key to the positioning and functionality of wearable devices. Fabric qualities vary enormously in terms of their characteristics such as stretch or rigidity and consequently the basic block will need to be verified and amended in relation to the fabric, or combination of materials, selected for a specified end-use. The assembly of garment components directly impinges on comfort, with modern joining technology offering possibilities for discrete smooth seams. Despite a breadth of design permutations, there are basic principles that may be taken into account in developing rules for standard blocks. An even squared grid drawn on the erect torso will demonstrate relative degrees of stretch required for different zones of the body (Watkins, 2007; see Figs 12.6a, b and c). In cutting for movement, base-layer garments should be designed to avoid a basic 'T' shape with seams placed in positions that may chafe or rub. In principle, a raglan cut will substitute the need for shoulder and set-in sleeve seams, reducing friction over the shoulder area for the carrying of backpacks. Underarm seams and body side seams should also be avoided as these will chafe during repetitive movement. The introduction of a body side panel will allow for garment fitting to be evenly dispersed around the torso. A two-piece sleeve, similar to the concept of a tailored sleeve, but with a flat sleeve crown, will enable sleeve shaping that follows the natural bend of the arm. The hem line may be ergonomically

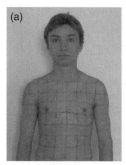 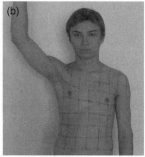 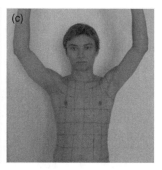

12.6 Natural skin stretch with movement of the body. (a) Normal view. (b) Right arm raised. (c) Both arms raised.

shaped as for the mid and outer layers. The base-layer block rules provide a starting point for further development.

12.6   Three-dimensional pattern making

12.6.1 Working on the body form

In the traditional three-dimensional pattern-making process, the pattern pieces are developed on a stand, also called a tailor's 'dummy' or 'mannequin'. Modeling, or 'draping', on a body form is one of the oldest traditional methods of creating a pattern. Reference lines in fine woven or adhesive tape are pinned or stuck to the stand to represent bust, waist, hip and arm holes, neck line, etc., as relevant. Design or style lines are also marked on the stand using fine tape. One half of the stand is used for symmetric styles and then the information is mirrored to create a full pattern. Traditionally, calico (toile), or a fabric with comparable characteristics to that specified within the design, is pinned onto the stand, with reference to appropriate grain lines. The information from the style lines on the stand, and all references to construction detail, are systematically transferred to the 'toile'. The toile/fabric pieces are then carefully removed, flattened, and pattern shapes traced off onto pattern paper. The pattern is checked for accuracy between components and details are added, such as grain lines, notches, buttonholes and correct seam and hem allowances.

12.6.2 Technical drawings

Initial design drawings should promote realistic proportions with a visual explanation of the front, back and side views to provide essential information on garment balance, pitch and articulation. In the design-driven performance sportswear sector, designers are becoming more creative in their presentation of technical garments and the detail of their functionality. The depiction of front, back and side views, with articulation in terms of 'arm-lift', and added movement at knees and elbows, is quite demanding in terms of 2D representation. There may also be the added complications of fold-away hoods, ventilation, storm flaps, interior features, etc. This type of 'illustration' is needed to clearly explain the design concept throughout the design development process as well as, in a more graphic format, to effec-tively sell the product in point-of-sale material. Technical working drawings are required to inform both the two- and three-dimensional pattern making process, see Fig. 12.2. There is now an additional demand for the design team to resolve in describing the incorporation of wearable electronics!

12.6.3 Designing in the round

The success of the design development process will largely depend on the suitability of the design brief and the designer's inherent understanding of the form and proportion of the human body and an ability to work 'hands on' in three dimensions. It prompts a return to designs being created in realistic proportions, and detailed specifications on materials and construc-tion techniques being recognized as a valued part of the creative process. Clothing design development that is 'fit for purpose' requires design lines and detail geared to the technical, aesthetic and cultural demands of the end use as opposed to styling that is driven by transient fashion with varying degrees of superficial embellishment. Recent textile innovation in tandem with novel garment joining techniques introduces a new aesthetic that blends form with function. When first introduced, materials made from made fibres such as nylon were made into waterproof garments with non-breathable coatings that had to be designed as loose fitting styles to provide ventilation; but, with the introduction of more breathable coatings and laminates, and elastomeric stretch attributes, outer layer functional gar-ments throughout the layering system have become increasingly close fitting and ergonomically cut.

The normal convention of starting from drawings, often with exaggerated proportions, that only depict front and back views, should be avoided. The designer should begin by observing the body in terms of three-dimensional size, shape and predominant posture. Standard measurement charts, with erect diagrammatic figures, may be consulted and compared but be amended to reflect the natural measurements of the target wearer or market. Addi-tional measurements may be taken directly from the body to calculate movement such as arm lift, elbow and knee articulation. Ideally, the techni-cal and aesthetic design process should blend working on the stand with flat pattern development, informed through liaison with users for the veri-fication of overall comfort, fit and functionality.

The designer can begin by working directly onto the tailor's dummy, making and refining decisions with regard to design lines and detail as the initial prototype evolves. Trial fabrics should be sourced that have comparable characteristics to those for the proposed design development. The traditional use of calico for the development of toiles (initial prototypes) will not be appropriate to give a true representation of the characteristics of technical and predominantly man-made technical materials. The dummy, mannequin or stand will provide an initial reference point as opposed to an accurate and comprehensive representation of the human body. The subject will have a particular size, shape and posture that will need to be mapped onto the pattern development. The process of designing in the round demands a continual sequence of cross referencing between development on the dummy, testing and amending the toile (initial prototype) on the subject, transferring information into a flat pattern and remaking the toile from the flat pattern for verification on the subject. This process will be repeated until the shape, fit and balance of the body and pitch and articulation of the sleeve are considered fit for purpose. A final prototype should be tested in the selected fabric(s). A visual, digital, record should be made of the design development process.

12.6.4 Body shape and size

There is no standard guidance for size, shape and fit. The last UK sizing survey in the public domain took place in 1952 when the standard size for women was found to be Size 12. The recent SizeUK survey, not in the public domain, has found that the average UK woman is now size 16 (Sizemic, 2007). Feedback from this survey has indicated that 60% of shoppers in the UK have difficulty in finding clothes that fit. Garments may vary in size specification depending on the style and ease allowance. Fashion collections and style conscious, active sportswear ranges seldom cater for larger sizes. There is a confusing mix in differences between size charts from different companies, and different countries, in addition to the changing needs of the body for different age groups, or degrees of fitness. 'Small', 'Medium' and 'Large' for one target group will be quite different from that of another category of end-user. [TC][2] state (2008) that their 3D Body Measurement System consists of four strategically placed sensors, which use white light to register more than 200000 data points on the body, scanning the whole body in less than six seconds and producing a true-to-scale 3D body model within minutes for the custom fitting clothing as well as other 3D product development. Three-dimensional scanning provides body shape analysis relevant to medical applications and, in health clubs, for fitness management. Body scanning is being used to directly inform the design and fit of customized compression hosiery

(William Lee Innovation Centre, 2007). Body scanners are now located in some retail outlets or may be self-operated by the scan subject, in a fully private environment, without the need for a 'scanner technician' to operate the scanner([TC]², 2008) .The Size UK data may be mined for the specific end-use of commercial customers.

12.7   Placing of materials

12.7.1 Textile selection

Technical textiles exist in diverse fabric constructions and finishes, with a range of contrasting characteristics and performance attributes. A knowledge of their structures and properties is essential for the interpretation of design requirements into the appropriate cutting lines and positioning of materials within the garment layering system. Choice of materials will be dependent on the demands of the environment and the end-use. The designer-pattern cutter should source textiles for their technical and aesthetic attributes in relation factors such as weight, handle, flexibility, insulation, breathability, absorbency, rigidity, abrasion resistance and protection. In designing in the round, materials should be selected to take advantage of their stability or degree of stretch, with both style and cutting lines positioned to work in harmony to support and/or move with the body. Both elastomeric and mechanical stretch textiles will have one- or two-way stretch characteristics. The direction of stretch or bias of the fabric, chosen to enhance movement, must be clearly indicated on the resulting pattern pieces.

Addressing predominant posture, repetitive and extreme movement, thermal regulation, moisture management, impact protection and safety are some of the key concerns in three-dimensional design development. Mesh inserts or strategic laser cut patterns can increase air flow for increased moisture wicking in high sweat areas; mixed fabric weights and variable stretch characteristics may be ergonomically placed to enhance mobility and support. Fibre choice is key to technical textile development with a mix of generic and brand names promoted. Polypropylene is relatively inert and transports moisture. Polyester is used in lightweight easy-care constructions throughout the layering system from base layer wicking textiles to lofty insulating fleece and fibre pile fabrics. Polyester is also used in softer handle outerwear fabrics. By comparison, nylon fibre is relatively stronger with High Tenacity nylon providing light-weight protective shell materials and nylon Cordura providing abrasion resistance. Three-dimensional spacer knitted or woven constructions provide impact protection and thermal regulation (Heathcoat, 2008). Fabrics must be considered in combination with the selection of appropriate garment assembly techniques.

12.7.2 Zoning and mapping

Strategic 'zone construction' is current terminology used to describe the positioning of carefully selected textile structures, sympathetic to the needs of different areas of the body, in order to increase garment performance. Innovation, in terms of fibre development and textile engineering, often takes inspiration from bio-mimicry, a study of plants and animals. Terminology such as 'second skin' is used for base-layer garments, where textiles are designed to enhance moisture wicking and ventilation or provide variable stretch in areas or 'zones', mimicking the function of the human skin. Fleece pile fabrics are being engineered to incorporate variable patterning, in developing fabric structures replicating animal fur, to provide areas of added protection and/or ventilation, relative to the demands of the body (Patagonia, 2008). 'Body mapping' is also current terminology used to market outer layer garments, where features such as breathability, abrasion resistance and thermal regulation have been addressed through fabric selection, in combination with garment cut and construction (Gore, 2008b). The development of these engineered knitted and woven structures, with their zones of contrasting fabric and yarn properties, provides garment functionality that is integral to aesthetic appearance.

A functional design attribute of modern clothing, now often taken for granted, is stretch. This provides enhanced movement and fit in close-fitted or restrictive cuts, but is also used to provide enhanced comfort in everyday clothing. Stretch, in combination with ergonomic cut, enables the putting on and taking off of garments throughout the layering system. Stretch is a key attribute in the fit of intimate apparel and base-layer garments for sport and medical applications. Variable stretch textiles are used in compression garments that offer targeted support in maintaining muscle alignment and a reduction in the loss of energy in athletic performance. Additional reinforced taping, outlining and supporting muscle groups, contributes to the aesthetic appearance of the garment. Stretch fibres contribute directly to the placing of textile sensors and embedded wearable electronics that must be held in appropriate locations within the garment system. Recent sports applications for wearable electronics in base-layer garments have adopted seam-free knit technology. Strategic zone construction is integral to seamless knitting techniques. Machinery producers have established terminology such as 'Seamfree' (prevalent in the structuring and shaping of intimate apparel and base-layer garments; Santoni, 2008), 'Scan to Knit' (in medical hosiery; Stoll, 2008; William Lee Innovation Centre, 2007) and heavier gauge 'Whole garment' knitwear (Sheima Seiki, 2008). These technologies offer the potential for the wearer's individual size and shape to inform the digital development of personalized patterns.

12.7.3 A new aesthetic

A new aesthetic 'look' has become prevalent, predominantly in the performance sportswear and intimate apparel sectors, early adopters of technical textiles. In these technical textile driven areas with predominant use of synthetic and man made-fibres, novel garment design engineering techniques are emerging for garment cutting and the stitch-free joining of materials. The selection of fibre properties and textile structures has a direct influence on the choice of garment manufacturing methods and finishes. Innovations in garment heat bonding and laser welding are particularly suitable for technical textiles made from a high percentage of synthetic fibres (TWI, 2006; Prolas, 2008). For intimate apparel, heat seam bonding in combination with heat-moulded components produces 'sewfree' garments with clean design lines that enhance comfort through reducing friction in key sensitive areas (Sew Systems, 2008). Functional design details, such as zip openings, garment edges and perforations for ventilation, may be laser cut to provide a clean and non-fraying finish. The Gore Airvantage jacket has been developed with the strategic 'body mapping' of multilayer bonded channels which may be inflated or deflated to regulate insulation (Gore, 2008b). Waterproof zips, for main closures and pockets, sleeve tabs and hood reinforcements, may be bonded without stitching (Ardmel, 2008; Wilson, 2004). Sonic stitching is used primarily for disposable garments such as forensic suits, medical end use and clean room garments. Moulding of components in bulkier materials is used for protective elements in garments for contact sport and occupational wear.

12.7.4 Incorporation of smart attributes and embedded technologies

This process of designing in the round becomes more complex with the introduction of wearable technology. Engineered knit, laser finishes and bonding techniques enable textile sensors and miniaturized electronic devices, and their controls, to be dispersed throughout the layering system. Textile-based electronics demand accurate positioning and attaching, embedding or encapsulation in garments constructed from knitted, woven and non-woven textile assemblies. Seam-free engineered knit, for example the Numetrix sports bra (Numetrex, 2008), may incorporate conductive yarns positioned for the monitoring of vital signs. Mid-layer conductive polymeric textiles may heat up when a power source is applied, either from a battery or mains power: for example, in the Berghaus Heat Cell jacket (EXO², 2008). Smart, shock-absorbing textiles provide impact protection, such as 'd3o' Laboratory's materials (d3o, 2008), with a range of garment applications that include skiwear with the protective fabric integrated at key contact areas such as knees and elbows. In the event of a sudden impact, molecules instantaneously lock together to provide a protective barrier. Dow Corning's 'Active protection' is another smart impact protection impregnated textile adopted by Rukka for extreme body protection in the area of motorbiking. Conductive metallic fibres and polymers are used in light-weight textile assemblies for soft keyboards and as controls for wearable devices. Outer shell garments incorporate moulded textile buttons for electronic switches, as soft control systems for electronic devices such as iPods, MP3, Mini Disc players and cell phones. France Telecom introduced a fibre optic weave as an early example for flexible displays used on a sports backpack. Flexible solar panels may now be positioned on outerwear jackets for the harvesting of power. The Finnish company, Reima, in cooperation with Clothing+, has successfully used optical fibre in a children's winter all-in-one, the Robotec, for safety purposes (Risikko and Marttila-Vesalainen, 2006).

12.8   Future trends

In the emerging sector of smart clothes and wearable technology, a new breed of designer is needed to collaborate within a multi-disciplinary team in approaching clothing design from a product design perspective. A user-centred design process must be applied beyond the current boundaries of fashion design and be inclusive in bringing the attributes of new textile related wearable technologies to a breadth of end-users and, in particular, the rapidly growing 'New Consumer Majority', the ageing community (Wolfe and Snyder, 2003). Clothing designers must have an inherent sense of three-dimensional form and enjoy adopting a practical hands-on approach to designing in the round. They must develop an understanding of advanced materials to inform textile sourcing and selection in relation to the demands of the body and the end-use. Universal design that promotes comfort for everyday clothing will embrace a range of postures and environments beyond the predictable requirement of the fashion cat walk. A practical and logical understanding of designing in the round, and cutting for movement, will support an iterative creative process. Design development must be continuously evaluated and amended, and the detail of the wearable technology verified or rejected, and new solutions discussed in collaboration with the design team and the end-user.

Technical designers require an understanding of the integral components utilized within technical clothing development and an awareness of the key terminology. They need an overview of the various emerging manufacturing technologies within the industry to drive new design concepts. The design team must co-ordinate the selection of appropriate materials and trims, corresponding to correctly shaped and sized patterns, in combination with the specified machinery and techniques required to assemble a quality finished garment. The smart clothing design team must adapt to addressing the everyday clothing demands of a range of age groups and figure types with their individual physiological, cultural and style requirements. [TC]² have led research in the use of body scanners, utilized in the SizeUK survey in optimizing size and fit, linked to techniques for 3D product development. The use of body scanners for an analysis of body shape, to inform the placement of advanced material structures and their particular functional attributes, is evident in the elite performance sportswear sector.

Speedo has been a leader of integrated textile and garment development for competitive swimming, with design enhanced by a study of biomimicry. The 'Fastskin' suit, emulating shark skin, was launched for the Sydney Olympics in 2000, with subsequent designs progressing innovation in the cutting and placing of sophisticated textile structures and finishes. In March 2008, the Speedo LZR Racer suit was introduced 'made from an ultra-lightweight, low drag, water repellent, fast-drying fabric, unique to Speedo, called LZR Pulse', claimed to be 'the world's first fully bonded swimsuit that is ultrasonically welded'. The suit is 'cut from a unique 3 piece pattern and constructed in 3D shape to fit like a second skin.' LZR Panels are embedded into the base fabric to create a 'Hydro Form Compression system', helping to compress the entire swimmer's body into a more streamlined shape (Editorial, 2008). The Speedo 'Aqualab' global research and development team used 3D body scans to help 'understand the precise shape of swimmers' bodies, allowing them to revolutionize pattern design and anatomical fit.' The use of Computer Fluid Dynamics (CFD) software enabled Aqualab to 'identify the water flow and hotspots around a virtual swimmer's body, allowing them to eliminate drag most effectively.' Aqualab 'scanned the bodies of more than 400 elite swimmers and held technical tests involving more than 100 different fabrics and suit designs.'

Dupont originally identified the concept of proprioception as a marketing argument in the promotion of Lycra® power, with evidence to show that a degree of compression enabled athletes to be more accurate in their body movements. The Adidas Innovation Team (AIT) has carried out extensive research into the biomechanics of top athletes in relation to the effect of movement on muscle compression in key areas. AIT, as an early adopter of the concept of cutting and placing of performance materials through the use of garment bonding technology, has recently announced the development of the 'TechFit Powerweb' body-suit, with bonding technology developed by US extrusion specialists Bemis. As Berthold Krabbe, Team Leader apparel of AIT says, reported by Unsworth (2008), the suit has been developed to harness the intrinsic power in the human body and improve the athletes' performance. It provides 'dynamic support by containing and amplifying muscle movements and linking key joint and energy systems together' (p. 6). The suit has a combination of an elastic base fabric with a 'two-layer Thermoplastic Polyurethane (TPU) film coating', claimed to provide 'excellent directional stretch and energy return to support athletes' movements, maximizing the use of energy produced by the body during exercise'. The kinetic energy of the athlete's movements is 'constantly stored and released through the elastic behaviour of the TPU.' Results are said to provide athletes with increased joint power, better posture, improved feedback for more accurate movement and faster recovery times due to increased blood circulation.

Performance sportswear design has led the way in the adoption of textile-based wearable technologies that will filter down to everyday clothing. The Speedo Aqualab and Adidas Innovation Team are initial examples of future cross-disciplinary design research and development teams embracing fabric-driven innovation in wearable technology. This chapter aims to contribute to the development of a shared language that will enable textile and clothing designers to liaise with electronics and communications providers, as well as with potential end-users. In the development of wearable technology, a knowledge of the human body is the fundamental starting point for designing in the round. Developments in body scanning linked to 3D computer aided design and the subsequent digital garment chain are discussed in more detail elsewhere within this book.

12.9   References

ANDERSON, K. (2005) Patternmaking: Past to Present. [TC]^2 Techexchange.com [URI http://www.techexchange.com/thelibrary/patternmaking.html, accessed May 27th 2008]

ANON. (2005) Sports bra monitors heart rate with built-in monitor. *Gizmag Emerging Technology Magazine*, December 23rd. [URI http://www.gizmag.com/go/4952/, accessed May 27th 2008]

ARDMEL (2008) Ardmel automation [URI http://www.ardmel-group.co.uk/, accessed June 2nd 2008]

ARMSTRONG, H. J. (1987) *Patternmaking for Fashion Design*. New York: HarperCollins Publishers.

BRIATORE, V. (2004) *Dainese: Il Design Salva la Vita*. Milano: Abitare Segesta.

D3O (2008) Intelligent shock absorption. [URI http://www.d3o.com/, accessed May 19th 2008]

EDITORIAL (2008) NOW – Space Age. *Textiles: The Quarterly Magazine of the Textile Institute*, (1): 19–20.

EXO^2 (2008) Berghaus heated clothing. [URI http://www.exo2.co.uk/berghaus.html, accessed May 19th 2008]

GORE (2008a) Airvantage™ Adjustable insulation. [URI http://www.gore.com/en_xx/products/consumer/airvantage/index.html, accessed May 19th 2008]

GORE (2008b) Gore-Tex Fabrics – Gore Comfort mapping technology. [URI http://www.gore-tex.com.au/www/348/1001412/displayarticle/1024369.html, accessed May 27th 2008]

GOTTFRIED MEDICAL (2007) Burn garments – Compression therapy for hypertrophic scars, contractures and keloids. [URI http://www.gottfriedmedical.com/conditions. treated/burn.therapy.garments.html, accessed May 27th 2008]

HEATHCOAT (2008) About us, Heathcoat Fabrics Ltd. [URI http://www.heathcoat. co.uk/, accessed May 27th 2008]

HIBBERT, R. (2004) *Textile Innovation*. London: Line, 2nd ed. [URI http://www. textileinnovation.com, accessed May 27th 2008]

KETTUNEN, I. 2001. *Muodon Palapeli*. WSOY: Helsinki. pages 35–36; 62–64.

MCCANN, J. (2000) *Identification of Requirements for the Design Development of Performance Sportswear*. M.Phil Thesis. Derby: University of Derby.

MCCANN, J. (2005) 'Material requirements for the design of performance sportswear'. In: Shishoo, R. (ed) *Textiles in Sport*. Cambridge: Woodhead Publishing Ltd.

NOBLE BIOMATERIALS (2008) X-STATIC® – Noble Biomaterials. [URI http://www. noblebiomaterials.com/page_xstatic.asp?itemid=52, accessed May 27th 2008]

NUMETREX (2008) Strapless heart monitor clothes by Numetrex. [URI http://www. numetrex.com/, accessed May 19th 2008]

O'MAHONY, M. and BRADDOCK, S. (2002) *Sportstech – Revolutionary Fabrics, Fashion and Design*. London: Thames & Hudson.

OUTLAST (2008) Outlast: Advantages. [URI http://www.outlast.com/index. php?id=71&L=0, accessed May 27th 2008]

PATAGONIA (2008) Outdoor clothing, technical apparel and gear. [URI http:// www.patagonia.com/web/eu/home/index.jsp?OPTION=HOME_PAGE& assetid=1704, accessed May 19th 2008]

POLARTEC (2008) [URI http://www.polartec.com/, accessed May 26th 2008]

PROLAS (2008) The Laser Welding Company [http://www.prolas.de/download/ ProLas2005.pdf, accessed June 2nd 2008]

RISIKKO, T. and MARTTILA-VESALAINEN, R. (2006) *Vaatteet ja Haasteet*. WSOY: Helsinki. pp.12–15.

ROEPERT, A. (2006) Wearable Technologies. *World Sports Activewear*, November/ December pp.16,17.

RUKKA (2007) Avantex Innovation Prize for Rukka SRO Anatomic motorcycling outfit. [URI http://www.rukka.com/lfashion/rukka/rukkawww.nsf/vwpages/ E1EC0D24B183F57EC22572F000419C74?OpenDocument&Expand=1.2.1, accessed 10th June 2008]

SANTONI (2008) Circular electronic seamless knitting machines. [URI http://www. santoni.com/en-index.asp, accessed June 2nd 2008]

SEW SYSTEMS (2008) Designers and developers of sewing systems and machinery for the textile industry. [URI http://www.sewsystems.com/, accessed May 19th 2008]

SHEIMA SEIKI (2008) Wholegarment knitwear. [URI http://www.shimaseiki.co.jp/ wholegarmente.html, accessed June 2nd 2008]

SIZEMIC. (2007) SizeUK Data Set. [URI http://www.humanics-s.com/uk_natl_anthro_ sizing_info.pdf, accessed 8th Nov 2007].

STOLL (2008) [URI http://www.stoll.com/, accessed June 2nd 2008]

[TC]² (2008) 3D Body Scanner and 3D scanning software. [URI http://www.tc2.com/ what/bodyscan/index.html, accessed June 2nd 2008]

TILKE, M. (1982) *Pukujen Kuosit ja Leikkaukset*. Lehtikirjakauppa: Turku.

TWI (2006) 'ALTEX' – automated laser welding for textiles [URI http://www.twi. co.uk/j32k/protected/band_3/crks3.html, accessed June 2nd 2008]

TWIGG, J. (2007) Personal communication. University of Kent, September.

UNSWORTH, R. (2008) Adidas maximises athletic performance. *Textiles: The Quarterly Magazine of the Textile Institute*, Industry News (1): 6.

WATKINS, P. (2007) Brands ignore special qualities of stretch-garments. *WSA* July/ August 2007, 18.

WILLIAM LEE INNOVATION CENTRE (2007) Scan2Knit [URI http://www.k4i.org.uk/ about/Scan2Knit_Case_Study_20_08_07.pdf, accessed June 2nd 2008]

WILSON, A. (2004) Sewfree – Bemis, 100 Innovations. *Future Materials*, Issue 6. [URI http://www.inteletex.com/adminfiles/PDF/fmBemis.pdf, accessed June 2nd 2008]

WOLFE, D.B. and SNYDER, R.E. (2003) *Ageless Marketing*. Chicago: Dearborn Trade Publishing.

13
Developments in fabric joining for smart clothing

I. C. AGNUSDEI, University of Wales Newport, UK

Abstract: New technologies and methods being used in the garment manufacturing include bonding and welding. Bonding uses an adhesive and the two types of welding are ultrasonic and laser. The traditional method of sewing is been added too by these new technologies, not supplanted. This chapter looks at these bonding methods and includes information about the components, machinery used, the physical join, the type of applications and the advantages and disadvantages of using each method.

Key words: textile bonding, textile sewing, textile welding, textile joining.

13.1   Introduction

One of the first methods of manufacturing a garment was to use a needle and thread. The needle would have been made from bone and the thread from animal sinew (Issenman, 1997). The first sewing machines, as we know them today, were produced on a mass scale around the 1850s and this event sped up the sewing process, allowing clothing to be joined at a faster and more accurate pace. Fabrics are still most commonly joined using stitching but the range of methods available for making seams in fabrics is increasing as the need to join new fabric types arises, as new joining techniques are developed, and as a result of a drive towards greater automation. In contrast to stitching, welding or bonding typically provide a continuous seam suitable for fluid or gas sealing. Adhesive bonding is widely used in laminating of fabrics, sealing previously stitched seams, and is being used increasingly in various forms to provide a seaming method in its own right. Adhesive methods are applicable to natural and synthetic fabrics. Welding methods are used for synthetic fabrics, woven or non-woven, coated and uncoated. Welding methods such as ultrasonic and laser techniques have become possible partly due to the application of synthetic fibres in increasingly sophisticated textile constructions. These fabrics enable the new methods to be implemented because of their thermoplastic properties, i.e. their ability to soften or fuse when heated, and subsequently harden in a remoulded shape when cooled. This chapter looks at the two methods with which the author has had most experience – bonding and ultrasonic welding – and compares these with traditional sewing. The experience is from a garment manufacturing point of view, and includes information about the components, machinery used, the physical join, the type of applications and the advantages and disadvantages of using each method. The development of laser welding for the joining of textiles is also introduced.

13.2   Traditional manufacturing methods: the sewing machine

13.2.1 Components

Fabrics of all kinds can be joined using this traditional method. A sewing machine incorporates a needle and thread to join the fabric together. Thread is spun in different fibres that include cotton, silk, rayon, and polyester, with the last as the most popular and economical. Thread is produced in a number of thicknesses, on a number of different size reels and in a multitude of colours. Needles come in a number of sizes with different profiles for different purposes, e.g. ball point for silks or flat edge for leather. Threads have to be sourced that are compatible with the fabric. For example, a fabric with an elastic content will require a bulked thread. The appropriate choice of thread, in conjunction with the correct setting of the machine, will allow the seam to be flexible in relation to the fabric composition and structure.

13.2.2 Machinery

There are a number of brands of both domestic and industrial sewing machines. Within a production environment, a variety of machinery may be used to produce a garment. A lockstitch machine will make a straight seam while an overlocker will crop and neaten the seam edge. A coverstitch machine, mainly employed for knit constructed fabrics, will flatten and protect a seam and is also used to finish a hem. This machinery may have attachments to enable additional processes to be performed.

13.2.3 Physical join

To make a join, initial processes must be completed. The first process is to wind thread onto a bobbin using the mechanism attached to the sewing machine. This is then placed into a case and together put underneath into the base of machine in the area known as the rotary hook. The remaining reel of thread is placed on top of the machine and the thread is wound down through a number of tension discs and eyes, to the needle. The hand wheel at the side of the machine is then turned to enable the needle to descend into the machine and pick up the thread from the bobbin. This has to be in place to allow for an untangled start to sewing. The machine is now ready to commence sewing. Fabric is placed onto the bed of the machine and the foot of the sewing machine is lowered to hold the fabric in place. The needle pushes its way through the fabric to hook up with the thread from the bobbin case. The thread is pulled up to form a chain through the fabric. This process is continued to the length that is required as the machinist feeds the fabric pieces through, aided by the motion of the teeth on the base plate.

13.2.4 Applications

Applications for this traditional sewing method are many including products related to clothing, interiors, automotive and medical end-uses.

13.2.5 Advantages and disadvantages

Advantages of traditional sewing methods for the production of garments include:

- Employment of a number of people
- Skills in the use the machinery can be learnt relatively quickly
- The technology can be used by a wide range of industries
- Products are relatively cheap to produce
- Industry and retailers understand the process.

Disadvantages include:

- The relatively cheap process provides availability but resulting competition
- Quality control can be inconsistent
- Demand on skills and experience to work with the different types of fabric and range of manufacturing processes required
- Sewing joins can be weak at certain points.

13.3   Bonding

13.3.1 Components

Adhesive bonding utilises a third material at the interface between the fabrics to be joined. This binds either chemically or mechanically with the fabric surface and typically infiltrates the fibrous materials to generate strength. In the bonding process, the components include the machine, the adhesive and the fabric. The machine for the bonding of garments is not visually dissimilar to a traditional sewing machine. Where it differs is in the thread being replaced with adhesive and the traditional bobbin and case being replaced by heating and air elements. Adhesives can be selected for application to most synthetic fabrics and may be used to make joints between dissimilar materials.

Hot-melt adhesives are selected such that they melt at a lower temperature than the fabrics to be bonded. The fabric outer surfaces are therefore unaffected by the process. An adhesive may be applied in powder form, as a melt pre-applied or applied when the parts are brought together, as a liquid or as a continuous film or web at the interface. Adhesive tapes are also applied to seams externally. The adhesive for bonded garments may be produced in the form of a film with a paper backing that can either be cut into sheets for specific areas of lamination, or produced as tape for seaming and hemming. The adhesive film may also be applied to elastic components for their attachment. Bemis is a prominent producer of suck films for Europe, USA and the Far East (www.bemisworldwide.com). There are a number of qualities of soft melt films with standard thicknesses, from 1 mil (25 micron) to 7 mil (175 micron), which can be cut into different width tapes. A stretch tape has also been developed with a layer of elastic between two layers of 2 mil (50 micron) film. Bemis provide a selector guide for their range of films, although each combination must be tested with the chosen fabrics before a final specification. Bemis also develops specialist types of film for specific applications on request.

Other adhesive bonding products include:

- Pin dotting glue, with rows of pin dots applied to the fabric. As this is not a continuous adhesive, it has the capability to be relatively breathable although it will result in a weaker join than that provided by continuous film.
- Glue thread, under the brand name 'Grillon', is produced by the worldwide producer of performance polymers EMS Griltech, a subsidiary of The EMS- Chemie Holding AG. It has the appearance of a normal sewing thread and can be applied by means of a lockstitch machine or in the looper of an overlocker. Heat is then applied to the join resulting in the glue thread melting, thus bonding the seam.
- 'Stretchline' produces elastics, with adhesives applied primarily for the lingerie market (www.stretchline.com).

Bonding techniques have specific fabric requirements with important factors that include:

- Silicones, Teflon or other slippery finishes will prevent absorption of the glue.

- Hydrophilic properties (water loving) that enable the glue to penetrate into the fibres enable a successful bond. A hydrophobic fabric can be used but it will take longer for the glue to penetrate the fibres.
- Thermoplastic properties decrease the potential for burn marks on the garment. The average melting temperature is between 180 and 200 degrees Celsius.
- Dyeing may have an effect on the bond: stronger colours tend to have a poorer bonds than weaker colours, due to highter dye contents.
- Weight and thickness of fabric will affect the transference of heat, which in turn may affect the bond of the seam or lead to burning of the fabric.

All garment production tests must be considered prior to finalising design specifications. Polyester and nylon are suitable fibre content for bonding because they are both hydrophilic and thermoplastic in their properties. Other factors to take into consideration with fabrics;

- In fabric types such as lace and mesh, the glue will go through the holes within the structure, thus not guaranteeing a successful bond.
- Curling fabric will also not achieve a good bond and will look unsightly.
- Inconsistent quality of bulk purchase fabric, compared to the sample fabric, will give inconsistent bonding
- All trims and components must also be heat resistant and Teflon free including, for example, end of zip tapes.

13.3.2 Machinery

The purpose of the bonding machinery is comparable with a lockstitch sewing machine in the joining of layers of fabric together to form seams. There are two methods that may be employed to heat the film to the required melting point; hot air and sonic frictional heat. The purpose of both types of machinery is to heat the glue/film to the required melting point to create a successful bond. The AT720 bonding machine is an example of the hot air process that incorporates a side knife (Fig. 13.1). This has been developed by a UK company, Sew Systems of Leicester (Sew Systems, 2008). This company produces various bonding machine types for different applications, initially developed for the lingerie market. As in the case of a traditional sewing machine, attachments can be added to allow for a range of manufacturing processes.

Other equipment needed within the manufacturing sequence includes:

- Flat bed heat press: for small and large area lamination and attaching zips and small components such as hook and loop tabs.

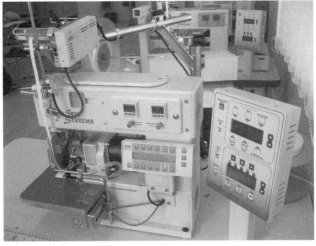

(a)

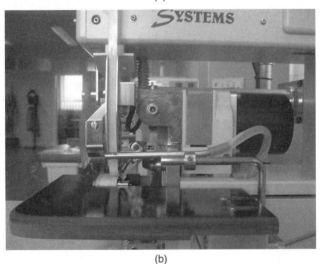

(b)

13.1 (a) AT720 bonding machine with side knife from the Leicester based company Sew Systems Ltd, (b) Close-up of the foot of the AT720 bonding machine showing tape in place ready for bonding.

- Small tacking machines; to perform small spot welds to hold pieces together prior to using the bonding machine. This is equivalent to using tacking thread or pins in the traditional sewing method.

A bonding machine (see in Fig. 13.1) is very similar in appearance to a sewing machine. The most comprehensive bonding machine is one that enables both the upper nozzle, providing hot air, and lower base plate to be heated at the same time or each to be activated individually. This is the best solution for most fabrics as heat may be applied evenly, softening the glue evenly thus providing a good bond. Also, certain fabrics and manufacturing processes may require heat from above or below thus allowing the choice, especially for the joining of different qualities of fabric requiring varying degrees of heat on the upper and lower sides of the join.

13.3.3 Physical join

The joining process is, simplistically, sticking two pieces of fabric together using glue/adhesive. Important factors in achieving the correct bond are:

- Accurate heat setting to melt the film to the softening point to achieve a good bond.
- Appropriate glue for the fabric, selected to ensure a bond will be formed and the fabric will not lose its original handling properties.
- Correct speed setting at which the fabric proceeds through the machine.
- Appropriate length of time and correct pressure in the machine; in the case of the flat bed, to prevent burning of the fabric and transference of the glue to the face side of the fabric

A bond is successful when appropriate amounts of glue penetrate both layers of fabric and exist between the two layers. Each glue/film has its own softening point and glue line temperature (GLT); this GLT is measured using a temperature strip, which is fed through the machine to test when the correct temperature is achieved. The appropriate installation and use of the machine is also important. A flat bed must be placed on a level surface to ensure even pressure. Bonding must be done in the middle of the machine bed as the edge of the flat bed is cooler, preventing the film from melting and bonding correctly. For example, the AT720 hot air machine must have correct hot air pressure, perfect positioning of the air nozzle, and alignment of the foot plate and guide or folder.

As part of the design development process, it is wise to test the fabrics prior to any garment sampling being initiated, in order to save time and money. When the successful sampling of a garment has been achieved and, prior to ordering materials in bulk, further tests need to be carried out. These include:

- Peel test: this should be done 24 hours after the bonding process has taken place to allow the glue to set. The test involves attempting to peel apart the two layers of fabric. This test can be performed by the supplier of the glue/film, such as Bemis.

- Shear bond strength: this is tested as above on overlap seams.
- Colour shade: this is done on the thickest part of the garment on dyed fabrics an hour after the bond.
- Shrinkage test: this is done on each batch of fabric, one hour after the bonding process, on a piece measuring 50 cm by 50 cm. Measurements need to be recorded and subsequently used as a reference in production for quality control.
- Wash test: this is important, done 24 hours after the process.

As well as fabric and film choice, thought has to be given to the type of seams and their positions on the garment, and the types of components to complete the garment. Points to consider when deciding upon these factors are:

- Points at which a number of seams come together must be taken into consideration so that even pressure is applied to ensure a successful bond.
- Too tight a curve in a seam can influence the regain of that part of the seam causing pucker, thus affecting the look, partly dependent on the fabric choice.
- Varying fabric thicknesses will result in uneven and inconsistent bonds, for example, if a textured fabric is used.
- A minimum of 10 mm overlap is necessary as anything smaller will result in a weak bond. This is also dependent on the garment type and the position of the seam and the degree of stress subjected to the garment when worn.

The process for producing a simple 10 mm overlap seam:

(i)   As an example, the film in the form of a 10 mm tape may be applied using the AT720k machine (with side knife) to both layers of fabric, taking care to apply the tape to the face of the fabric on one layer and onto the reverse side of the other fabric. The side knife cuts the fabric level with tape. Care should be taken not to cut off more than 2 mm.

(ii)   The paper backing should be peeled off the tape and the layer of fabric with the film on the reverse side of the fabric placed over the layer with the film on the face. If necessary, a spot welder should be used to tack the two layers together. Care must be taken not to allow the adhesive film to attach itself to itself. (Some of the softer films that have a low melting point remain tacky, which is helpful for tacking without the spot welding machine but risks the adhesive sticking to itself.)

(iii) An AT720 machine (a similar generic machine, from Sew Systems, minus the knife) may be used to bond the two layers together.

13.2 Soft shell jacket completely bonded. All seams are lapped seamed using the AT720. A flat bed heat press was used for both the hem and cuff facings, the collar and the pocket bags (garment courtesy of Jane McCann).

Figure 13.2 shows an outerwear jacket that has been manufactured entirely by bonding processes using 'soft shell' water-repellent three-layer laminated fabric. All seams have been lapped-seamed using the Sew Systems AT720 machine. A flat bed heat press was used for attaching components such as hem and cuff facings, collar and the pocket bags.

13.3.4 Applications

Bonding applications are becoming widespread both in the sports and fashion markets and, in particular, lingerie. Bonding has been used for a number of years in the automotive industry and also in the outdoor clothing market for laminating fabrics and waterproofing stitched seams. Multi-layer fabrics can be created by bonding materials together to maximise the function of a garment. Encapsulation of components such as bra wires can eliminate processes within the traditional garment manufacturing sequence. Currently, seams are taped after having been ultrasonically welded. Elastic with adhesive film applied is used to bind the edges of lingerie. Other applications within a garment include the attachment of brand logos, embroidered logos and components.

13.3.5 Advantages and disadvantages of bonding processes

Advantages include:

- Bulky stitched seams are replaced with smooth, flatter and almost invisible seams that may be less intrusive and abrasive to the wearer.
- Garments are lighter in weight.
- Garments have a cleaner, more 'technical' appearance.
- A reduction in stitches, or no stitching at all, can improve waterproofness.
- Breathability may be increased by eliminating layers of fabrics within a garment.
- Components, such as zips, can be attached without sewing, eliminating processes.
- Bonding enables a seam to stretch with good recovery.
- The basic machine, cited as an example, has interchangeable attachments for different processes.

Disadvantages include:

- Initial financial layout for the machines and time and expertise required to train staff can be both time consuming and expensive.
- A greater level of accuracy is needed from the machinist as once a seam has been bonded it is permanent, unlike a sewn seam that can be unpicked.
- Currently, it is a more costly production process per garment, although this will be reduced in future.
- The thread industry will be affected.
- Bonding machines must be placed in a separate area because the dust produced by traditional sewing machines will affect the quality of the bond.
- Garment repair or alteration work will be impossible other than by the manufacturers.

13.4   Ultrasonic welding

13.4.1 Components

Ultrasonic welding uses heat generated as a result of high frequency (20–40 kHz), low amplitude mechanical vibrations applied under compression to the joint line. The ultrasonic sound waves pass through the layers of fabric to fuse the fibres together. This method can also incorporate a cutting edge to enable 'cut and seal' processing. The ability to ultrasonically weld textiles depends again on their thermoplastic content and the desired end result. As a minimum, the material must have uniform thickness and a thermoplastic content of 65%. Yarn density, tightness of weave, elasticity of material and style of knit are all factors that can influence the weldability.

13.4.2 Machinery

There are two types of machinery used to ultrasonically weld textiles. One is referred to as a continuous welding machine. In rotary/continuous mode, the machine is similar in appearance to a traditional sewing machine with a configuration where the traditional needle is replaced with a horn or wheel and the bobbin and case with a drum on the base of the machine. The fabric is fed between the fixed vibrating horn and a rotating wheel to provide a continuous seaming process. The wheel is normally patterned to provide local intensification of the ultrasonic energy and to provide a variety of patterned seams. Seaming rates of at least 30 m/min have been described. The second type of sonic welding machine is an intermittent machine with the joining of fabrics carried out in one action. In this 'plunge mode welding' the fabric is held under pressure onto a patterned anvil by a vibrating horn, the process typically taking around 1 second to complete. The shape and size of the join will be defined by the tool. A variety of applications will require different tools. Applications include strapping, belt loops, filters and vertical blinds. In both cases, the fabrics are melted throughout their thickness. Both types of sonic welding machinery will produce a cut and sealed edge.

13.4.3 Physical join

The continuous welding machine may be used to join seams of between 1 mm and 10 mm, dependent on the size and shape of the wheel. If a 1 mm join is required, it will have to be reinforced with a tape backing to make it secure. Alternatively, a 10 mm join would be made on overlapped seams, using a patterned wheel. As in the case of a traditional sewing machine, a pedal at the base of the machine is depressed. The layer or layers of fabric are passed between the horn/wheel and drum. When the machine is in motion, pressure is applied. With this combination, frictional heat is produced and the fibres in the fabric melt where the horn/wheel meets the raised area on the drum. It is at this point the join or cut is made.

Intermittent welding works in the same way, with applied pressure and heat generation, but in one specific area only, with one shape. This is used mainly to attach components within a garment, e.g. bra strap. Care must be taken to have the correct settings. As with the bonding machines, when passing the fabric through, speed and heat generation must support each other to prevent stalling. Stalling, or going too slowly, may result in the fabric puckering or burning, or both.

13.4.4 Applications

Applications for continuous welded products include branded sportswear and disposable garments for the forensic, medical, food and clean room industries. Lingerie, packaging, tents, tablecloths, and lace making can also be manufactured using this technology. Applications for intermittent welds include, toys, components in the manufacture of garments and disposable products.

13.4.5 Advantages and disadvantages

Advantages include:

- Bulky stitched seams are replaced with smooth, flatter almost invisible seams that may be less intrusive and abrasive to the end-user of a product.
- Garments are lighter in weight.
- A relatively cleaner, more 'technical' appearance.
- A reduction in stitches, or no stitching at all, can improve the water-proofness of a garment.
- For seams of 1 mm reinforced with tape, this type of join enables a seam to stretch, with good recovery.
- Basic machinery comes with interchangeable attachments for different processes.

Disadvantages include:

- Initial financial layout for the machinery and the time and expertise to train staff can be both time consuming and expensive.
- A greater level of accuracy is needed from the machinist, as once a seam has been welded and taped it is permanent unlike a sewn seam that can be unpicked.
- It is a more costly production process per garment, although this will come down in the future.
- Thread industry will be affected.
- Any machine repair or alteration work will be impossible other than by the manufacturers.

13.5 'New' technologies

13.5.1 Laser welding

'Welding of textiles is often associated with the production of a rigid brittle seam that needs to be hidden in use. Laser welding has been developed to offer an alternative, providing soft seams, with the opportunity to seal against air or fluids and increase production automation.' (Jones, 2008b).

Welding is a thermal process requiring melting of material at the fabric surfaces that are being joined. It is applicable to fully or partially synthetic fabrics with thermoplastic components (e.g. nylon, polyester or polypropylene yarns and PVC or polyurethane coatings), which are compatible when melted together. As a general rule, the fabrics to be welded must be of the same thermoplastic. The material to be melted may be the fibres of the fabric, a thermoplastic coating or a film added at the joint in combination with the fabric fibres (Jones, 2008a).

Leading research and development in the textile area is being carried out by The Welding Institute Limited (TWI, 2008). TWI is a UK independent research and technology centre, working with collaborators such as Clearweld, the Gentex Corporation, Prolas (laser technology) and Pfaff (providing the generic sewing machine body) in developing and patenting methods for the laser joining of textiles. This collaboration has developed a technique for the welding of thermoplastic textiles so that they only melt at the interface between materials, rather than through the full thickness. This results in a join that has a greater flexibility and softer feel than one made with other welding methods. The outer texture of the fabric is also retained. Again, for successful joins, materials are required to have a minimum percentage of thermoplastic properties in woven or non-woven, coated and uncoated assemblies. This more flexible welding solution uses transmission laser welding to apply a well-controlled amount of heat, just to the contacting surface of the fabrics. The selective heating is achieved by introducing a low visibility laser-absorbing fluid onto one or both of the fabric surfaces, or to a polymer film, which is then placed between the surfaces to be joined. An infrared laser beam is then directed along the seam line. The beam passes through the fabric, heats the absorber and generates a weld which seals the interface (TWI, 2008). The resulting aim is to cut or join single or multi-layers of materials. Resulting beneficial features of the process include:

• Control of melt volume and hence seam flexibility.
• Sealed seams in one operation, avoiding use of tapes – curved seams become possible.
• Potential for high speed seaming and automation.
• A novel appearance to the seam – new design opportunities.

- Welding may be achieved through several layers, so closed products with internal welded structures are possible (Jones, 2008b).

Equipment specific to the laser welding of textiles is being developed and prototypes are being used for evaluation purposes. A partnership exists between TWI and Coleg-sir-gar, a higher education institution in Carmarthern, South Wales. Within the 'Shared Technical Resource Centre', a prototype machine has been set up with the aim of developing different joining methods for garment manufacture using this technology for the benefit of small manufacturing businesses interested in the adoption of this novel joining technique. The technology is currently being developed for use in the textile industry to provide leak-proof seams, such as in personal protective clothing, containment bags or, waterproof jackets and tents, where traditional stitching methods leave a presence of holes with the potential of weakening the fabric in the region of the seam. Seam-sealing compound or tape has often been added to stitched seams to give a seal, which adds cost to the joining operation. Recent developments in both laser and materials technology now allow the benefits of laser welding to be realised in fields such as garments, shoes, protective wear, inflatable structures, airbags, sails and many other applications within the automotive, electrical and medical industries. With further development, laser welding, which needs access to only one side of the fabrics, is offering the potential to automate the seaming process for garments (Jones, 2008a).

13.5.2 Rapid prototyping

Rapid Prototyping is also known as Additive Fabrication, Laser Sintering, and Stereolithography. It is a computer-aided design technology that has been used in the engineering field for many years. An STL file is the format used by the software of this technology to generate the information needed to produce the 3D models on the stereolithography machines. A thermoplastic polymer material is the most commonly used, with the emergence of metal and other organic materials. Basically, the machine produces the product by extruding the material from a nozzle. It is laid down in a similar movement to that of a photocopier. A laser is used to sinter the materials – to fuse them together using heat. This produces the product. When a product proceeds from the development stage, the process is known as Rapid Manufacturing. Its applications in the textile area have included a football boot (a collaborative project between prior2lever and freedom of creation) and a handbag in its own packaging. The advantages of rapid protyping include elimination of warehouse stock and the assembly process, huge reduction in transport costs and the ability to provide just-in-time production. Freedom of creation (2007) is a design and research company specialising in rapid manufacturing and research and development with industrial partners. They produce products using this technology available in retail outlets or via their website.

Loughborough University also has a research and development group for rapid prototyping and is researching into the development of manufacturing technologies using new, emerging materials, design protocols and working methodologies within the supply and demand chain of manufacturing.

13.6 Conclusion

As with all garment design development, a confirmation of the correct choice of fabric, components and manufacturing technology should be considered as a whole to achieve a garment that is functional, aesthetic and economical to produce, for the intended end-user. The new technologies discussed have given garment designers and manufacturers additional scope for innovation in the design development and production processes. A mixture of both traditional and new technologies can be employed. A designer's existing understanding of the traditional use of fabrics, garment pattern cutting and manufacturing techniques will continue to provide background knowledge for the utilization of these new technologies. To investigate these technologies initially, the designer and technicians may take existing garments and experiment in making them using the new methods. The experience gained will inform subsequent projects.

At present, these novel processes are emerging primarily in the outdoor and performance sportswear garment categories. DIM is a brand that designs and produces lingerie using both heat bonding and ultrasonic welding. Branded outdoor sportswear such as 'The North Face' and 'Patagonia', also use ultrasonic welding and heat bonding processes. These companies use terminology such as 'magic seam' and 'Composite Seam System', respectively. Freedom of creation utilise the rapid manufacturing process to produce textile products that come in their own packaging. Although these new technologies are expensive to implement and utilise in the current manufacturing environment, the techniques will eventually become cheaper. Development is needed in the transference from the sample to the production methodologies, with the necessary financial commitment.

Overseas factories can now produce garments using some of these technologies. The present technologies described lend themselves, primarily, to garments that are simple, with as few components as possible or specified for specific areas of a garment. Additional restrictions for the designer include the current limitation in fabric choice, based on the use of synthetics such as polyester and/or nylon. The psyche of the customer is another consideration. Do people want glued garments? Do women know they are already wearing glued garments – for example, lingerie? It may be a job for the marketing department to explain the new technologies to the public in terms of what is available to meet customer needs. Customers are looking towards more bespoke, limited editions of garment collections, with quality and that stand out from the crowd. Could this technology find its way into the fashion market via this route? Is it a way for the United Kingdom to compete with the Far East?

13.7   Sources of further information and advice

The International Textile Machinery Association, ITMA, is the world's largest international textile machinery exhibition, taking place every four years. It was last held in 2007 in Munich, Germany and the next will be in 2011 in Barcelona, Spain (www.itma.com).

ISPO is the International Trade Fair for Sports Equipment and Fashion, held in Munich, Germany, now focused on an annual winter show (January). Brands for sports, including ski and snowboarding, performance sports and technical innovations and processes, showcase their products.

Bemis: manufacturer and supplier of thermoplastic film for bonding also provide a free analysis service to test your product to see if it is suitable to use with their film.

Sew Systems are a supplier of machinery for the bonding technology, www.sewsystems.co.uk.

13.8   References

FREEDOM OF CREATION (2007) www.freedomofcreation.com
ISSENMAN, B.K. (1997) *Sinews of Survival: Living Legacy of Inuit Clothing*. Columbia: University of British Columbia Press.
JONES, I. (2008a) *Welding and adhesive bonding of textiles*. TWI Ltd. [URI http://www.twi.co.uk/, accessed July 21st 2008]
JONES, I. (2008b) *Textile welding using lasers*. Technology update. TWI Ltd. [URI http://www.twi.co.uk/, accessed July 21st 2008]
SEW SYSTEMS (2008) Sew Systems Ltd, 53 Iliffe Avenue, Oadby, Leicester, LE2 5LH, UK E-mail: info@sewsystems.co.uk. www.sewsystems.co.uk
TWI (2008), www.twi.co.uk

13.9   Useful internet addresses

www.stretchline.com
www.bemisworldwide.com
www.clearweld.com
www.technical-textiles.com

www.emschem.com
www.thenorthface.com
www.patagonia.com
www.twi.co.uk
www.fkm-sintrtechnik.de
www.freedomofcreation.com
www.lboro.ac.uk/mrg

14
Digital embroidery techniques for smart clothing

A. TAYLOR, University of the West of England, UK

Abstract: The chapter gives a background into computerised embroidery, providing a context for current developments in the field. The chapter aims to demonstrate the exciting and innovative ways in which cross-disciplinary collaborations between science, technology, medicine, art and design are shaping the future of digital embroidery. Although an overview of the process and materials relating to digital embroidery has been touched upon, it is hoped that the references will lead to more detailed and specific information. The chapter should be seen as an informative research tool and a springboard for those seeking further information about the subject.

Key words: digital embroidery, e-broidery, computerised stitching, smart textiles, computer embroidery.

14.1 Introduction

Digital embroidery is no longer the domain of large manufacturing companies, stitching large quantities of fabrics for mass production. The wide availability of the automatic computerised embroidery machine, from domestic machines to a range of professional and industrial, and with all major sewing machine companies offering digital embroidery machines and software packages within their range, has resulted in increased competitiveness and more affordable prices. This has led to an increase in smaller companies offering customisation services. Companies, sports teams, schools organisations and clubs all require logos on garments as part of their corporate ID. Embroidery is an effective means of adding customised text and image directly onto garments or in badge form, and adds value to the product. Demands of corporate health and safety regulations also require staff to wear uniforms, and embroidery is an ideal solution as a means of identification.

Customisation of garments and products within the domestic sewing machine market has led to a rise in sales of digital embroidery machines, particularly in the older population who, at retirement, have upgraded their machines. This has also encouraged an increase in computer literacy amongst the adult stitching fraternity.

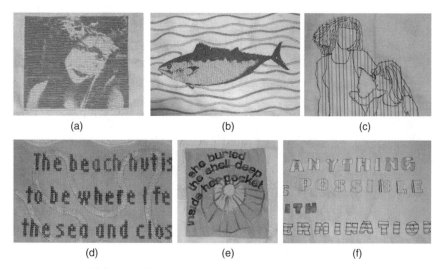

14.1 (a) Scanned photograph, (b) Photographic and digitisation, (c) Appliqué, (d) Cross-stitch text, (e) Photographic, digitisation and text, (f) Appliqué text (designs – Alison Taylor, photographs – David Bryson).

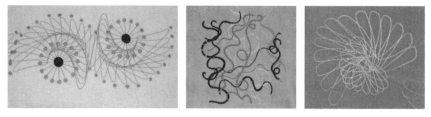

14.2 Digital embroidery designs that could incorporate wires, switches, sensors or other electronics (designs – Alison Taylor, photographs – David Bryson).

The variety and versatility of new digital machines and software mean that embroideries can be stitched in large multiples or as individual bespoke pieces, allowing for a broader range of applications and innovative developments. Digital embroidery has been tailored to meet the demands of the surgeon in the operating theatre, designing custom-built ligaments for reconstructive surgery. It has been introduced as a solution to the dilemma of damage to aircraft wings caused by collision with flocks of birds, and to the accurate placement of electronics into cloth towards the development of the wearable computer.

14.2   The embroidery machine

14.2.1 Schiffli

Invented by Isaak Groebli in Switzerland using the combination of needle and shuttle to form a stitch, Schiffli, meaning 'little boat,' describes the shape of the shuttle.

This method of embroidery uses rows of needles held on a horizontal rack with a substrate material mounted in a vertically held pantograph. The primary yarn runs partially through the substrate, held in place by a second interlocking yarn at the rear.

Original Schifflis used large bobbins to hold maximum amounts of thread. Needles had to be close together to get the embroidery on a single piece of fabric and, as a result, smaller bobbins were required. The shuttle was modified so that the thread would unwind within it. The first machines were slow to run, at speeds of 28 stitches per minute, and time-consuming to set up, with smaller machines having approximately 350 needles and larger models up to 1400, requiring the rethreading of each individual needle when changing colour.

Today's machines have come a long way. Sewing speeds have increased nearly 10 times and automatic colour change has been added. Computerisation has also improved the accuracy and production of the machines. However, Schifflis are still very large, with some nearly as big as a two-storey building, resulting in a vast outlay for any company; they are also heavy and expensive to operate.

The machine is ideal for stitching sheer fabrics and is renowned for the embroidery of lace, furnishings and fashion fabrics where a large metreage is required. The fabric is reversible and does not pucker, whilst remaining soft and pliable.

The cost factors, the introduction of the multi-head machine and the decision to transfer production to the Far East has led to a decline of the industry in the States and the UK.[1]

14.2.2 The multi-head

The market for embroidery evolved as orders grew smaller, with the need for customisation and a quicker turnaround. The Singer Sewing Company introduced the first multihead embroidery machine in 1911, featuring six sewing heads and a pantograph attachment, but although better for stitching smaller jobs, they were still slow and inefficient. In the 1950s, more sophisticated machines employing the latest technology revolutionised the industry.

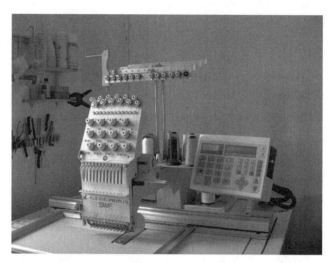

14.3 Multihead embroidery machine, Ellis Developments Ltd
(photograph – Alison Taylor).

New developments have allowed for major improvements, including
computerisation and automatic colour-change and thread trimming. Frames
have been developed to handle a variety of sized embroideries and machines
have become smaller and more affordable. The multihead embroidery
machine has opened up an entirely new market, making it a viable alternative to the larger Schiffli.[2]

14.2.3 Domestic market

Although multiheads are able to hold 30 sewing heads, demands by the
domestic and small business markets have allowed for the development of
smaller, cheaper machines from single to 6 heads. These lightweight, portable and easy to operate machines require only one person for the job, and
fast sewing speeds of over 2000 stitches per minute allow for greater
customisation. Unlike the Schiffli that can stitch only on flat fabric, the
multiheads are capable of stitching on finished garments and three-
dimensional surfaces such as baseball caps.

Many domestic sewing machine companies have developed single-head
embroidery machines. The competitiveness of this market has brought the
prices down to £2000 upwards, plus £1000 for the software.[3]

14.2.4 Cornely

Cornely embroidery uses a single-needle head with a substrate material
held in a pantograph, which is moved under computer control. Decorative

chain stitch and surface corded effects can be applied to the fabric as well as the tacking of heavy cord to the substrate by coiling the cord in a wrapping yarn, then stitching the wrapping yarn to the substrate with a chain stitch.[4]

14.2.5 Irish

The Irish machine is able to produce a straight, side and satin stitch with variations in length and width, allowing the skilled operator to almost paint with thread. This effective method of embroidery can cover large areas quickly, using multiple colours or directional changing of threads to alter light reflection. Satin stitch appliqué, padded appliqué, cut work and cut edges such as scalloping are most successful.

14.3 Production methods

14.3.1 Digitising

Digitising is the process of adapting artwork in the form of image and text into data that can be read by a digital embroidery machine. The data is translated via a computer software programme and saved as a file format such as HUS, DST, PSC or PEC.

Digitising is a very specialised job, requiring training and practice. The quality of the embroidered outcome depends on the skill and expertise of the embroidery digitiser. The design quality of the image is important but so is the smooth stitching of the embroidery. The organisation of the running order of thread colours and the stitch length, density, directions and jumps, and planning the path or route are important design factors in creating a successfully stitched outcome.

The image can be downloaded from a camera photo file, drawn using a digital drawing pad and pen or drawn and scanned then reworked by design programmes such as Photoshop and imported into the embroidery software. The image must be cleaned and prepared prior to digitising. The stitched circle needs to be designed as an oval to take into account the pull of the fabric and size of the stitch. Too small and the image and text can lose impact or suffer in the stitching out process.

The digitiser understands the process of the embroidery machine and the software programme involved. Stitching output requires the machine to move the hoop under the needle directionally. Consideration to the fabric type, the choice of backing or stabiliser, threads, registration of design, compensation and stitch density according to stitch type will affect the quality of the outcome. Washing the garment can cause distortion if the digitising has not been carried out properly.

The designer can choose from a variety of stitch types that include running stitch, double and triple running stitch, satin stitch, broken edge satin stitch, cross stitch, chain stitch, appliqué or fill stitch. The fill stitches allow the designer to choose from a variety of pre-determined textured stitch options. Stitches can be altered in size, density and direction, allowing for a multitude of design options. Before stitching, the software will programme a network of stabilising stitches or underlay, which helps to prevent distortion. At the beginning of the stitching, the machine will automatically find the centre of the hoop and will start with a few stitches that can be cut off at a later stage. Jump stitches will move the needle and thread to a different area to be stitched and will always occur at the start of a new colour. Machines with single needles require threads to be cut and changed and needles rethreaded manually every time there is a colour change. Commercial machines are able to run multiple pre-threaded colours that are automatically stitched out in turn. Three-dimensional effects can be achieved by the grading of colours. Depending on the complexity and size of a design and the number of colour changes, stitching out a design can take from a few minutes to a few hours.

Some software programmes allow for auto-digitising but the aesthetic quality is often poorer and can often result in technical problems during the stitching out process. Hand digitising can be more economical in the long run, as the digitiser can control the optimum number, density and order of stitching.

Ready designed, digitised embroidery files can be downloaded free or purchased from internet websites priced per stitch. Formats differ, depending on the make of machine, and conversion programmes are available to translate one stitch file format to another. Many companies offer customised digitising as a service for embroiderers and designs can be sent electronically to the customer, offering a rapid response service.

It is important to know the copyright policy of the company, as some designs can be used only for personal reproduction and cannot be resold as a design (see copyright issues).

14.3.2 Editing or customising

Once the design has been digitised, it is saved as a stitch file and can be edited by converting it to a different stitch file format, changed in size to fit different formats or hoop dimensions, repeated, mirror imaged, rotated, stretched or distorted, added to another design, split or arranged to fit a template. This stage in the design process also allows for text to be added. Colours can be altered and sorted and the order which the design is stitched can also be changed. It is also possible for individual stitches to be cut out of the design.

14.3.3 Loading the design

After digitising and editing, the design is saved to the specific file format of the machine to be used. Common design file formats for the domestic embroidery machine include PES, HIS, VIP, ART, JEF, SEW and DST. The saved file is then transferred to the sewing machine, using a variety of methods applicable to the machine. When the memory device is placed in the sewing machines, many machines allow further editing options within the computerisation inherent in the machine itself.

14.3.4 Embroidery hoops

Embroidery hoops vary in size according to the make and manufacture of the embroidery machine. If stitch files are converted from one machine to another, they may not fit the dimensions of the hoop unless the design is edited. The embroidery hoop is a clamping device designed to give stability and tension to the fabric and to hold the backing and fabric in place under the needle in the sewing machine. The hoop can be loosened and tightened to attach and remove the fabric to and from the hoop. The hoop defines the area to be stitched and is an important tool in securing the fabric and ensuring the registration of the design. Registration is vital if the design is to be repeated or continued into another design. The warp and weft of the fabric must be straight, parallel down and across the hoop. Loose tension can cause wrinkles, while too tight a tension can also cause distortion of the warp and weft when the fabric relaxes out of the hoop. This can cause more delicate fabrics such as organdie or knitted jersey to be distorted or marked by the hoop and can be avoided by placing the fabric on top of a hoop stretched with backing fabric using an adhesive bonding agent.

The size of the hoop depends on the size of the area to be stitched; for example, a small logo requires a small hoop. The shape and dimension of the hoop allows for a complex shape to be stitched, e.g. a baseball cap. The hoop is then attached to the machine which moves the embroidery hoop under the needle.

14.3.5 Stabilising

In order to prevent stretching, puckering, pulling and distorting the design and the fabric whilst embroidering, a backing or stabiliser is required. This is applied to the back of the fabric to be stitched and is secured by sandwiching and stretching the fabric and stabiliser in the hoop. The stabiliser holds the stitches in place without damaging the outer fabric and also acts as a shock absorber, counteracting the needle as it perforates the upper fabric. Choosing the correct weight and appropriate backing is one of the most important factors in successful embroidery. The most common types of backing are non-woven or bonded. Backings come in a variety of forms. The cut-away is cut from the back of the outer fabric after stitching, the tear-away is torn from the back of the embroidery and the dissolvable is either removed by putting in hot or cold water or by ironing.

Fabric stretch, flexibility, 'give' and tolerance are important factors to consider when choosing a backing, which has to be stable to prevent movement during the stitching process. The structure of the weave or the knit pattern must be analysed for stability. A large design with a heavy stitch count on a loose knit or weak fabric will require stronger or heavier backing to retain and stabilise the stitch pattern than a heavier weight woven fabric.

Once the embroidery has been completed, as much backing as possible needs to be carefully removed. Putting too much strain on the top fabric when pulling the backing away can cause damage to the thread stitching and top fabric. This may also disturb the embroidery, causing puckering and warping, and change of direction to warp and weft. Tear away products are easier to remove, as cutting can cause mistakes by accidentally cutting into the embroidery thread and top layer fabric. Using several layers of lighter weight tear away rather than one heavier weight can make it easier to tear away and less likely to damage the embroidery, though it can be more time consuming. Dissolvable backing is ideal for transparent fabrics or when the fabric is to be seen on both sides.

The success of the embroidery can be measured only after the hoop has been removed and the backing torn away, and after the fabric has relaxed to its normal position.[5]

14.3.6 Threads

The main threads used for digital embroidery are rayon, polyester, metallic and cotton. Silk and linen are also used, but rayon and polyester make up 90% of the embroidery market. Embroidery thread is used on the top only for most embroidery, unless the item is to be reversible such as quilts, towels, organdie and transparent fabrics, or lace patterns where both sides need to be viewed. If this is the case, then embroidery thread of matching colour and weight must also be wound onto the bobbin.

Normally, specialist bobbin thread should be used, which is very light in weight and can be nylon, polyester or cotton. This can be purchased pre-wound on bobbins or wound from a spool onto the machine's compatible bobbins. The weight of thread or thread count determines the thickness of the thread. 40wt thread is the most popular used for embroidery: 35wt or 30wt may be used if a heavier effect is required, particularly for a straight outline stitch or for effects such as blackwork or redwork where a hand-stitched effect is required. The heavier the thread, the quicker an area can be covered. A 50wt thread is very fine, can be used for delicate work and is particularly good for blending colours.

Rayon, with its strength, sheen, cost effectiveness and availability of colours, is the most popular thread. It is more elastic than other threads making it more suitable for lighter weight fabrics. Rayon is dyed using an absorption process by placing the thread in a dye vat where the solution is absorbed. The thread is removed from the dye and washed twice to remove the excess dye. There may, however, be a residue of dye left in the thread which should disappear after the first wash of the garment. The recommended wash for the first time is 60 °C, but many garments are washed at 40 °C.

When embroidering workwear, polyester is recommended for its hard-wearing qualities. It can withstand heavy laundering and chemical treatments and is abrasion resistant. Polyester is dyed under pressure in sealed containers filled with heated dyestuff which is forced into the thread, becoming part of the thread itself. It can withstand higher washing temperatures and there is less chance of a colour bleed during laundering. Polyester also has greater tensile strength than rayon, making it harder to break.[6]

Cotton is also dyed using the absorption process. It has a much duller appearance and is not used as much as rayon.

Metallic threads consist of a metallic polyester film wrapped around a carrier thread such as polyester or nylon. This thread is more sensitive and problematic, running better off a vertical spool due to the twisting action caused by the flat thread kinking and twisting back on itself. It is essential to choose the correct needle size of between 12/80 and 14/90. Metallics react badly to abrasion and the thread does not bend well, making small stitches unsuitable. The garment can be dry cleaned or washed in cool to warm water and dried on a low heat setting and ironed over a cloth.

Phosphorescent glow-in-the-dark thread, emitting light stored in the yarn after exposure to light can be washed up to 30 °C and up to 50 times without loss of luminosity. It is also light fast and dry cleanable. Such thread is useful for safety wear as well for its aesthetic qualities.

Solar active colour changing embroidery threads react to absorbed light from sources such as ultra-violet light, sun or artificial UV light. The white thread will change colour after exposure and a miraculous transformation will take place. The polyester thread will need to be ironed over a cloth and can be laundered in the usual way.[7]

Prism hologram thread can give a unique reflective multi-tone quality. The stitches reflect the lights around them to create a shimmering 3D effect and are available in a range of different colours.

Fusible thread is fine, like dental floss, and can be positioned accurately, which is an advantage for detailed work. It creates a soft, flexible bond giving less stiffness and bulk than a fusible bonding web. It can be used for positioning appliqué and bindings and hold other fabric pieces and threads down before stitching on top. Puckered, undulating and bubbled textures can also be created. The fusible thread can be used in the bobbin or as a top thread, or be hand-stitched through a needle.[8]

Dissolvable thread can be very useful in digital embroidery and is manufactured using PVAlc (polyvinyl alcohol) by the dry spinning method. The threads are easily dissolvable in plain warm or hot water without the aid of chemicals. Using these threads enables interesting surface qualities to be achieved, particularly through stitching, gathering then baking polyester fabrics to cause shrinkage and puckering.[9]

Conductive threads are either spun or twisted on a nylon or Kevlar® core and contain conductive material, such as silver, stainless steel, copper, tinned copper or gold that enables conductivity. These yarns have electro-mechanical properties and can be embroidered together with a non-conductive thread to create the substrate for a textile that will enable an electrical current to pass through. Fabrics using conductive threads or yarns, woven, knitted or embroidered to form a circuit can be powered by a battery source in order to create warmth. Conductive threads have been used in the aerospace, medical and industrial applications. Some conductive threads also have magnetic properties. These threads have also been used in the manufacturing of fencing garments – when touched by the opponent's foil, this activates the electronic connection, enabling the scoring system.[10]

Antimicrobial threads have been developed to capitalise on the demand for performance fabrics used in active sportswear and the health and well-being industries. Silver is used in many healthcare products due to its anti-microbial properties. X-Static is a silver fibre produced by Noble Fiber Technologies, Pennsylvania and has been developed for use in active sports-wear in order to maintain freshness and odour control. X-Static can also act as an anti-static as well as a thermal regulator. NanoHorizons using nanotechnology has developed an antimicrobial additive called E47 with the same antimicrobial qualities as silver, integrating nanoscale engineered particles into fibres on such a small scale that they cannot be detected by human senses.[11]

14.3.7 Needles

The choice of the correct needle for the job is essential, otherwise faults can occur such as poor tension, puckered fabric, broken or shredded thread, large holes in the fabric, perforated fabric, inferior stitching or even damage to the machine, the bobbin or throwing off the timing. The type of sewing machine will determine the variety of needle system.

The needle size can be determined by the performance of the fabric type, such as delicate or medium weight, open weave or knit, felted or leather, velvet or terry towelling. Lightweight fabrics such as georgette or organdie need a small needle size, 60/8. Mid weight woven fabrics such as jersey, lycra, linen or calf leather use an average needle size of 80/12 to 70/10. Heavy fabrics such as denim, vinyl, canvas, require a 90/14 to 100/16 and for very heavy fabrics 110/18 or 120/19. Thread types will also have an impact on size, particularly metallic or holographic, and varying weights such as 40 or 30 thread counts. The type of stitching whether decorative or functional also plays an important factor. Thread should be able to pass freely through the needles eye to produce even regular stitches. The fabric's type determines the shape of the needle's point.

The needle must be changed regularly, as a blunt needle can also cause problems.

14.3.8 Waterproofing the embroidery

Waterproof and/or breathable coated fabrics, when embroidered, lose their performance qualities on the area that has been decorated or embroidered. Seams will have been bonded rather than stitched to prevent the needle perforations letting in water. Waterproofing products are available that will not only re-proof the area that has been decorated but will also waterproof and protect the actual embroidery, enhancing the performance and wear.

14.4   Engineering applications

14.4.1 MASCET Project

MASCET (The Manufacture of Structural Composites using Embroidery Techniques) was a project set up to investigate the use of embroidery techniques for the manufacture of reinforcement preforms for composite structures produced by liquid moulding. Such composites are made when a liquid resin and a fibrous reinforcement are used together, resulting in a composite material that has very different physical and performance properties from the original materials used.

The project was a collaboration between a number of partners, including University of Nottingham Department of Mechanical Engineering, Ford Motor Company, Essex and Ellis Developments Ltd, Nottingham. It involved the experimentation of fibre placement onto an embroidery substrate material. This aimed to save weight over the conventional alloy part and reduce waste and assembly time for the automotive industry.[12]

14.4.2 Through thickness reinforcement of composites for improved impact resistance and delamination resistance

The Mascet project was completed in 1997 and, since then, Ellis Developments have examined the use of stitching technologies to hold together different layers of textiles, similarly for use in structural composites. Another project, 'Through-thickness Reinforcement of Composites for Improved Impact Resistance and Delamination Resistence' (through stitching) has the potential to reduce the splitting apart of the layers which form many textile reinforced composites. If the finished part is subjected to a heavy impact, there is a danger that the layers may delaminate, and the composite will fail in its task. Through stitching has been found to improve impact resistance, and may have particular application in aircraft which are subject to bird strikes. 'Our work on through stitching is continuing at Ellis Developments in conjunction with QinetiQ (formerly DERA) at Farnborough, England, Airbus UK and others. Through stitching is proving to be a rapid, flexible and inexpensive method of achieving the desired results' (Ellis Developments Ltd[13]).

Techniques with composite fabrics are used to develop loudspeakers, parts for the automotive industry and aircraft wings.

14.4.3 The 3D stitching robot

Carbon fibre composites are used in aircraft production and other manufactured outcomes such as wind turbine blades, and motor and marine sports where lightweight and high performance qualities are essential. The high cost of composites has been a disadvantage in the past but with new production techniques the costs have decreased and the performance has improved. QinetiQ describes its new production methods – 'The latest technique involves using dry fibre, formed in the shape of the final component, with liquid resin.' 'QinetiQ has installed a 3D stitching robot that makes this process even more cost effective by creating dry fibre performs which reduce handling and lay-up time. The robot then selectively stitches the net-shape preform in order to optimise its strength and weight to the precise requirements of its end application.'[14]

The robotic stitcher is housed within QinetiQ's Cody Technology Park, Farnborough and is teamed with three different sewing heads, offering three different sewing methods, which are interchangeable during the manufacture of any single perform. The stitching techniques are tufting providing through thickness reinforcement, blind stitching using a curved needle and two-needle single-sided stitching. 'The robot possesses a "7th axis" because it is mounted on a 3 metre long rail. When combined with the natural sweep of the robot arm, it can operate over a very large area, enabling large components to be produced, typical of those required by the aircraft industry.'[14]

14.5   Medical applications

14.5.1 Ellis Developments Ltd

As well as the development of textiles for engineering applications, Ellis Developments, the Nottingham based textile company also specialises in the design and development of surgical implants. The surgical implants using embroidery techniques (SITE) project was set up with partners Ellis Developments, Pearsalls Sutures, and the Division of Vascular Surgery, University of Nottingham led by Brian Hopkinson. Mr Hopkinson has an international reputation for his work in endovascular surgery. Anson Medical Ltd advised on shape memory alloy. The objective of the project was to investigate the use of embroidery techniques for the manufacture of surgical implants. The project developed a graft stent for the repair of abdominal aortic aneurisms using endovascular techniques. The project has demonstrated that fibre placement can be applied to textile fibres using the CAD/CAM techniques of modern embroidery systems with the placement of metallic shape memory alloy wire.

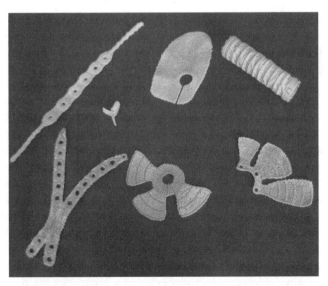

14.4 Ellis Developments Ltd, a textile technology company specialising in surgical implants and reinforcement for plastics (composite materials) http://www.ellisdev.co.uk/. Examples of embroidered implants (photograph – Alison Taylor).

Ten thousand people die per year due to ruptured aortic aneurisms. The vascular stent is designed to line the aorta and can be inserted using endoscopic techniques. Open surgery puts patients at a significantly greater risk and there had been a 25% failure rate for current endoscopic techniques. The implant was designed to incorporate radio-opaque markers and be thin enough to pass through a 7.3 mm catheter, with densely packed fibre walls providing low porosity. Polyester strip springs at the top and bottom of the stents allowed for a seal to be made against the artery wall.

The Cornely, Schiffli and Multihead lockstitch machines were investigated but the Multihead was chosen on grounds of cost, the suitability of the software, and portability, with single to 36 heads allowing for individual customisation to future mass production.

The flexibility of digital embroidery techniques allows for rapid turnaround from customisation of a design from an email request for the surgical implant to manufacture and transfer to the operating table.

Conventional vascular repair textiles are reinforced using a crimping method to concertina the fabric, but this produces a high bulk fabric that is unsuitable for insertion through a catheter. Therefore, the fabric has to be straight-walled with a reinforcement material. High columnar strength is essential to prevent the graft collapsing under the pressure of pulsating blood. Experiments were carried out with threads such as conventional braided polyester suture thread, nylon, and polyethylene and polypropylene monofilament. Eventually a super elastic alloy was used. (Nitinol alloy has been successfully used in other surgical implants but its biocompatibility is being questioned.) A total of 153 designs for reinforcement patterns were stitched out and sampled, and a ladder pattern was chosen that provided sufficient reinforcement to prevent collapse, and also columnar strength. The nitinol wire was reshaped, set, and then precisely stitched into place on the base cloth using conventional lockstitch embroidery methods. By stopping the embroidery at an appropriate stage, the nitinol wire is fed through the base cloth to emerge externally before continuing the stitching process. The process is finished by hand-stitching as no machine method is yet available to form the narrow tube required. After the graft stent is closed into a tube, the device is scoured and sterilised and is then ready for loading into the implantation device. Successful operations have been carried out on sheep and are still going through the process of clinical trials with other embroidered implants in various stages of development.

Shoulder repair devices have been successfully used on a number of patients requiring reconstructive surgery and a prosthesis for the replacement of an invertebral disc of the cervical spine. This embroidered shape fits around a cervical spinal disc prosthesis. The textile is designed to replace the spinal ligaments and hold the disc prosthesis in place.[15]

14.6 Art

14.6.1 Grayson Perry

The art of digital embroidery was brought to public attention when Grayson Perry won the Turner Prize in 2003 for his ceramics. The image of Grayson wearing his digitally embroidered dress at the awards ceremony has been reproduced internationally.

Although better known for his pots, embroidery has always been an outlet for his expression. Grayson sees embroidery as a form of semaphore and uses it as a contradiction to the lack of acceptability of the medium by the Art Establishment. His imagery has a universal resonance and is often disturbing, provocative and erotic, with symbols of anti-war, anti-violence, sex, the ageing and the disabled, with the message even more poignant when beautifully stitched onto duchess satin and made into pretty girlish frocks worn by his transvestite alter ego, Claire. 'When I wear my dresses I say to people "I'm wearing the heraldry of my subconscious".'[16]

14.6.2 Michael Brennand-Wood

Trained as a painter but influenced by painters who used textiles as part of their visual vocabulary such as Rauschenberg and Tapies, Wood has developed a body of work using digital embroidery as a result of his 'Field of Centres' residency at the Harley Foundation, Nottingham. He was influenced by lace made on the Schiffli machines and by gaining access to digital embroidery machines whilst on a residency at the University of Ulster.[17]

14.7 New technology and e-broidery

14.7.1 Maggie Orth

Maggie Orth is responsible for the term e-broidery, which describes the digital stitching of electro-conductive threads into textiles. She currently works at International Fashion Machines and has a PhD from MIT Massachusetts Institute of Technology. Her background is in fine arts performance and environmental art and she has worked for a time at the media lab in MIT where she discovered textiles. Following in the path of Steve Mann,[18] a pioneer in wearable electronics and wearable computing and co-founder of the Wearable Computing Project at MIT, Orth's approach couldn't be more different: 'Most wearable computers still take an awkward form that is dictated by the materials and processes traditionally used in electronic fabrication. The design principle of packing electronics in hard plastic boxes (no matter how small) is pervasive and alternatives are difficult to imagine. As a result, most wearable computing equipment is not truly wearable in the sense that it fits into a pocket or straps onto the body. What is needed is a way to integrate technology directly into textiles and clothing.'[19]

Maggie Orth's e-broidery projects are described in detail in a joint paper that she shares with E.R. Post, P.R. Russo and N. Gershenfield. The introduction to the paper sets the scene: 'Highly durable, flexible and even washable multilayer electronic circuitry can be constructed on textile substrates, using conductive yarns and suitably packaged components. In this paper we describe the development of e-broidery (electronic embroidery, i.e. the patterning of conductive textiles by numerically controlled sewing or weaving processes) as a means of creating computationally active textiles. We compare textiles to existing flexible circuit substrates with regard to durability, conformability and wearability. We also report on: some unique applications enabled by our work; the construction of sensors and user interface elements in textiles; and a complete process for creating flexible multilayer circuits of fabric substrates. This process maintains close compatibility with existing electronic components and design tools, while optimising design techniques and component packages for use in textiles.'[20]

The paper describes the development of a range of textile projects:

• The Firefly Dress made from silk, metallic silk organza, conductive threads, beads, LEDs and conductive Velcro, incorporates stitched electrical circuits using conductive fabric to transport power throughout the dress. As the wearer moves, LEDs attached to fuzzy conductive pads make contact with the conductive fabric layers, causing lights in the skirt and necklace to flicker and change colour.

• The Musical Jacket 1997 was an early experiment in the integration of computing into clothing. The Levi jacket is adapted to become a musical instrument. The digitally embroidered keypad over the left pocket is sewn with mildly conductive thread that, when touched, sends a signal to another processor which runs a midi synthesiser. Sound is projected through two mini speakers in the pocket and the system is battery powered.

• The digitally embroidered, soft-to-handle Musical Ball uses conductive threads as pressure sensors as opposed to hard buttons or traditional keypads. This instrument allows children and novices to interact with and explore music in a spontaneous, expressive way, which more traditional instruments do not allow.

• The Electronic Tablecloth was created for guests at a cocktail party to play a game of Jeopardy, communicating personal information by touching the tablecloth. Each tablecloth contains a decorative embroidered tag-reader and a decorative embroidered keypad, coaster, ID tags and centrepiece displays. These electronic textiles were connected to a large computer and database.

- The Pom-Pom dimmer is a soft, whimsical pompom that connects to an electrical light source and controls light levels with just a gentle tap. International Fashion Machines' patented electronic-textile technology has developed conductive yarns that, combined with colourful recycled carpet fibres, are touch sensitive to make up the soft switch, which is available to buy on the IMF website.[21]

14.7.2 Rachel Wingfield

Rachel Wingfield, artist turned engineer, creates reactive luminous surfaces by exploiting electroluminescence to help seasonal affective disorder (SAD) sufferers. Light has a profound influence on the emotional and physiological state. The absence of light can cause severe medical problems for sufferers. Her designs that place light generation into bedding help to alleviate the symptoms. The Light Sleeper collection of bedding embroiders electroluminescent wire and silk. The reactive light emission simulates sunrise and helps to reset the user's body clock.[22]

14.7.3 Joanna Berzowska

Joanna Berzowska is an Associate Professor of Design and Computation Arts at Concordia University, with a Master of Science from MIT. Her career has been devoted to the integration of new technology within fashion and textiles. Although she experiments with a broad range of textile techniques in order to resolve often complex electronic innovations, digital embroidery has been a major focus. She co-founded International Fashion Machines in Boston, is a member of the Hexagram Research Institute in Montreal and is the founder and research director of XS Labs.[25] Berzowska explains 'XS Labs is a design research studio with a focus on innovation in the fields of electronic textiles and reactive garments, where we try to break down the traditional boundaries between disciplines. Many of our electronic textile innovations come from the fact that we look at the technical but also cultural history of how textiles (and garments) have been made for generations (weaving, stitching, embroidery, knitting, beading, quilting, tailoring) but use materials with different electro-mechanical properties, which enables us to construct more complex textiles and garments with electronic properties.'[26]

14.7.4 Others

The spacesuit glove with embedded robotic controls and developed by NASA, features an embroidered keypad with quantum tunnelling composites embedded in the spacesuit activity glove. The Softswitch keypad, developed by Softswitch Ltd, is a flexible embroidered keypad on cotton. Composites that use multiple layers of conductive textiles and an elastomer loaded with fine metallic particles, allow electron tunnelling between them when brought closer together by the pressure of a person's touch.[23]

The Softswitch technology is being used for jackets to hold MP3 players such as the iPod by Apple. This technology is also incorporated into jackets designed by Burton Snowboards that allow the freedom to activate the control by touching the sleeve. A similar product by O'Neil is on the market, also in the form of a snowboarding jacket. Tactile fabric keyboards been developed by Elexon and Gorix. The collaboration between Philips and Levi has developed smart garments with integrated mobile phones and MP3 players operated by embroidered switches.[24]

14.8   Future trends: what does the future hold?

With an increase in the ageing population and the shift in customer spending power from the younger market to the 'Baby Boomers', there is a greater demand for garments to fit a broader range of sizes and fulfil a greater variety of user needs. The older we get, the more different we become from each other. Sizing is an issue as body shapes change as we age and with conditions such as osteoporosis and arthritis more prevalent. New fabrics that offer wicking, cooling and heating, antimicrobial qualities and other medical and wellbeing properties can be applied to resolving problems and offering a better quality of life. Branding is less important and functionality, comfort, weight, washabilty and fit are of greater consequence. That is not to say that all of these should take priority over style, taste and design. The demand for mass customisation is on the horizon and the need for a tailored-to-fit garment that suits needs and tastes at mass market prices with a rapid response turnaround from point of sale to delivery must be met.

Body sizing scanners can accurately measure individual body shapes and that measurement can be transferred to the retailer from the customer who would be able to choose a preferred design and appropriate fabric(s) for the garment, which could be customised individually using a combination of digital print, laser etching, laser cutting and digital stitch. The customer would be able to try the garment on their own personal computerised 3D avatar and even take advice from their online personal shopper. Once chosen, the details would be electronically sent to the manufacturer who would then make the garment and deliver directly to the customer. This process would have a rapid turnaround.

Embroidered implants would speak to embroidered circuits on garments that in turn would converse with the smart environment, communicating medical information monitored from the body, allowing the garment and environment to react and adapt. The implants could also act as a means of identification and tracking of the individual.

Embroidered composite materials that are lightweight, yet strong, could be incorporated into specific parts of a garment that would provide localised protection for vulnerable parts of the body whether under attack, taking part in an extreme activity or weak due to medical conditions such as osteoporosis.

Robotic 3D stitching could produce large-scale composite materials, lightweight but strong, that could be used for ecological building projects such as housing and cheap, easy to erect shelters for homeless refugees.

The increase in world population and the exhaustion of renewable resources will increase the need for recycled products. The embellishment of recycled textiles through digital embroidery would give a new life to worn garments. As technology gets smaller and cheaper, the digital embroidery machine may become as commonplace as a mobile phone. Designs for stitch could be shared and downloaded like ring tones, allowing for designer logos to be added. Individuals will be able to communicate ideas and identity through text and image displayed on the garment.

Asked about the future, Maggie Orth says 'there is no doubt we will have computer screens decorating textiles, but I hope people will explore textiles as a new art form, not as computer screen art but as smart materials interacting with computation. It is a new creative space'.

14.9   Sources of further information and advice

BERZOWSKA J and BROMLEY M, *Soft Computation Through Conductive Textiles*, XS Labs 2007

BRADDOCK CLARKE S E and O'MAHONEY M, *Techno Textiles 2, Revolutionary Fabrics for Fashion and Design*, London, Thames and Hudson, 2005.

BRADDOCK CLARKE S E, O'MAHONEY M, *Sportstech, Revolutionary Fabrics, Fashion and Design*, London, Thames and Hudson, 2002.

JEFFRIES S, 'Top of the Pots', *Guardian* article, Friday, November 21st 2003. http://arts.guardian.co.uk/turnerprize2003/story/0,,1090056,00.html

MCQUAID M, *Extreme Textiles, Designing For High Performance*, London, Thames and Hudson, 2005.

NET COMPOSITES 12 6 2007 http://www.netcomposites.com/news.asp?4425

POST E R, ORTH M, 'Smart Fabric, or Washable Computing' *The Digest First IEEE International Symposium on Wearable Computers*, Cambridge, MA, (1998).

RUDGLEY K, 'Sew Cool, Illustrations Embroidered in Style', *Embroidery*, Vol 54 No 6.

SMITH E, 'Soft wear', *Selvedge*, edition 00, page 27, May/June 04.

http://www.embroiderydesigns.com

http://www.interrogatingfashion.org/

http://www.textilefutures.co.uk/RSA/exhibitors/rachel-wingfield.htm

http://www.loop.ph/new/wingfieldbio.html

http://www.maggieorth.com/
http://www.maggieorth.com/art_pompom.html
http://ifmachines.com/
http://freeembroiderystuff.embroiderydesigns.com/magazines.aspx
http://www.eurostitch.com/home.php?taal=eng&ed=200408
http://loop.ph/bin/view/Loop/DigitalDawn
http://www.husqvarnaviking.com/
http://www.abc-embroidery-designs.com/Embroidery_Threads/
http://www.gs-uk.com/
http://www.annetcouwenberg.com/index.htmlww.cooperhewitt.org/
    EXHIBITIONS/extreme_textiles/index.asp
http://www.chillingeffects.org/dmca512/notice.cgi?NoticeID=1747
http://www.fiber-lineinc.com/index.php?id=2
http://www.windstarembroidery.com/embroidery-information.cfm
http://www.taunton.com/threads/
http://www.etn-net.org/shop/magazines/2007_3e.htm
http://www.swfamerica.com/
http://www.hand-embroidery.co.uk/
http://layersofmeaning.org/wp/?p=200
http://www.1-art-1.com/michael_brennan_wood.html
http://www.caa.org.uk/exhibitions/exhibition-archive/2007/coming-soon/
    michael-brennand-wood.html
http://www.clothandculturenow.com/Michael_Brennand-Wood.html
http://www.craftscouncil.org.uk/
http://embroidery.embroiderersguild.com/
http://www.selvedge.org/default.aspx
http://www.berzowska.com/
http://hybrid.concordia.ca/~joey/
http://www.xslabs.net/work.html
http://www.qinetiq.com/
http://www.maggieorth.com/bio.html
http://www.brother.co.uk/
http://www.annatextiles.ch/publications/fraefel/2machi.htm

14.10 References

1   http://www.annatextiles.ch/publications/fraefel/2machi.htm
2   http://www.gs-uk.com/php/brother_industrial_embroidery_machines.php
3   http://www.brother.co.uk/
4   http://www.industrialsewmachine.com/webdoc3/application/Embroidery/cme.htm
5   PURBA J S, *The Embroidery Book*, https://www.etc-embroidery.co.uk/catalog/index.php
6   http://www.abc-embroidery-designs.com/Embroidery_Threads/
7   http://www.gs-uk.com/
8   http://www.quiltersthread.com/store/special.html
9   http://www.nitivy.co.jp/english/yarn.html

10 http://www.bekaert.com/bft/Products/Innovative%20textiles/
    Base%20materials/Yarns.htm
11 http://www.x-staticfiber.com/index3.htm
12 http://www.ellisdev.co.uk/mascetproj.html
13 http://www.ellisdev.co.uk/site.html
14 http://www.qinetiq.com/
15 http://www.ellisdev.co.uk/site.html
16 HOGGARD L, 'Grayson Perry, The Heraldry of the Subconscious,' *Selvedge*, page
    22, edition 00, May/June 04.
17 Schoeser M, *Bucking the Trend, Essay for Michael Brennand-Wood, Catalogue
    for Field of Centres Exhibition*, The Harley Gallery and The Gallery Ruthin
    Craft Centre, 2004
18 http://www.wearcam.org/mann.html
19 http://www.maggieorth.com/bio.html
20 POST E R, ORTH M, RUSSO P R and GERSHENFELD N, E-broidery: Design and Fabrica-
    tion of Textile-based Computing *IBM Systems Journal*, Vol. 39, Nos 3&4,
    Armonk, NY, IBM Corporation, (2000).
21 http://www.maggieorth.com/art_pompom.html
22 http://www.loop.ph/new/wingfieldbio.html
23 LEE S, *Fashioning the Future, Tomorrow's Wardrobe*, p. 26, London, Thames and
    Hudson, 2005.
24 http://www.softswitch.co.uk/
25 http://www.xslabs.net/work.html
26 http://www.berzowska.com/

15
Developments in digital print technology for smart textiles

C. TREADAWAY, University of Wales Institute Cardiff, UK

Abstract: This chapter outlines developments in digital print technology and highlights the ways in which these are significantly influencing the design and manufacture of printed textiles. Direct and non-direct methods of digital ink-jet printing, including dye sublimation and pigment printing, are explained along with the various types of print heads used. Three-dimensional digital print technologies are also described including: Stereolithography (SLA) Selective Laser Sintering (SLS) Three Dimensional printing (3D) and Fused Deposition Modelling (FDM). The final section of this chapter looks ahead to emerging technologies in which decoration becomes integrated within the construction of the whole garment using dynamic colour effects and moving images.

Key words: textile digital ink-jet printing, dye sublimation ink-jet printing, printed textile design, three dimensional printing, light emitting textiles.

15.1 Introduction

The integration of wearable technology looks set to significantly enhance apparel in the near future; however, functionality alone is not enough to sell a product in a world in which image, style and fashion predominate in marketing strategies. The appearance of clothes, their colour, shape and embellishment, communicate personal identity and enhance both desirability and economic value. Surface decoration and colour convey social and cultural significance, as well as communicating the individual values of the wearer. The processes that are used to translate surface pattern designs onto the surface of textiles influence the visual characteristics of apparel fabrics. An appreciation of the technical constraints and limitations imposed by these technologies, as well as the design opportunities they provide, are essential if designers are to successfully create functional and appealing designs in the future.

New methods of printing colour and pattern on fabric surfaces are now possible as a result of digital technologies. This chapter outlines some of the developments in print technology and highlights the ways in which these influence the design and manufacture of the end product. The ubiquity and availability of digital and communication technology is changing the design process and challenging our preconceptions of printed textile design. In the past, manufacturing processes dictated the types of designs that would be printed, including the numbers of colours used and the visual appearance of the pattern; for example, the style and scale of repeat structure used (Bunce, 1999). Analogue manufacturing process are labour intensive at the preparatory stage; making, sampling and prototyping of textile designs expensive and time consuming. The advent of digital textile ink-jet technology looks set to radically affect manufacturing processes. Already many printed textile companies use digital printing for prototyping and sampling; and the introduction of production speed ink-jet printers looks set to challenge traditional methods of mass-produced printed textiles in the near future (http://www.dupont.com/inkjet/en/artistri/index.html).

Developments in three-dimensional printing may provide us with new methods of decorating textiles concurrently with garment creation, patterning and forming the product in a single process (Delamore, 2004). The final section of this chapter looks ahead to emerging technologies that offer the potential for a completely new manufacturing paradigm in which decoration becomes integrated within the construction of the whole garment. Future pattern on clothing may be created using dynamic colour effects and moving images as a result of nanotechnology, luminescent fibres and flexible textile displays. These evolving methods of creating surface pattern also have the potential for dynamic change and interaction. In the future it may be possible to download surface patterns for garments from the Internet in the same way as we currently download music and video files (Farren, 2004).

15.2   Digital imaging and printed textile design

Computers have been used since the 1970s as a tool to assist the preproduction of artwork for printed textiles for mass production (Leak, 1998). These processes involve the development of repeating units of pattern, adjusted to fit around rollers or onto flat screens in order to sequentially print the individual colours in a design. Artwork is colour reduced using CAD software so that positives can be made for each roller or screen. The introduction of digital ink-jet printing as a manufacturing process eliminates many of these technical restrictions that have constrained creative artwork development; direct translation of digital imagery is now possible onto textile substrates. For the textile designer, this means that digital imaging can be used to innovate concepts that no longer adhere to the conventions of traditional printed textile designs (Briggs, 1995). Patterns can be non-repeating along the width of the cloth, and garment pattern shapes can be printed containing surface pattern that is engineered to fit, accommodating the flow of the design across seams, fastenings and openings (Campbell,

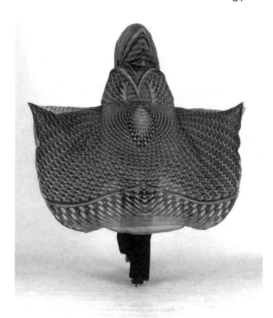

15.1 'Transformation Icarus' digitally printed garment designed and printed by J. R. Campbell and Jean Parsons (2005).

2005) (see Fig. 15.1). Garments can become customizable, personalized, and unique, with each print potentially an individual rendition of the design concept. Unlike traditional analogue printed textile processes there are no economies of scale; it is equally cost effective to produce a single garment as many metres of a design. As manufacturers embrace this technology for mass production, they are beginning to require more adventurous design solutions that offer a new visual approach, and take advantage of the potential of the digital print process (Treadaway, 2006).

The economic benefits of global manufacturing have meant that many companies now operate with factories located many thousands of miles from their design and marketing headquarters. Digital imaging enables designers to create and communicate design data with rapidity, accuracy and ease, both 'in house' and to suppliers and manufacturers around the world. With increasing 'time to market' pressures and 24/7 working strategies, digital artwork development is crucial. The potential of sharing digital data also enables a variety of expertises to be combined. Textile and garment designers along with marketing and retail personnel can collaborate in the development of a product. The Internet is providing greater variety in consumer choice between products, and some companies now offer customer participation within the design process itself (www.cybercouture.com). Research into mass-customization using the Internet has shown that there is economic potential for this method of co-design collaboration with the customer, to create personalized digitally printed products (Campbell, 2005). This type of production has a number of commercial advantages, including the reduction of waste, and transportation and warehousing costs, since only goods that are specifically required for a customer are manufactured.

There are significant advantages for textile designers working with digital print technology. Internet access enables digital image files to be emailed directly to digital textile print bureaux and printed fabric can be returned to the designer by mail. Individual art to wear, or designer-maker craft products, can be created without the need for costly equipment or workshop space. Fashion designers are able to print one-of-a-kind fabric for unique garments. Many top couturiers include digitally printed garments within their collections in order to extend the thematic content of a collection and to push the creative boundaries of their work. Notable designers who are using this technology include Issey Miyake, Julian Roberts, Hussein Chalayan, Jonathan Saunders, Basso & Brooke and Siv Stoldal (http://www.dazeddigital.com/eyespy/item.aspx?a=331). Hamish Morrow, working in collaboration with textile designer Philip Delamore, have together exploited the potential of the technology to the extreme by using video images of patterns projected onto garments on the runway as surface decoration for limited-edition garments capturing both time and motion within the garment decoration.

15.3   Digital ink-jet printing

Textile ink-jet printers resemble their counterparts developed specifically for paper printing. They comprise a structure through which a roll of fabric or paper can be fed, beneath a movable electronic print head that sprays a controlled amount of colour onto the fabric/paper surface (see Fig. 15.2). Raster Image Processing (RIP) software controls the translation of the image on-screen to that created on the fabric surface, determining the exact amount of each colour required to recreate those identifiable as pixels in the digital design. Colours are contained either in individual printer cartridges or bulk feed units, with the choice of substrate predetermining the type of dyestuff required. Digital print processes used to manufacture and sample printed textiles include indirect (dye sublimation or heat transfer ink-jet printing) and direct ink-jet printing methods, and use either pigment or dyestuff.

15.3.1 Dye sublimation (indirect) ink-jet printing

Dye sublimation printing is widely used for printing onto both woven and knitted white polyester or polymer-coated substrates. The process involves

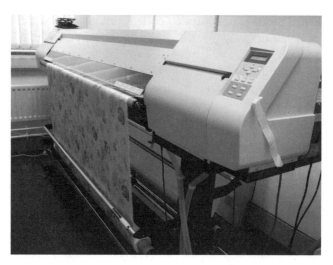

15.2 Mimaki TX2 textile ink-jet printer.

printing the dyestuff onto paper using a conventional large format ink-jet printer. The colorant is then transferred onto fabric via a sublistatic (heat transfer) process in which the dyestuff sublimes, i.e. turns from a solid deposit on the paper surface, into a gas. This penetrates and colours the fibres of the substrate. Sublimation dyes are able to produce a wide colour gamut on fabric and use a four-colour print process: cyan, yellow, magenta and key (black) CMYK. Occasionally, light magenta and light cyan are also included for printing pastel tones.

One of the recent major breakthroughs in dye sublimation ink-jet printing has involved the development of papers capable of absorbing the carrier liquid (usually water) whilst fixing the dye particles onto the paper surface. The transfer paper is designed to release as much of the colorant as possible during the sublimation process as variance in the amount of colour discharged directly affects the colour yield on the fabric (van Houtum, 2005). Dye sublimation is widely regarded as one of the least environmentally polluting print processes since the main waste emission consists of the paper produced after printing. There are no final post-printing wash-out treatments required and the paper can be recycled or reused; it is frequently used by florists to wrap flowers or to make paper bags. British designer Rebecca Earley, whose garments are sold under the Be Earley designer label, has worked extensively with this printing technique. Her concerns with the environmental damage caused by traditional textile print processes led her to explore eco design techniques. Recycled polyester garments are embellished and given a second lease of life using exhaustive print processes including dye sublimation (http://www.tedresearch.net/docs/intro/intro.htm). Like Earley, Kate Goldsworthy also makes extensive use of

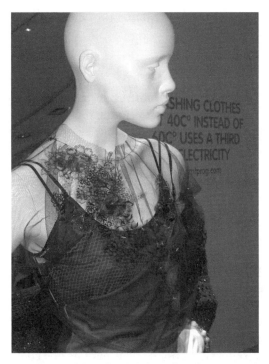

15.3 Dye sublimation ink-jet printed garment designed by Kate Goldsworthy.

digital dye sublimation processes to develop environmentally friendly textiles (see Fig. 15.3).

15.3.2 Textile digital (direct) ink-jet printing

For economic reasons, the majority of printed textiles today continue to be printed using traditional analogue roller processes; nevertheless, there is a continual increase in quantities of fabrics that are being produced in the textile industry using direct ink-jet digital printers. Unlike dye sublimation printing, which is limited to patterning of synthetic fabrics, this process is suitable for many different textile substrates. The main advantages of direct digital ink-jet printing are its flexibility and low set-up costs. Very short production runs are economic as no costly tooling or screen preparation is entailed as in conventional analogue processes. There are few limitations on colour; millions can be reproduced at a similar cost as a design containing only one or two. The major disadvantages of this technology concerns the initial economic investment required in purchasing the equipment and the slow printing speed. Advances are constantly being made with the technology and machines capable of production printing speeds now available; it is likely that, in the not too distant future, digital ink-jet printing will predominate (Dehghani, 2004).

Print heads

Colour is applied to the surface of the substrate from a print head that is controlled by a computer in a non-contact printing process. The print head is located inside a carriage which travels along and slightly above the width of the fabric, depositing small droplets of colour as it moves. The droplets are emitted from the print head in several ways depending on the type of mechanism used in the device, each producing varying characteristics. There are three main types of print heads:

• Thermal drop on demand (DOD). This technology is widely used in bubble-jet home computer printers. An electrical current is used to heat a plate inside the print head containing a dye reservoir. The liquid vaporizes and forms a droplet which is then released onto the fabric. This is the most inexpensive type of print head available.
• Piezoelectric drop on demand. This print head operates via an electronically controlled mechanism in which an electrical current is used to charge a piezoelectric material. This causes it to change shape and allows colour to be squeezed through a nozzle. This permits fine control of the quantities of colour released. Very small droplets are created, resulting in fine print definition from this type of head.
• Continuous. Continuous ink-jet print heads operate via electronic control of statically charged droplets of colorant. Some of these droplets are deflected from a continuous stream onto the substrate through a nozzle. Those droplets that are not deflected are collected in a gutter for reuse. This type of head is less prone to clogging and so can be used for printing more difficult colorants that may block other types of print heads (Fig. 15.4).

Colorants

The type of colorant used for digital ink-jet printing is determined by the fibre content of the substrate and the end purpose of the print. The most widely used colorants for digital ink-jet printing are reactive dyes, used to print cellulose fibres including cotton and linen, and acid dyes used on protein fibres such as silk and wool.

Substrate preparation

It is necessary to prepare the substrate, prior to printing, with chemicals that enable colour to bond with the fibres (in the case of reactive dyes) and

15.4 Mimaki TX2 textile ink-jet printer – detail of print head.

to inhibit the wicking effect that occurs through the capillary movement of dye along the fibres of a woven fabric.

Reactive dyes require alkali, moisture and heat in order for the chemical colour bonding reaction to occur to make the colour fast. It is not possible to apply the alkaline at the same time as the colour, since heat and moisture are required in the operation of the print head and this would jeopardize the required chemical reaction on the substrate. Fabrics for digital ink-jet printing can be purchased ready prepared and need to be stored and treated carefully as the applied chemicals are not stable and have a limited shelf-life. Once printed, the substrate is dried and steamed to enable the chemical reaction to occur, giving stability to the printed colours.

15.3.3 Colour management

There are many variables that affect colour in digital ink-jet printing, and successful colour management remains one of the key determinants of successful adoption of this technology by industry (Xin, 2006). The chemicals used to prepare the substrate prior to printing, the finishing process, as well as the print process itself, introduce variables and complicate the colour calibration process. By developing consistency in substrate suppliers, environmental conditions in which the fabric is stored, types of inks and finishing processes used, it is possible to accurately reproduce colours in a digital design. The expertise of those providing the print service is a crucial factor in the equation, since appreciation of the subtleties in colour variance is a complicated business. The perception of colour relies on subjective experience and has been shown to be influenced not only by physical factors such as ambient light and chemical constituents of the dyestuff, but also on psychological factors including memory (Beau Lotto, 2004). Colour management software is used extensively to calibrate what is viewed on the computer monitor with the printed output and RIP software accommodates necessary adjustments in colour application according to the type and weight of the substrate to be printed. Tools such as spectrophotometers and light booths provide constancy in viewing of colour and its accurate calibration. Colour profiles specific to the printer, ink, substrate and post-printing processing used are developed within the RIP software to enable the operator to select from a menu the appropriate quantity and hue of colour to be applied during printing.

For designers working outside of industry and without access to spectrophotometers and colour management tools, digital colour translation remains a difficulty. Research has shown that empirical iterative trial-and-error techniques in which colours are tweaked on a computer monitor prior to sampling on fabric can produce satisfactory results (Polvinen, 2005). Some practitioners find that the best approach is to work with the colours and use colour anomalies as stimulants for further creative exploitation, including embellishment with other textile techniques (Treadaway, 2004a).

The post-printing treatments, which include steaming to fix the colorant and washing to remove the pre-treatment chemicals, greatly complicate the colour management process. Accurate testing of colours is possible only after processing and the resulting chemical changes are stabilized. Each stage in the digital print process introduces new variables and highlights the need for consistency at every step.

15.3.4 Printing with pigments

Digital ink-jet printing with pigments eliminates some of the complications described in colour management using textile dyes. Unlike reactive, acid and disperse dyes, pigment colours are fine particle size solids and adhere to the printed textile surface through the use of external polymeric binders. These colorants require heat or air drying to fix them and unlike dyes their colours remain unchanged through the process. Nevertheless, printing pigment inks with ink-jet technology also poses technical problems concerning the application of colour through the print head. The pigments are crystalline particles, varying in size between 50–150 nm, which have the potential to block the heads and affect precision in the delivery of colour, whereas dye inks comprise uniform dye molecule solutions (Fu, 2006). Research is currently ongoing to develop enhanced print heads capable of consistent ink-jet application of textile pigment colours. Pigment colours provide a reduced colour gamut and currently lack the colour vibrancy and density of hue that are possible to print using textile dyes.

15.3.5 Special effects

Future developments in textile digital ink-jet colorants are likely to be influenced by advances in nanotechnology. Chemical colour constituents and additives can now be digitally ink-jet printed onto textiles due to the manufacturing of ingredients at molecular scale. Future nano-engineered ingredients are likely to include chemicals with luminescent and photochromic properties, making possible ink-jet printed colours with the potential to emit light in darkened environments and change colour in sunlight. These new colorants have many potential safety applications as well as providing further innovative decorative surface pattern options (Lee, 2005).

The European Union funded 'Digitex' project is developing digital ink-jet technology to apply nano-scale textile surface finishing processes, such as fire resistance and self-cleaning properties to the surface of fabrics (http://www.dappolonia-research.com/digitex-eu.com/doceboCms/).

15.3.6 Hybrid craft

The fragmented dot matrix of digitally printed colour contrasts visibly with the uniform spot colour strength produced in the analogue silk screen process. These different visual characteristics and the relative ease with which it is possible to digitally print textiles have encouraged designers to respond to the printed output in a variety of ways. Some practitioners develop the printed surface further using hand painting, embroidery and other traditional textile print processes whilst others refine and adjust the digital image in order to maximize the colour effects in a print. Rapid print production provides greater time for further embellishment using hand techniques leading to hybrid processes that combine digital and hand crafting (Treadaway, 2004b). Techniques such as embroidery, appliqué, beading and application of colour effects by hand are used by designers to enhance and add economic value to digitally printed fabrics.

15.4   The future: printing new structures, patterns and colours

Printing technologies have traditionally been used to apply decoration or pattern embellishment onto the surface of a textile substrate. In the future, it looks increasingly likely that print technology will also be involved in garment construction. Integration of garment design and surface pattern will place new demands on designers who will need to understand how pattern and colour work in both two and three dimensions, and be able to negotiate sophisticated design software to enable them to develop their ideas.

15.4.1 3D printing and rapid prototyping

Developments in three-dimensional printing, currently used for product rapid prototyping, have the potential to radically alter the way garments of the future will be made. Current research at the London College of Fashion is investigating how this technology might be used to develop individually tailored garments, integrating data obtained from three dimensional body scanning and using Three Dimensional (3D) computer software. Working in conjunction with Freedom of Creation (http://www.freedomofcreation. com/) and Prior2Lever, designer Philip Delamore has developed prototype textile structures and shoes (see Figs 15.5 and 15.6). This technology has

15.5 Prototype boot.

15.6 3D printed sole unit developed by Philip Delamore (2004).

15.7 3D body scanner.

huge implications for the fashion industry and suggests a future in which it is possible to download design data from the Internet in order to print unique garments at home, tailored to the customers' exact specification, with data acquired from 3D body scanning processes (see Fig. 15.7). Although the integration of complex surface pattern and colour into the product is not yet possible, the potential exists for designers to create garments in which shape, textural surfaces and embellishments can be printed as one form.

There are a number of technologies currently available that enable three-dimensional structures to be digitally printed using a variety of different materials including thermoplastic polymers, ceramic powders and metals. The processes are broadly similar and are used to print three-dimensional objects by layering materials one on top of another using files that have been created in Computer Aided Design (CAD) or 3D modelling software. Although the term rapid prototyping implies this is a speedy process, in reality these are slow, expensive and precise technologies that provide the opportunity to recreate digital files as 'one of a kind' physical artefacts.

Rapid prototyping technology is developing quickly and the surfaces of the resulting printed forms are becoming increasingly refined, needing less finishing treatments. It is also possible to print electronic components simultaneously within the construction of a product via this technology. The following are four digital print technologies used to create three-dimensional forms:

- Stereolithography (SLA). This process involves the use of a laser beam, which is traced onto a vat of liquid photopolymer; the liquid solidifies wherever the laser hits the surface. The build process is repeated layer by layer and the evolving object lowered sequentially into the vat so that additional layers can be printed on top of the first. Support structures are required for overhangs and undercuts, depending on the nature of the design being printed. These are fabricated concurrently with the printed object and, when the printing process is completed, are cut or broken away to reveal the finished artefact.
- Selective laser sintering (SLS). A laser is used to sinter or fuse a thermoplastic powder material, which is first spread and compacted via a roller onto the bed of the device. The computer-directed laser selectively melts and bonds the powder to form each layer of the object. The process continues, layer by layer, until the complete object is fabricated. Unlike SLA, no support is required for overhangs and undercuts, since the powder bed provides a scaffold during fabrication. On completion, excess powder is brushed away to reveal the finished printed object. Various thermoplastic materials including nylon, glass-filled nylon and polystyrene can be printed using this technology, along with metals and ceramic substances.
- Three-dimensional printing. This process is similar to SLS but instead of using a laser to sinter material together, an adhesive is used for bonding. Ink-jet technology is used to deposit a computer-directed quantity of liquid adhesive in a two-dimensional pattern onto a compressed and flattened powder bed. The powder becomes bonded in the areas of deposited adhesive to create a layer of the printed object. Layers are printed sequentially and there is no need to build support for overhangs since this is provided by the powder bed (see photographs illustrating this process in Fig. 15.8).
- Fused deposition modelling (FDM). This process uses molten plastic or wax extruded through a computer-controlled ink-jet print head. Thermal and Photopolymer Phase Change Ink-jet (PCI) printers use a single jet to print both a plastic material for the item to be created and a support material. Minute droplets of these materials, kept in a molten state in a connected reservoir, are forced through the heated print head. The

15.8 3D printing process. (a) Data is communicated to the 3D printer from the computer. (b) Detail of printer carriage and print head. (c) The powder bed is levelled and compressed prior to printing. (d) The computer directed print head travels across the powder bed, printing layers of liquid adhesive onto the powder surface to build the form. (e) When the print is completed, excess powder is deposited into half the powder bed, which is lowered to form a recycling unit. (f) The printed form is exposed by carefully brushing away excess powder. (g) The fragile printed form is removed with care from the powder bed. (h) Excess powder trapped inside the form is gently brushed away. (i) Any remaining powder is removed from the internal and external surfaces of the form using pressurized air in a vacuum unit. The form is hardened with a layer of adhesive.

droplets cool as they are printed and harden to form a solid layer. Once a layer has been printed, it is levelled to make it a uniform thickness and the printing process resumes until the entire form has been created. When complete, the support material is melted, broken or dissolved away. The photopolymer PCI uses an ink-jet print head to deposit both build and support materials. Each layer of polymer is cured (hardened) following deposition using an ultraviolet light mounted on the print head. The support material is washed away using pressurized water, leaving the printed three-dimensional structure (see Fig. 15.9).

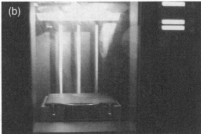

15.9 Fused Deposition Modelling (FDM) machine, (a) general view, (b) internal detail, (c) final printed form.

15.4.2 Light-emitting textiles and flexible displays

Textiles in which pattern may comprise moving images and dynamic changing colours may be possible as a result of technological advances in flexible textile displays using polymer light-emitting diodes (PLEDs). This emissive technology consists of a polymer material, printed onto a substrate of glass or plastic, which does not require the backlighting or filters that are currently used in liquid crystal displays. PLED technology is very energy efficient and is currently being developed into ultra-thin displays. This technology can also be used in reverse to create photovoltaic (PV) devices, which convert light into electricity and are used to make polymer solar cells (Lee, 2005). These will provide bendable, disposable solar power sources for numerous applications and have the potential to power flexible PLED textile displays, capable of displaying pattern and moving video images. Applications requiring dynamic patterns, which may be responsive to light or emotion, will provide new creative challenges to designers in the future. Surface patterns that evolve in real time will bring a new meaning to the concept of 'design collection' and will require agility in creative thinking and ability to visualize complex sequences of changes. Design work will exist as software to be downloaded and updated, casting the designer in a new role, as creator of digital rather than physical artefacts (Farren, 2004).

15.5   Digital printing for smart clothes and wearable technologies

This chapter presents a variety of digital printing technologies that are in the process of transforming the ways apparel textiles are designed, manufactured and patterned. New processes challenge old ideas and designers need to be familiar with the potential of these emerging technologies to appreciate how they may influence the visual appearance of design solutions and how they can be used to embellish textile surfaces (Ujiie, 2006).

As digital textile ink-jet printing becomes more readily available, consumers will demand products that are customized to their specific needs and requirements, or may wish to participate more actively in the design process themselves. Our increasingly connected world offers potential for greater collaboration between designers, manufacturers and the consumer via the Internet. Distributed working methods enable designers based in one country to work with printed textile manufactures on the other side of the world. Technology that enables teams of designers, with a variety of expertise, to collaborate in ways that would have been unimaginable a few years ago, are already in place in a world in which 'time to market', 'print on demand' and '24/7 working' are vital elements in global textile manufacture.

New printing technologies will be useful for visually enhancing clothing in which electronic devices have been incorporated. Wearable technology needs to be both functional and aesthetically pleasing and may require specific areas of a garment to contain printed decoration. Digital ink-jet printing enables a flexible and cost-effective method of producing specific short-run or customized printed garments in which decorative elements can be strategically positioned to enhance or disguise electronic components or power sources (Hynek, 2004) (see Figs 15.10 and 15.11).

Current developments in rapid prototyping and three-dimensional printing suggest that, in the future, it may be possible to print entire garments with integrated electronics and power supplies. Designers using these technologies will be required to navigate complex 3D software design systems and work collaboratively with teams of other designers, utilizing multiple areas of specific expertise, in order to realize their designs. Digital printing technology will play an essential role in enabling these new kinds of textile design concepts to be developed, communicated and reproduced as garments. 3D printers designed for desktop use have already been developed and enable small plastic objects to be printed at home. These machines heat powdered nylon under a halogen light source, which is then printed in microscopic layers to create a three-dimensional object (www.desktopfactory. com and www.fabathome.org).

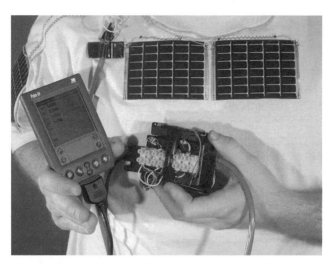

15.10 Garment incorporating photovoltaic cells (Campbell, 2005).

15.11 Digitally printed solar powered tie designed by J.R. Campbell (2005).

Research is under way into the development of self-replicating digital printers that will enable general-purpose manufacturing machines to be printed (www.reprap.org). These devices could potentially enable all kinds of artefacts, including whole garments, to be downloaded from the Internet and printed cheaply at home. The implications of this on retail and manufacturing industry are enormous. The future design process will inevitably have to change to meet the demands of the imminent digital printing revolution, in which the customer may wish to purchase software in preference to physical goods.

15.6   Acknowledgements

The author thanks J.R. Campbell, Jean Parsons, Philip Delamore, Kate Goldsworthy, Julian Sanders (Studio SDA), Peter Hathaway and Mal Bennett for their contributions.

15.7   References

BEAU LOTTO, R. and PURVES. D. (2004). 'Perceiving Colour.' *Review of Progress in Coloration* 34: 12–25.

BRIGGS A. and BUNCE, G.E. (1995). 'Breaking the rules: Innovatory uses of CAD in printed textiles.' *Ars Textrina* 24: 185–203.

BUNCE, G. (1999). 'CAD and the role of the printed textile design'. *Conference Proceedings, CADE 1999*. University of Teesside: Computers in Art and Design Education.

CAMPBELL, J.R. and PARSONS, J. (2005). 'Taking advantage of the design potential of digital printing technology for apparel.' *Journal of Textile and Apparel, Technology and Management* 4(3).

DEHGHANI, A. (2004). 'Design and engineering challenges for digital ink-jet printing on textiles.' *International Journal of Clothing Science and Technology* 16(1/2): 262–273.

DELAMORE, P. (2004). 3D printed textiles and clothing on demand. *RMIT Intermesh Symposium*, Melbourne, Australia.

FARREN, A. and HUTCHINSON, A. (2004). 'Digital Clothes: Active, dynamic and virtual textiles and garments.' *Textile, the Journal of Cloth and Culture* 2(3): 290–307.

FU, Z. (2006). 'Pigmented Ink Formation.' In Ujiie H. *Digital Printing of Textiles*. Cambridge: Woodhead Publishing Ltd.

HYNEK, J.S., CAMPBELL, J.R. *et al.* (2004). 'Application of digital textile printing technology to integrate photovoltaic thin film cells into wearables.' *Journal of Textile and Apparel, Technology and Management* 4(3). http://www.tx.ncsu.edu/jtatm/volume4issue3/digital_printing.htm (acc.24.04.08)

LEAK, A. (1998). *A Practical Investigation of Colour and CAD in Printed Textile Design*. Nottingham: Nottingham Trent University.

LEE, S. (2005). *Fashioning the Future: Tomorrow's Wardrobe*. London: Thames and Hudson.

POLVINEN, E.M. (2005). 'International collaborative digital decorative design project.' *Textile: The Journal of Cloth and Culture* 3(1): 36–57.

TREADAWAY, C. (2004a). 'Digital Imagination: the impact of digital imaging on printed textiles.' *Textile, The Journal of Cloth and Culture* 2(3): 256–273.

TREADAWAY, C. (2004b). 'Digital Reflection.' *The Design Journal* 7(2): 17.

TREADAWAY, C. (2006). Digital imaging: Its current and future influence upon the creative practice of textile and surface pattern designers. PhD thesis in Art and Design. Cardiff: University of Wales Institute Cardiff.

UJIIE, H. (2006). *Digital Printing of Textiles*. Cambridge: Woodhead Publishing Ltd.

VAN HOUTUM, A. and EERLINGEN, M. (2005). 'Transfer Yield: A key success factor for inkjet dye sublimation.' *SGIA Journal* 9(2): 3–4.

XIN, J. (Ed) (2006). *Total Colour Management in Textiles*. Cambridge: Woodhead Publishing Ltd.

16

Environmental and waste issues concerning the production of smart clothes and wearable technology

M. TIMMINS, UK

Abstract: This chapter identifies some of the sources of textile waste, and examines some of the environmental and social effects of their manufacture, use and disposal in a situation where landfill and incineration, the traditional methods of waste disposal, are becoming less and less available. New legislation (Waste Electrical and Electronic Equipment (WEEE) Regulations 2004) concerning electrical and electronic waste disposal is outlined, with particular reference to the topic of Smart Clothes and Wearable Technology. In the latter part of the chapter, some broad guidelines for designers are offered.

Key words: textile waste, textile waste disposal, Waste Electrical and Electronics Equipment (WEEE) Regulations, end-of-life disposal of items of Smart Clothes & Wearable Technology (SCWT), life cycle considerations for design of items of Smart Clothes & Wearable Technology (SCWT).

16.1 Introduction

Any manufacturing process produces waste, and any product eventually, at the end of its life, becomes waste, too. Textile-based products are no exception, and over the ages ways have been found of minimising textile waste in the first place, and where possible recycling and reusing the waste, perhaps in very different products. These procedures have never been seen as being a particularly vibrant side of textile manufacture, but factors are emerging that bring the topic into sharper focus. One factor is the increasing difficulty of disposing of any kind of waste, not just textile-based items. Landfill and incineration, two main routes for disposal in the past, are increasingly difficult to use. Environmentally safe and socially acceptable landfill sites are becoming very scarce, and bulk incineration is largely unavailable because of air pollution problems. With the advent of smart clothes and wearable technology (SCWT) products, an additional set of new factors comes into play. Although, at present, production of these items is much lower than for traditional textile-based items, the production of SCWT involves incorporating non-traditional materials into, for example, items of clothing. These materials, most usually electronic systems, energy supplies, and interconnecting wires or fibre optic cables, represent a collection of materials now seen to have a particularly bad impact on the environment when landfilled or incinerated. From the middle of 2007, Europe-wide legislation demands that many items of waste electrical and electronic equipment must be collected and disposed of separately, and this legislation is likely to apply to at least some items of SCWT.

This chapter, then, identifies some of the sources of textile waste, and examines some of the environmental and social effects of their manufacture, use and disposal. The new legislation concerning electrical and electronic waste disposal is outlined, with particular reference to the topic of SCWT. In the latter part of the chapter, some broad guidelines for designers are offered.

16.2   Textile waste

Waste from textile manufacture arises from many sources, some of which are inherent in the nature of the raw materials used and some associated with inefficiencies in the manufacturing process. Some waste, particularly in the area of clothing and fashion, is caused because a flat two-dimensional fabric needs to be cut into irregular shapes so that the finished garment can be fitted to the three-dimensional human form. In addition to this manufacturing waste, there is a considerable problem with textile items, usually clothes, that have come to the end of their natural life. Whether this is because of accidental damage during use, garments becoming worn out through long use, or simply discarded because they are no longer in fashion, a mountain of waste textiles arises over surprisingly short times. It is estimated that some 10 million tonnes of textile waste from all sources is sent to landfill each year in the USA and Europe.[1] Amounts of waste of this magnitude represent a considerable wasted economic resource, and in addition may have a serious negative effect on the environment.

As far as fibre manufacture is concerned, if waste arising from inefficient use of energy and water is disregarded, there are two principal forms of manufacturing waste. The first is direct fibre waste, generated, for example from machine start-up or shut-down. This waste can sometimes be reintegrated into the fibre production stream. The second type is material that cannot be assimilated in this way. Under the best circumstances it can be raw material for another product – possibly by another manufacturer. If, however, these by-products cannot be used in this way, they usually end up as landfill.

For end-of-life textiles, most usually clothing, items with some residual life can be cleaned and used as clothing again. Some may be cut up and used as cleaning rags, or further shredded to make fibre-based fillings (and in some cases high grade paper). The remainder will once again be sent to landfill. Figure 16.1 shows the main sources and destinations of textile waste.

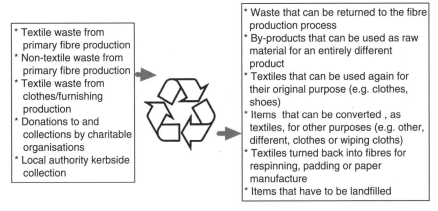

* Textile waste from primary fibre production
* Non-textile waste from primary fibre production
* Textile waste from clothes/furnishing production
* Donations to and collections by charitable organisations
* Local authority kerbside collection

* Waste that can be returned to the fibre production process
* By-products that can be used as raw material for an entirely different product
* Textiles that can be used again for their original purpose (e.g. clothes, shoes)
* Items that can be converted , as textiles, for other purposes (e.g. other, different, clothes or wiping cloths)
* Textiles turned back into fibres for respinning, padding or paper manufacture
* Items that have to be landfilled

16.1 Waste sources and destinations in textile production and use.

16.3   Environmental effects of the textile manufacturing process

Throughout the world, there is growing concern about the effects of human activities on the environment. Changing weather patterns, depletion of natural resources such as oil (doubly relevant to the textile industry as a source of fuel and as the raw material for most artificial fibres), rising sea levels and many other manifestations become more and more apparent. There is growing realisation that using natural resources at current rates is not sustainable, and that all processes need to be scrutinised to make sure they are as resource-efficient as possible; nowhere is this more apparent than in reducing waste. Large quantities of textile – or indeed any – waste cause many environmental problems in attempting to dispose of them. The options are few. Some textile items, as shown in Fig. 16.2, may be reused in one way or another. Some may be burned, but much textile waste is, in practice, put into landfill sites. Suitable sites are increasingly difficult to find, and soon become filled. In theory, many textiles will break down when buried in a landfill site – they are biodegradable, and so should eventually be turned back into their component chemicals. Unfortunately, however, the conditions inside most landfill sites are not very suitable for this bio-degradation. The required moisture and oxygen from the air are scarce deep inside a landfill site, and so the textile item just sits there, often for many years.

Any textile-related process, such as the manufacture of garments, has effects on the environment. Some are more marked than others, and these effects occur at every stage of the process. Some of them are fairly obvious, but others may be less so, as Table 16.1 shows. The same general

Table 16.1 Effects of the various stages of garment manufacture on the environment

Production stage	Environmental effect
Natural fibre growing (e.g. non-organic cotton)	Herbicides, pesticides and fertilisers damaging to watercourses and wildlife routinely used. Crop residues burned leading to air pollution.
Artificial fibre production (e.g. polyamide, polyester)	Production based on oil, a non-renewable resource. Accidental release of chemical intermediates can result in short- and long-term health problems.
Spinning into yarn	For industrial level production, machinery largely driven by electricity, the production of which uses non-renewable fuel, and produces gases contributing to global warming.
Conversion into fabric	Similar considerations to the spinning process apply here.
Dyeing and finishing fabric	Production of dyestuffs, which in industrial quantities are often poisonous, carries risks to humans and wildlife. Requires considerable amounts of water, with a risk of the used water polluting streams and rivers.
Garment production	Modern production methods can involve shipping part-finished garments across continents, using non-renewable fossil fuel.
Retailing	Garments are often transported over very long distances by transport using non-renewable fossil fuels and producing air-polluting traffic fumes.
Cleaning during the useful life of the garment	Garments are often washed more often than is strictly necessary, and using higher temperature water than is really needed. The wash water drained into the waste system needs to be purified, but may escape into watercourses. Increased wash temperatures mean more energy is used, adding to air pollution during its generation.
Disposal at the end of the garment life	Most garments end up as mixed waste in landfill. Possible landfill sites are becoming very scarce, and there is a considerable danger that rain washing through a landfill site can dissolve substances dangerous to the environment into watercourses.

considerations apply to textiles used for any other purpose, such as furnishing, upholstery, bedding, industrial fabric items and many others. Environmental effects in these other cases will vary to a certain extent depending on the item under consideration, but generally the descriptions in Table 16.1 will apply.

In addition to these effects on the environment, such industrial processes have effects on the human population. For example, standard cotton production methods may involve the use of land that a local population might use better for growing food. Frequent spraying of the cotton crop[2] to provide high enough yields can affect the health of local people as well as those directly involved in growing and harvesting the product, and there may be a temptation on the part of the growers to use child labour for harvesting. Growers of 'organic' cotton attempt to overcome these problems, but production levels are low at present.

Garment production may take place in circumstances where extremely low wages are paid, and at all stages of the process, careless handling of waste products can pollute the air as well as the ground and watercourses. This can mean health is affected, and in addition food sources (such as fish stocks) may become depleted or unusable.

Because of these issues, and because of some well-publicised cases of textile manufacturers being involved in major environmental and social scandals, many organisations have adopted procedures to minimise their effects on people and the environment. Some adopt rigorous regimes of accrediting their suppliers under a self-developed 'Corporate Social Responsibility' (CSR) programme, and others adopt an 'Environmental Management Scheme' (EMS) such as ISO14001, where performance is externally verified. These supply chain issues can assume considerable importance. For example, a garment producer may wish, for commercial reasons, to be able to reassure their clients that the products they sell have high environmental and social credentials. They will be unable to do this unless the item in question has been designed and produced to similarly high standards. For the producer to be able to give this guarantee of high standards, they in turn must place similar demands on their own suppliers. In some cases these guarantees will need to be backed by these formally recognised EMS schemes.

16.4 Particular issues for smart clothes and wearable technology

Other than the traditional recovery routes for textile waste outlined previously, scant regard is often given to the issues of waste and recycling for textiles reaching the end of their life. The usual route for disposal is landfill. Part of the problem is the nature of textile-based items, particularly (but not exclusively), garments. Garments are often constructed from many substances, not all of them textile in origin, and even where the components are textile-based they can be of very different fibres. A sportswear over-jacket could have a polyamide outer shell, an acrylic fleece insulating layer and a cotton lining. It is highly likely to include zip fasteners, either plastic or metal, buttons of similar materials, drawcords of yet another fibre, and even wooden toggles. It could have leather trim, and transparent plastic pockets to hold a ski pass. All of these need to be firmly fixed or sewn together, and although in principle each component could be recycled and reused, in practice, disassembly is a difficult process.

Figure 16.2 shows the hierarchy of ease of reuse/recycling of a selection of textile-related products. Smart clothes and wearable technology (SCWT) items, largely based as they are on textile foundations, are no different in these respects, but there is the extra complication of the electronic systems incorporated in them.

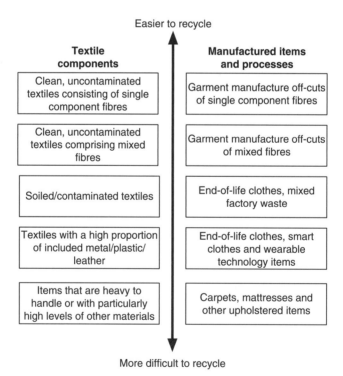

16.2 Ease of recycling of different textile products.

Other considerations may also apply; for example, heavily dyed textiles are sometimes more difficult to recycle, and waste containing fibreglass nearly impossible to use again. SCWT items occupy a place low in the this hierarchy; that is, they are more difficult to recycle, because of three inter-related issues. The first is simply that such items tend to be in garment form and garments are in the category of more difficult items to reuse/recycle anyway. The second is that where SCWT items contain electrical and elec-tronic systems – and most of them do – the metals incorporated add to the overall metal content of the item, making it more difficult to reuse/recycle the item. The third issue is that electronic-system based SCWT items should be considered in the light of new legislation (Europe-wide, but with inter-national implications) concerning the safe disposal of electrical and elec-tronic equipment. This legislation, and some of its possible effects on SCWT, will be outlined in Section 16.5.

One way of understanding the problems associated with disposal of these mixed substance items – whether they represent those with a SCWT content or not – is to consider how firmly the 'contaminant' is attached to the item. Those that are loosely attached can be removed – and perhaps reused – but those that are firmly attached and closely integrated into the item become difficult-to-remove. Some 'contaminants', of course, *have* to be closely integrated. Zips and other fastenings need to stay firmly in place, but others, such as garment trims, can be more loosely integrated. For items of SCWT, however, some choices may be possible at the design stage about the degree of integration that an electronic system is incorpo-rated into a garment. Consideration of these choices at this early stage can have a marked effect in helping to reuse, recycle and dispose of the item at the end of its life-cycle. As an example, within a garment, system interconnections such as wires could be woven or knitted into the fabric structure before the garment is made up. This would be regarded as a very high degree of integration, and would thus represent a situation where reuse/recycling would be made more difficult. A design where such interconnections could be added after a garment is constructed, possibly under an openable flap designed into the seams, would help eventual recycling because they could be more easily removed. This would repre-sent a much lower degree of integration. Power supplies are a particular case. Although it might be possible to closely integrate such items into a wearable item, and provide suitable recharging systems built in, at some stage they will need to be replaced. The additional fact that laundering batteries is inadvisable serves to point up the case for looser integration of power supplies.

As is often the case in the design process, there are some conflicts that need to be resolved in the matter of close/loose integration of technology items into SCWT. Some of the main factors are given in Table 16.2.

Table 16.2 Main factors in the integration of technology items into smart clothes and wearable technology

Factors tending to suggest closer integration	Factors tending to suggest looser integration
Retention of traditional appearance.	Where recycling/disposal of the item would be made easier.
Likelihood of casual interference from the user with the electronics system.	Frequent washing/cleaning required.
Likelihood of loss or damage to the device.	Power supplies needing to be recharged/replaced.
Appearance and/or functioning of the item would be unacceptably affected by attaching an electronic system.	Rapidly changing technology requiring electronic modules to be upgraded.

16.5   Waste electrical and electronic equipment from smart clothes and wearable technology

Implicit in the definitions of SCWT is the direct incorporation of electronic or electrical items into the very structure of the fabrics and other raw materials from which products are made. Whether these are closely integrated with the structure of the item itself or more loosely integrated, SCWT products will include substances that previously have not been used to any great extent in the textile field. Such substances provide desirable properties such as conducting (or resisting) electrical current, heat retention or cooling, allowing wire-free communications, or enabling electrical power to be routed from one part of a garment to another. They may, in the form of flexible solar panels, actually generate energy to provide power for more loosely integrated equipment such as mobile phones or medical monitoring apparatus.

Incorporating new materials into existing textile processes is, however, notoriously difficult. Textile machinery is finely tuned to existing raw material supplies – and even here different batches of materials may need significantly different machine settings. The incorporation of even small amounts of a non-traditional textile material can have a very great effect on process settings, quality and general manufacturing system productivity. But it is these non-traditional materials that have to be incorporated into SCWT products in order to meet the required characteristics.

In addition to these considerations, some of the substances used in traditional electronic and electrical manufacture have a clearly damaging effect on the environment, and EC-wide attempts are being made to deal with their safe use and disposal. The kind of electrical and electronic items that might be incorporated in SCWT products are also the type of article that is currently under the scrutiny of the EC, and hence by manufacturers in the UK.

New legislation designed firstly to restrict the use of certain traditionally used substances, and secondly to control legacy substances in already-manufactured items came into force in the UK in the middle of 2007. These Regulations ('Waste Electrical and Electronic Equipment (Producer Responsibility) Regulations 2004' and 'Prohibition of the use of Certain Hazardous Substances in Electrical and Electronic Equipment Regulations 2004') will bring into force EU Directives first published in 2002.[3] The first regulation (the WEEE regulation) is the one that has received most publicity, containing as it does a requirement for producers or importers of consumer equipment to arrange to take back end-of life electrical and electronic equipment for environmentally-sound and safe disposal. The second (usually called the RoHS regulations after the title used in the EU Directive) has received less attention in the media, since it can be seen as applying prohibitions to manufacturers making or importing electrical and electronic items, rather than the public buying products.

For WEEE, the Regulations apply to ten product types (see the full listing in Table 16.3), covering a wide range of consumer equipment. Product Types 3 (IT and Telecommunications), 4 (Consumer Equipment), 7 (Toys, Leisure and Sports Equipment) and 8 (Medical Devices) would appear to relate to the types of item that constitute SCWT, but product type 9 (Monitoring and Controlling equipment) could be relevant as well. If these five areas are relevant and the Regulations apply, then manufacturers or importers of SCWT items will have to make arrangements to recover at least 70% of end-of life product and then recycle at least 50%. This 50% is not half of the amount recovered, incidentally, but half of that originally produced. This need could well influence the initial design of the item, in just the same way that vehicle manufacturers need to design-in end-of-life considerations when producing cars. It should be emphasised, however, that the actual amounts of WEEE relevant SCWT items are likely, at least for the forseeable future, to be small compared with the manufacture of non-smart items.

As far as RoHS is concerned, the situation for SCWT is likely to be somewhat different. The RoHS regulations are designed to prevent the use of six substances recognised as particularly hazardous to the environment. These include lead – traditionally used in solder – as well as mercury and cadmium, often found in miniature power supplies. The hexavalent form of the element chromium and two fire retardants are on the list, and they are less likely to be found in items of SCWT. Their position may, however, also need to be considered in the context of the fact that chromium is used to make chrome leather, and can be used as a mordant for some dyes. The named fire retardants could be encountered in foam plastic. It is presently far from clear how chromium and the named fire retardants, in a non-electronic context, will relate to RoHS considerations, since RoHS applies to electrical and electronic equipment, not ordinary textiles. Some guidance may, however, be available through the EU Directive on end-of-life of motor vehicles, where chromium and certain named fire retardants are given consideration, and are being phased out.

The effect of RoHS on SCWT will be that manufacturers will not be able to supply electrical or electronic items containing substances on the list, and substitutes will have to be found by those manufacturers. The items affected are those included in the WEEE Regulations, but there are exceptions. One of these is 8 (Medical Devices), where, for example, lead-shielded garments for protecting cancer patients undergoing radiation treatment would be allowable.

As previously noted, the legislation is designed to prevent polluting substances being released into the environment. Since the legislation in the UK is derived from European legislation, similar legislation must be enacted throughout Europe, by mechanisms that will vary in different member states. The consequences, however, extend beyond the EC borders. Any relevant components and finished products imported from the rest of the world will also have to conform, or they will be refused import licences – and manufacturers across the world are unlikely to set up different production lines just for imports to the EC. It is worth noting in this context that China has introduced similar legislation to RoHS, but with even more stringent restrictions.

There is considerable discussion (and dissent) as to the scope and applicability of the Regulations, especially the WEEE Regulations, and clarification will take some considerable time. Although the Regulations are quite specific in many ways, borderline cases abound, and SCWT may be a case in point. A complicating factor in judging whether the WEEE Regulations apply in the area of SCWT is the concept in the Regulations of an item's 'primary purpose'. If the primary purpose of a smart item such as, for instance, a jacket, is to keep someone warm but has built-in entertainment hardware, then this might not be regarded as a WEEE item. This is because the primary purpose is to keep the wearer warm, and it would do so even if the embedded electronics ceased to work. On the other hand, a vest that has been designed to monitor bodily functions – heartbeat, breathing, temperature and so on – would have these measurements as its primary purpose, and consequently be regarded as WEEE at the end of its life. It is very difficult at present to predict how SCWT programmes and products will be affected. The effect may be minimal, but becoming informed of the issues could prevent problems in the future. Table 16.3 details the WEEE product categories, and Table 16.4 the RoHS prohibited substances.

Table 16.3 WEEE product categories

1. Large household appliances
2. Small household appliances
3. IT and telecommunications appliances
4. Consumer equipment
5. Lighting equipment
6. Electrical and electronic tools
7. Toys, leisure and sports equipment
8. Medical devices
9. Monitoring and controlling equipment
10. Automatic dispensers

Table 16.4 RoHS prohibited substances

Lead
Cadmium
Mercury
Hexavalent chromium
Polybrominated biphenyls (PBB)
Polybrominated diphenyl ethers (PDBE)

16.6 Developing a strategy for minimising environmental effects

As previously noted, all activities associated with manufacture of any product have an effect on the environment and the people in that environment – it is simply a matter of degree. For designers and others associated with SCWT, the aim should be to minimise these effects by taking action in two principal areas. These areas interconnect. The first is concerned with the specification and sourcing of the materials used in the product itself, and the other is in developing design and manufacturing strategies that allow for recycling, reuse and proper disposal of the items, or parts of the items. The influences to consider these topics will vary. Sometimes the pressures come from direct customer demands, and sometimes from personal beliefs and opinions of designers and manufacturers. In other cases, supply chain pressures will be the driver. In yet other cases, legislation demands better environmental performance. As previously noted, the influences of the WEEE and RoHS Directives may come into play when the electrical/electronic aspects of SCWT are considered.

In the first area, that of raw material acquisition, concerns will be around the sources of the materials to be used, both for the fabric and in the electronic systems incorporated in some way with the fabrics.

As far as the electrical/electronic systems are concerned, the RoHS directive will apply. This is the case since for all purposes, not just SCWT. Electrical/electronic systems and component producers seeking to import to Europe must supply 'RoHS-compliant' items, even though such producers may manufacture outside Europe.

For other parts of SCWT items, the question of raw material supplies is by no means as clear. Designers and manufacturers always need to take into account such traditional factors as appearance and wearability, both practical and aesthetic. The need for cleaning and possible garment repair are also important issues, and underpinning all of these is cost. But commercial and legal considerations may now have an additional effect for all wearables, not just SCWT items. Fibres from more sustainable sources, where environmental and social concerns are an important issue in their production, seem destined to move out of niche markets into mainstream ones, and designers and manufacturers will need to take on board this additional aspect.

The second principal area of concern for SCWT items is end-of-life considerations. This topic has traditionally been regarded as being of little concern to designers and manufacturers, except in those important cases where textiles are able to be shredded and reprocessed into other textiles (or different products). But disposal of waste in any form is now an international dilemma, and in general, end-of-life textile products make a significant contribution to the problem. Even for products lying outside the definition of SCWT, there is an imperative to consider how fabric-based items may be reused or recycled at the design stage, or at the very least have a minimal effect on the environment if they are finally landfilled or incinerated. Disposal of textile (and other) waste, whether before or after its use by consumers, is now in most countries a matter for increasingly stringent legal control. Design and manufacture to minimise both this pre- and post- consumer waste helps to minimise environmental problems – and, of course, can save costs. As far as SCWT items are concerned, there is the additional factor of the possibility of the item concerned being regarded as Waste Electrical and Electronic Equipment (WEEE). If this is so, under one of the provisions of the WEEE Directive in the UK (and similar provisions in other European states) manufacturers will have to set up schemes to receive back and recover SCWT items, and arrange for their safe disposal.

In order to begin to take into account this type of issue in the design of items of SCWT, possible actions that could be taken include:

- Develop an awareness not only of the physical and aesthetic properties of fibres and fabrics under consideration, but also of the environmental and social aspects of their production. Specify, where possible, those with the lowest effects on society and the environment, preferably those that have been independently certified as being so.
- Plan for end-of-life considerations for items of SCWT at the initial design phase, not as an afterthought at the end. Prefer solutions where electronic systems are loosely integrated into the item, so that disassembly, and hence disposal, is made easier.
- Keep abreast of emerging waste disposal legislation, particularly for SCWT items. Waste control legislation is likely to become more stringent as time goes on.
- Seek reassurances from suppliers/importers about compliance, particularly in terms of RoHS. Most electronic component and systems manufacturers have a RoHS compliant range, even if they manufacture outside Europe.
- Even where compliance with emerging requirements may be difficult or impossible, be able to show that the matter has been considered, and all possible steps have been taken taken to minimise the effects of producing the item.

16.7   References

1   Y YANG, 2006 *Recycling in Textiles*, Georgia Institute of Technology, Woodhead Publications, ISBN 1 85573 952 6.
2   UK ENVIRONMENT AGENCY, 2005 *Life Cycle Assessment of Disposable and Reusable Nappies in the UK*. Section 5.6 collects information from US sources about environmental impacts of cotton production.
3   UK DTI website http://www.dti.gov.uk/innovation/sustainability/weee/page 30269.html

Part IV
Smart clothing products

17

Smart clothes and wearable technology for the health and well-being market

D. BRYSON, University of Derby, UK

Abstract: This chapter looks at the range of smart technology that can be used to help support our personal control over our quality of life, health and well-being. The number of activity monitoring devices and systems is increasing at the same time as national and international bodies are placing more emphasis on well-being to reduce healthcare costs. This then leads on to examining how we will balance the relationship between the technology, social need, and acceptance by the end-user.

Key words: health and well-being, personal monitoring, activity, smart technology, healthy acceptability.

17.1 Introduction

What do we need to live or what do we need to live well or as we would like? Are we driven by need, necessity or market forces? This chapter presents an overview of the current state of smart clothing and wearable technology in the health and well-being market.

Smart clothing and wearable technology is at a transition point. A lot of things that were previously wishful thinking are now entirely feasible but have yet to become reality. The designer needs to be clear what is the target market for the technology and its level of user acceptability, taking into account that there may be different types of user from young to old, different levels of technical ability, different levels of dexterity for using the technology from the problems of large hands and fingers to arthritis.

The types of user are critical to understanding this market sector. Is the user the final recipient of the technology, the care giver or the organisation buying the technology? Is the technology being used by the end-user who has bought it for personal reasons or has it been issued or applied to the client or patient in a health or social care context? It has been assumed that the older recipient will accept whatever is applied to them or given to them; as this market goes on changing in ability to spend and they make personal lifestyle choices, so they are expecting more from technology and for their use to be specifically considered in the design.

This chapter looks at what we mean by quality of life, health and well-being, diagnostic and health monitoring, activity monitoring devices and systems, and the relationship between the technology, need, acceptability and use, leading into possibilities for the future. It represents very much a snapshot of current applications and thinking towards new developments; it is not exhaustive in terms of technologies or applications, but is looking at trends and thinking in many ways as we leave behind the 'Heath Robinson' visions of wearable technology for health and well-being (Roggen, 2007) and move towards technology that is available to everyone through retail outlets as part of this new and growing market.

17.2 Quality of life, health and well-being

There are a number of ways of talking about health. In analysing disability, medicine has moved from medical models to social models. There are technical specifications giving precise figures for blood pressure, heart rate, breathing rate, body mass index (BMI), temperature. More recently, words like wellness and well-being have become popular and also the idea that health is the absence of ill-health, disease or any other condition.

In many ways it is easier to define 'Quality of Life' in terms of someone who can have a disability, or a condition which means they are not healthy, yet have a good life and live independently, compared with someone who is 'Healthy' but has a very poor quality of life due to other social factors. In using these terms in relation to smart clothing and wearable technology, 'Quality of Life' and 'Independent Living', there is a general acceptance (at least in the design and technology communities) that design and technology are able to enhance the quality of life and extend or enhance the ability for all of us to live independently.

The parameters surrounding 'Quality of Life' have been addressed in government documents with a number of definitions, including that used by Meeberg – 'Happiness and a feeling of well-being will also result from Quality of Life. When one rates his or her life as having quality, one will concurrently have a sense of self-esteem and pride regarding his or her life. It must be noted that a confounding scenario seems to be apparent with each of these consequences of quality of life in that each can contribute to, as well as result from quality of life.' (1993, p. 32).

'A striking feature is that the notion of "independent living" is now being used for different age groups and for a broad range of problems (varying from young criminal offenders with complex backgrounds to mentally or physically disabled people). As regards the older age groups, the concept of *independence in later life* is freely and widely used, but not explicitly defined. Independence is a complex and subjective concept. It is generally regarded as having a positive impact on the quality of life of older people' (European Commission, 2006).

The World Health Organisation (WHO) (2002) defines independence as the ability to perform the activities of daily life with no or little help from others (living independently).

17.3   Diagnostic and health monitoring

The diagnostic and health monitoring sector has always been a heavy user of technology and has been a key driver in product development and use. More recently, however, with budget cuts and constraints across health and social care, there is a developing gap between what is possible and what is being used. An additional factor in this developing gap is the ability to fully utilise or understand the possibilities of technologies with governmental concerns looking at massive countrywide developments and infrastructure changes that are driven by information management needs rather than by the carer–patient interface.

Information and Communication Technology (ICT) can deliver great benefits but these benefits depend on the broad adoption of technology not just by super-users or the technologically aware. 'ICTs can have a massive impact on all aspects of healthcare, from delivering the information people need to lead a healthy lifestyle to providing new tools to design tomorrow's medicines; from making healthcare systems more efficient and responsive to providing "in the home" and mobile healthcare technologies' (European Commission, 2008).

There are a number of important distinctions in the use of smart clothing and wearable technology; their use in personal monitoring versus health or social care personnel or organizations, where monitoring may be subsumed by surveillance, the functional versus the entertainment or edutainment, life-saving versus life enhancing. Issues arise from the use of smart technologies as tools within healthcare, e.g. the use of GPS technology for monitoring clients in nursing homes (Parnes, 2003) or for children (Wherifone, 2008). Systems that inform the GP of a patient's health status require considering the issue of consent and freedom of the individual where the use of the technology may limit or curtail the freedom of the wearer. The opposite has been the suggestion of using technology to overcome older people's fear of crime by overcoming the information gap between fear and actuality but the paper by Blythe (2004) also recognises the role of 'the politics of surveillance'.

The use of smart clothing and wearable technology in actual healthcare settings is limited with conventional wired technology still being used. Within a High Dependency Unit (HDU) following surgery, there is a massive surfeit of wired devices that should be using wireless technology. For example, post-operative monitoring: blood pressure cuff with cable, pulse oximeter on finger with cable, electrocardiogram (ECG) with cable, all connecting into heavy, conventionally powered machines. These are only three devices, but when you then add in the other connections a patient may have – intravenous line, heart line, epidural – there are a lot of wires and tubes for a patient to move or be moved. When there are that many connections already, consideration for extras that may be useful, such as glucose (Lymberis, 2006) or pressure monitoring, are unlikely to be added. If these devices could become wireless, using bluetooth or radio frequency (RF) transmissions linked into a central control computer, the amount of technology required reduces rather than increases. An example of clothing incorporating this level of functionality is the Sens Vest (Knight, 2005) measuring pulse, breathing, ECG and heart rate.

Van Langenhove (2004) has examined the application of smart textiles including the measurement of heart rate, ECG, respiration and notes that: 'The development of smart textiles starts to come at cruise speed. A part of the new materials and structures has already reached the stage of commercialisation; a much larger part, however, is still in full development or still has to be invented even. This applies especially for the very smart textiles. This phase is to be reached by 2010, so at medium term' (p. 72).

Similarly, Chensfold (2006) talks about new developments in an article headed '"Smart" Clothes: Prelude to Surge?' 'Some day apparel designers may turn us all into Inspector Gadget, allowing us to take multitasking to the ultimate level by essentially making us walking computers, though at this early stage innovations are much simpler' (p. 1).

The cycle of development and implementation is longer in diagnostic and therapeutic areas due to regulation and cost but, as we can see in the following sections, the early stages of commercialisation have been reached and the next phase of mass production is commencing.

Similarly, Barnard (2004), writes, 'After four hundred years of delivering health care in hospitals, industrialized countries are now shifting towards treating patients at the "point of need". This trend will likely accelerate demand for, and adoption of, wearable computing and smart fabric and interactive textile (SFIT) solutions' (p. 49).

The stage has been reached where we do not need more research, we need more products.

17.4   Activity monitoring devices and systems

An aspect of life that is in the news regularly is the link between activity and health. This can be seen as a Europe-wide issue, with the WHO's document on Physical Activity and Health in Europe stating in its foreword: 'Physical activity is a fundamental means of improving physical and mental health. For too many people, however, it has been removed from everyday life, with dramatic effects for health and well-being.

Physical inactivity is estimated to account for nearly 600000 deaths per year in the WHO European Region. Tackling this leading risk factor would reduce the risks of cardiovascular diseases, non-insulin-dependent diabetes, hypertension, some forms of cancer, musculoskeletal diseases and psychological disorders. In addition, physical activity is one of the keys to counteracting the current epidemic of overweight and obesity that is posing a new global challenge to public health' (Cavill *et al.*, 2006, p. viii).

And nationally in the foreword of the UK Government's 'Choosing Health' White Paper – 'An active lifestyle is key to improving and maintaining health. However, at present only 37% of men and 24% of women are sufficiently active to gain any health benefit. Three in ten boys and four in ten girls aged 2 to 15 are not meeting the recommended levels of physical activity. The challenge we now face is to encourage more people to become more active' (Department of Health, 2005, p. 4).

However, neither of these papers looks at the application of smart clothing or wearable technology for promoting health or activity. They are setting out what needs to be done, not how it can be done. The means to encourage activity that they mention are just to say that more activity is needed, more sport in schools, not how to stimulate the growth in activity. Applying smart technology and clothing to supporting an active lifestyle is not a route for all, but it can support it by providing not only monitoring, but also feedback direct to the end-user.

This avenue to promote healthy living has been examined by Meinander (2004) and Dishman *et al.* (2004, p. 179). – 'Novel approaches range from implanted and wearable technology to distributed networks embedded in the living environment to in-home delivery of health services from remote locations. These technologies target several pressing health needs of older adults, including promoting physical function and social interaction, facilitating early diagnosis, enabling self-monitoring of health status, and assuring adequate treatment. At the personal level, they contribute to preserving older adults' health and well-being and support their oft-stated preference for remaining independent as long as possible' (p. 179).

Wearable devices that monitor activity have become very popular across the range of ages in the marketplace from research use for children to stimulate activity, e.g. Fizzees (Futurelab, 2007) and Tagaboo (Konkel, 2004); to Nike and iPod combinations in the sportswear market; to the use of pedometers and GPS in the older market for personal monitoring to support improvement in fitness.

'Fizzees (Physical Electronic Energisers) is a prototype project that enables young people to care for a "digital pet" through their own physical actions. In order to nurture their digital pet, keep it healthy and grow, young people must themselves act in physically healthy ways' (Lee, 2006b). For more information about fizzees there is an online presentation about its use on SlideShare at http://www.slideshare.net/Dannno/futurelab-fizzees-project/.

'Tagaboo is an interactive game for two or more children that is based upon wearable radio frequency identification (RFID) technology. Tagaboo combines aspects from traditional athletic children's games with tagged physical objects that are bound to different sounds and behaviors' (Konkel, 2004).

Mobile phone technology is linking into this monitoring trend following on from the Apple/Nike collaboration 'Nike + iPod' (Apple, 2008), as mobiles become a more all-in-one device for music, notes, calendars. For example, the Adidas Samsung miCoach linking with Adidas for 'My run, My rhythm, miCoach' (Samsung, 2008).

The key features of these devices are the integration of technology in promoting a lifestyle change:

(i)   Garment containing means of monitoring, whether heart rate or accelerometer, to detect movement with connection to mobile device, e.g. track shoes or recent Nike+ Sportband (Nike, 2008).

(ii)   Mobile device which, in many circumstances, is now a mini-computer, and

(iii)   Connection to your personal computer either via a program or, more likely, through a web interface using Flash, either using the phone's Bluetooth capability or a detachable universal serial bus (USB) cable (Nike, 2008; Samsung, 2008).

Accelerometer data that has traditionally been collected from wrist-mounted accelerometers (Knight, 2007) can now be collected from your iPhone or iPod touch, both of which have integrated accelerometers that detect when you rotate the screen from vertical to horizontal. Similarly, these devices, even without a Global Positioning System (GPS), have the capability through wifi (iPod touch) or cell tower locations and wifi (iPhone) to locate where you have been and therefore the effort required if you walked downtown to the office or up hill when you went out to lunch. These sensors, together with more advanced software possible with the advent of manufacturers software development kits (SDK), mean even your phone could be monitoring your daily activity without your having to wear your trainers or pedometer. This functionality will extend monitoring from a sport-related to an everyday activity. This will lead to such devices, as Ash-brook and Starner phrased it, to 'act as intelligent agents in everyday life and to assist the user in a variety of tasks, using context to determine how to act' (2003, p. 1).

The key change from early wearable technologies to the present developments are the difference in size and capability of the mobile device that links the technologies together. The limitation is becoming the software development, integration of technologies and the marketability, not the bulkiness of the computer or feasibility of using the technology.

17.5   Technology, need, acceptability and use

There is a key relationship between the 'technology', whether it is large or small, bulky, easy to use, its actual application, i.e. medical, health or personal, and the acceptability of the device(s). This relationship is represented in Fig. 17.1. In a medical context, a patient will accept large bulky technology as the need is great; in health surveillance, for example, 24-hour blood pressure and heart rate monitoring, a certain amount of bulk is acceptable, but when one gets into personal uses such as activity monitoring systems, or where they are being equally used for entertainment or edutainment, then the lighter the better.

Acceptability also has a relationship with appearance; if something looks old-fashioned and bulky, it is less likely to be used. If, however, the item is fun, has a purpose and can be used quickly with little need for learning how to use the technology, it will be used and, in medical terms, compliance will be greater.

The trend within technology acceptance and size is for devices to be made smaller and less bulky with time. Compare the Sony Walkman, which first appeared in July 1979 with its cassette tapes, with an iPod shuffle or iPod mini. The technology is smaller and yet its capability is greater. However,

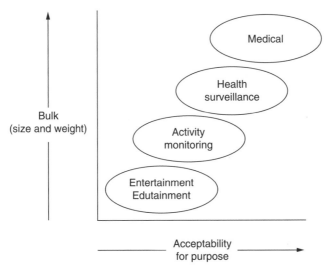

17.1 Relationship between bulk (size and weight) and the acceptability of the device to the end-user when meeting different needs.

there is a point at which devices requiring manual movements or input cannot be made smaller as the size of our fingers cannot decrease. In someone with large hands, the tip of the thumb can cover 6 digits on a mobile and the very tip of the finger has to be used to avoid hitting two numbers at once. This is where technology will come up with answers through alternative display and input systems, through heads-up displays (HUD), and input using the eyes. The research and applications are not likely to come from the medical research side but more likely via the games and entertainment industries, which are already using such devices to simulate complex activities as flying a plane or driving a car (Waters, 2008). The EmotivEPOC as their website explains, is 'based on the latest developments in neuro-technology, Emotiv has developed a revolutionary new personal interface for human computer interaction' (Emotiv, 2008).

The key features are all in place; it is now a question of systems integration leading to the creation of marketable products. An important part of this lies in software development and the ability of devices to work independently but also linked to networks or through a direct computer connection.

17.6   Future trends

Whilst everyone is looking for flexible screens or smaller/better screens for viewing, it may be the case that the next generation of devices need a HUD. You are about to go out for a run; your mobile/iPhone/PDA/iPodtouch is in your pocket and is linked by Bluetooth to your glasses with integral headphones, which allow you to see your planned route. As you run along, you can see details of how far you have gone from the start, overlaid on what you are looking at as you go through the streets or park. You can change music without having to use your hands via the built-in eye tracking software that knows what you are looking at in the heads-up display and scrolls to what you want to listen to.

You take the same equipment with you on a trip to go climbing, using the HUD for navigation whilst driving, then, before climbing, you switch to the climbing guide book, which overlays your route on the cliff face. As you are looking at the cliff, you can see overlaying the rock surface the climbs from very difficult through to severe and where they merge and separate; it also has marked recent changes in rock surface and dangerous areas, where grip is difficult due to water flowing down. Once you have completed your week's activities, everything is downloaded onto your personalised website and it tracks against your training plan.

Currently in the smart clothing and wearable technology market there is a problem with continuity as new ideas are seen as being fashion led rather than functional. There is a need for smart clothing to be around not just when fashion dictates but for the technology to be present long enough for it to mature and develop. Continuity is also a key factor in the healthcare market, which is reliant on a continuity of availability not fashion's here-today-gone-tomorrow way of working. As Lee emphasised in his talk in 2006 on Universal design, 'Non-trend based. We specifically didn't want to get into fashion aesthetics because we want our products to sell for 10/15 years after we put all the investments into it.' (Lee, 2006a).

Some areas are going nowhere very fast; fashion has got hold of quite a few technologies and used them creatively within an artistic context that garners publicity and funding; researchers use the technology to follow-up and promote its use and look closely at feasibility but always testing it in trials not taking products to market. Technologies like the 'Fizzees' should be widely available after they have proven effective. It is time for smart clothing and wearable technology to leave the laboratory and become mainstream.

The other trend should be for manufacturers to more widely examine inclusive design within the health and well-being market. Devices should work for everyone, not just the able bodied. Input devices like the EmotivEPOC and similar devices for input are useful to a wider population than just gaming. Other devices, such as 'Twiddler' (HandyKey, 2006), which has been around for 16 years, are great for the average user but could not be used by someone with rheumatoid arthritis. The next generation of technology has to be delivered using universal design principles (Lee, 2006a), so that all users can not only use them but will be willing to use them as part of their everyday lives.

17.7   Sources of further information and advice

Websites

http://www.innovation-for-extremes.org/
Recent publications on wearable and healthcare
http://www.seeingwithsound.com/newpubs/wearable_healthcare/

Podcasts

http://www.geekbrief.tv

17.8   References

APPLE (2008) *Nike + iPod Tune your run*. [URI http://www.apple.com/ipod/nike/, acccessed May 12th 2008]

ASHBROOK, D. and STARNER, T. (2003) Using GPS to learn significant locations and predict movement across multiple users. *Pers Ubiquit Comput* (2003) 7: 275–286.

BARNARD, R. and SHEA, T. (2004) 'How Wearable Technologies Will Impact the Future of Health Care.' In: Lymberis, A. and De Rossi, D. (eds) *Wearable eHealth Systems for Personalised Health Management. Studies in Health Technology and Informatics*, Amsterdam: IOS Press, 108: 49–55.

BLYTHE, M.A., WRIGHT, P.C. and MONK, A.F. (2004) Little brother: Could and should wearable computing technologies be applied to reducing older people's fear of crime? *Pers Ubiquit Comput* 8: 402–415.

CAVILL, N. SONJA KAHLMEIER, S. and RACIOPPI, F. (EDS) (2006) *Physical Activity and Health in Europe: Evidence for Action.* Copenhagen: World Health Organisation.

CHENSFOLD, C. (2006) 'Smart' Clothes: Prelude to Surge? *Apparel Magazine.* 47 (10): 1–3.

DEPARTMENT OF HEALTH (2005) *Choosing activity: A physical activity action plan.* London: Department of Health [URI http://www.dh.gov.uk/en/Publicationsand-statistics/Publications/PublicationsPolicyAndGuidance/DH_4105354, accessed May 12th 2008]

DISHMAN, E., MATTHEWS, J. and DUNBAR-JACOB, J. (2004) 'Everyday Health: Technology for Adaptive Ageing'. In: Pew. R. Van Hemel, S.B. (eds). *Technology for Adaptive Ageing.* Steering Committee for the Workshop on Technology for Adaptive Ageing, National Research Council National Academies Press.

EMOTIV (2008) EmotivEPOC [URI http://emotiv.com/INDS_3/inds_3.html, accessed May 19th 2008]

EUROPEAN COMMISSION (2006) *User Needs in ICT Research for Independent Living, with a Focus on Health Aspects.* Institute for Prospective Technological Studies, Technical Report EUR 22352 EN [URI http://ftp.jrc.es/eur22352en.pdf, accessed 25th March 2008]

EUROPEAN COMMISSION (2008) *Information can save your life. Europe's Information Society*, Information Portal. [URI http://ec.europa.eu/information_society/qualif/health/index_en.htm, accessed 25th March 2008]

FUTURELAB (2007) *Fizzees* [URI http://www.futurelab.org.uk/resources/documents/project_reports/mini_reports/fizzees_mini_report.pdf, accessed 25th March 2008]

HANDYKEY (2006) Twiddler 2 [URI http://www.handykey.com/site/twiddler2.html, accessed May 19th 2008]

KNIGHT, J.F. BRISTOW, H.W. ANASTASTOPOULOU, S. BABER, C. SCHWIRTZ, A. and ARVANITIS, T.N. (2007) Uses of accelerometer data collected from a wearable system. *Pers Ubiquit Comput* 11: 117–132.

KNIGHT, J.F. SCHWIRTZ, A., PSOMADELIS, F., BABER, C, BRISTOW, HW and ARVANITIS, TN (2005) The design of the SensVest. *Pers Ubiquit Comput* 9: 6–19.

KONKEL, M. LEUNG, V. and ULLMER, B. (2004) Tagaboo: A collaborative children's game based upon wearable RFID technology. *Pers Ubiquit Comput* 8: 382–384.

LEE, A. (2006a) What does universal design do for us as a company? Innovation through Inclusive Design – The annual symposium of the Helen Hamlyn Research Associates Programme 2006. [URI http://www.designcouncil.org.uk/en/Design-Council/Files/Podcast-Transcripts/Keynote-speaker—Alex-Lee/, accessed May 19th 2008]

LEE, T. (2006b) Physical inactivity and childhood obesity. Is technology to blame, or could it have some answers? FutureLab, Web articles. [URI http://www.futurelab. org.uk/resources/publications_reports_articles/web_articles/Web_Article474, accessed 25th March 2008]

LYMBERIS, A. and GATZOULIS, L. (2006) Wearable Health Systems: From smart technologies to real applications. *Conf Proc IEEE Engineering in Medicine and Biology Society.* 1 Suppl: 6789–92. [URI http://embc2006.njit.edu/pdf/1049_ Lymberis.pdf, accessed 25th March 2008]

MEEBERG, G.A. (1993) 'Quality of Life: A Concept Analysis.' *Journal of Advanced Nursing,* Vol. 18, No. 1, p. 32.

MEINANDER, H. and HONKALA, M. (2004) 'Potential Applications of Smart Clothing Solutions in Health Care and Personal Protection,' In Lymberis, A. and Rossi, D. (eds) *Wearable eHealth Systems for Personalised Health Management: State of the Art and Future Challenges.* Amsterdam: IOS Press.

NIKE (2008) Nike+ Sportband. [URI http://nikeplus.nike.com/nikeplus/, accessed May 12th 2008].

PARNES, R.B. (2003) *GPS Technology and Alzheimer's Disease: Novel Use for an Existing Technology.* Ageing and health. [URI http://www.cogsci.rochester.edu/u/ kautz/ac/NewsArticles/HealthGate/GPS%20Technology%20and%20Alzheimer' s%20Disease%20Novel%20Use%20for%20an%20Existing%20Technology%2 0CHOICE%20For%20HealthGate.htm, accessed 25th March 2008]

ROGGEN, D. ARNRICH, B. and TROSTER, G. (2007) *Wearable health and life-style assistants. Technology in support of well-being* [URI http://www.eurescom.eu/message/ messageMar2007/Wearable-health-and-life-style-assistants.asp, accessed May 16th]

SAMSUNG (2008) *Samsung Mobile.* [URI http://uk.samsungmobile.com/micoach/, accessed May 12th 2008]

VAN LANGENHOVE, L. and HERTLEER, C. (2004) Smart clothing: A new life. *International Journal of Clothing Science and Technology* 16 (1/2): 63–72.

WATERS, D. (2008) *Brain control headset for gamers* [URI http://news.bbc.co.uk/1/hi/ technology/7254078.stm, accessed May 19th 2008]

WHEREIFONE (2008) [URI http://www.wherify.com/wherifone/kids.html?page=kids, accessed 25th March 2008]

WHO (2002) *Active Ageing: A Policy Framework.* World Health Organization, 2002. [URI http://whqlibdoc.who.int/hq/2002/WHO_NMH_NPH_02.8.pdf, accessed 25th March 2008]

18

Smart clothing for the ageing population

J. McCANN, University of Wales Newport, UK

Abstract: This chapter looks at the potential for the application of wearable technologies in an area of the market that has been largely neglected by designers; the design of functional clothing for the rapidly growing ageing community. It looks at the development of smart functional clothing and wearable technology that may promote optimal, safe exercise and directly enhance the health and well being of the design-aware active ageing community.

Key words: active ageing, design, real-world needs, smart functional clothing, wearable technology, optimal exercise, well being.

18.1 Introduction

This chapter looks at the potential for the application of wearable technologies in an area of the market that has been largely neglected by designers; the design of functional clothing for the rapidly growing ageing community. Clothing is a major contributor to how people define and perceive themselves and is a necessary part of their everyday lives. Studies of the clothing needs of the ageing population suggest that stereotypes – the old as typically impoverished and unconcerned with appearance – are being challenged. As a group, the elderly have much in common with younger consumer groups, and their self-perceptions may offer a more reliable predictor of needs than chronological age (Lackner, 1998). Nevertheless, there is a tendency to target either 'everyone over 50 years' or the needs of those with identified 'medical conditions' (Williams, 2004). What is lacking is stylish and comfortable clothing for active members of older age groups who do not suffer from such conditions.

Transient fashion ranges are not generally geared to the physiological demands of the changing older body, resulting in clothing that is often uncomfortable due to factors such as inappropriate fit, styling, proportion and weight. Smart garment design is becoming increasingly sophisticated in the areas of performance sport, an early adopter of textile and technology innovation, but is targeted primarily at athletes and the youth market, and has not been adapted and designed in a suitable format for the benefit of older users. Medical devices have been developed for 'ill people' with little aesthetic appeal and with data feedback that may be difficult for the wearer to read and understand. Product development teams must embrace the design requirements of older wearers in terms of human physiology such as sizing and shape, the ergonomics of movement, predominant postures, thermal regulation, moisture management, protection and the psychological 'feel good factor'. Clothing may be difficult to take off and put on, fasten and have compromises in relation to quality and aftercare.

We are at the beginning of a new industrial revolution with the merging of technical textiles, wearable electronics and ICT. Wearable technologies, monitoring body signs such as heart rate, temperature, respiration, activity levels and location, may subsequently, over time, provide sufficient data to detect behavioural changes. Feedback may provide significant new understanding and alert the user, suggesting affirmative action, giving individuals the potential to ensure their own continued wellbeing (Nugent, 2007). Such advances could be utilised to promote health and wellbeing but may not be readily accepted by some older users due to badly designed user-interfaces that have small controls or displays that may prevent someone with a minor impairment from using them effectively. Little has been done to address physical and cognitive limitations when developing these new wearable products and related services to ensure that they are appropriate to the real-world needs of older people.

A new generation of older people, the 'Baby Boomers', has been accustomed to making choices in the design of their clothing and accessories throughout their lives. Within this period, fibre-driven textile development has embraced stretch, lightweight insulation, knitted fleece constructions and waterproof and 'breathable' protection with user expectations and requirements developing accordingly. There is potential for key cultural, social and behavioural limitations of the everyday lives of the active ageing community to be addressed by the use of a multifunctional technology-enabled garment layering system. This concept demands that design practitioners develop clothing, enhanced with unobtrusive assistive technology and appropriately designed technology interfaces, for older users who will willingly wear and enjoy them (Armitage, 2007). Set in the demographic context, where overall numbers of the over 60s are predicted to double to around 30% over the next decades, the development of core garment products for active ageing is now being recognised by the 'high street' market leaders as a key factor for serious consideration in their future market share.

18.2   Identification of older user needs

18.2.1 Collaborative design

'Some great brands have come about by companies treating their customers as people, not as data sets' (Wolfe, 2003, p. 31). This has been the culture in the performance sportswear sector which has adopted more of a product design approach in addressing user needs. As the ability to integrate innovative technologies into older people's lives becomes more prevalent, there is a danger that developments become both problem focused and technology led (Ryan, 2007). The alternative perspective is that innovations in design and technology should be led by the aspirations, desires and everyday needs of the end-user. A collaborative design approach involves users in requirements capture, understanding and specifying the context of use, design specification, and design and prototype development and evaluation. The psycho-social, lifestyle, health criteria and motivational issues that influence personal choice in clothing, connectivity and daily life must be established in consultation with an identified individual or target group (Ryan, 2007). Contextual information must be gathered on older people's existing daily living environments, including support systems, with a complementary assessment of literature, documents and other publicly available information sources (Ryan, 2007). Older-user design requirements must be assessed in terms of their expectations of the intelligent functioning of the whole 'clothing system' (Ryan, 2007; Nugent, 2007). Older users should be consulted throughout the technical design process in specifying and assessing the prototypes being developed. Evaluation should involve users in assessing the clothing and its functionality in their own preferred environments. The outcome of user evaluation will inform subsequent stages in design and technical development, resulting in the functionality of the system being improved. Collaborative engagement with older users will centre on matching individual needs, desires and expectations with a set of non-restrictive, supportive and wearable lifestyle-enhancing and monitoring facilities and associated services (Ryan, 2007).

18.2.2 Requirements capture

An existing user-needs driven design methodology may be elaborated to suit the needs of the active ageing. A hierarchal process 'tree', originating in the author's previous work in her 'Identification of Requirements for the design development of Performance Sportswear' (McCann, 2000), addresses a breadth of technical, functional, physiological, social, cultural and aesthetic considerations that impinge on the design of clothing that is intended to be attractive, comfortable and fit for purpose (see Chapter 3, Fig. 3.2). This tree of requirements uncovers a breadth of topics and sub-issues to guide the design development process for performance sportswear. A new layer of issues may be added to this tree to embrace the emergence of smart materials and wearable technology. Major topics have been organised under the areas of 'Form', that embraces aesthetic concerns and the importance of respecting the culture of the end-user, and 'Function' that embraces the generic demands of the body and the particular demands of the activity.

While the areas of investigation are organised under the main headings of 'Form' and 'Function', 'Commercial realities' is recognised as a topic with major impact on bringing innovative products to market for the emerging active ageing consumer. New strategies in routes to market are needed for the promotion and sales, customisation, maintenance, servicing and disposal of a hybrid mix of emerging wearable technologies for the changing demographic of older consumers.

18.3   Commercial realities

18.3.1 The new consumer majority

The 'New Consumer Majority' emerged in 1989, as adults of 40 and older became the adult majority for the first time in history (Wolfe, 2003, p. 8). Now in their 60s, this community encompasses a breadth of users with contrasting lifestyle requirements. Some are retired or semi-retired and some are setting up new businesses. Some are grandparents and/or are continuing to care for children, and many are carers of elderly relations. These 'Baby Boomers' will add about 2 million extra older people to those aged 70 and over in the 2030s (Metz, 2005, p. 40). The 1946–1950 Baby Boomers, who were born in post-war austerity with rationing and selective education, benefited from opportunities in education opening up in the 1960s and entered the labour market as the economy was beginning to prosper. Baby Boomers are now beginning their retirement and many hope to continue to lead an active lifestyle, with increased time for leisure activities. 'This is the only adult market with realistic prospects for significant sales growth in dozens of product lines for thousands of companies' (Wolfe, 2003, p. 21). However, 'thus far, marketing remains rooted primarily in the materialistic values that generally hold the most sway over people in the pre-middle-age years of adulthood. Because of this, many members of the new Consumer Majority feel marginalised by companies and their marketers' (Wolfe, 2003, p. xv). 'As the centre of gravity of society shifts away from the young, and the new old become included, the prospects of the continued dominance of the youth culture are beginning to look rather uncertain. Marketing orientated towards the young may therefore be based on increasingly shaky foundations' (Metz, 2005, p. 32). 'Older markets are not so much additional markets in the total scheme of things as they are replacement markets' (Wolfe, 2003, p. 210).

18.3.2 Segmentation

'The older you are, the more you become you. As you become your own person, you become less classifiable as being one type of customer than

another' (Wolfe, 2003, p. 37). The correlation between age-based segmentation and behaviour is by no means straightforward; the 'National Health
Service Framework for Older People' (Metz, 2005, p. 52) identifies three
stages within the life course defined by health status, the first as 'Entering
old age; completed career in paid employment and/or child rearing but still
active and independent'(Metz, 2005, p. 56/57). Another model depicts five
segments of old age as: early mature years (50–54); full bloomers (55–59),
pre-retirees (60–64), prime retirees (65–75), seniors (74+). In relation to the
phenomenon of cognitive age, 'older people see themselves as 10 to 15 years
younger than their chronological age' (Wolfe, 2003, p. 325). 'Meaningful
segmentation is more challenging because they are more 'individuated'. In
one shopping context an older person may fall into one context while falling
into another segment in another shopping context' (Wolfe, 2003, p. 36).
Close ties between degrees of life satisfaction and cognitive age have been
defined as the 'Feel-age', the age one feels independent of physiological
conditions, the 'Look-age', the age one believes they physically appear to
be, the 'Do-age', that corresponds to associations of certain activities with
certain age groups and the 'Interest-age', that corresponds to associations
of certain interests with certain age groups (Wolfe, 2003, p. 325). 'Lifestyle
segmentation is particularly relevant for leisure activities, holidays and
travel, where the service offered can respond to the positive attributes of
individuals, their elective choices as opposed to their unavoidable needs'
(Metz, 2005, p. 62).

18.3.3 The performance sports sector

In January 2008, the international sports trade event, ISPO, presented 'Best
Ager', a market study that has identified the real potential for the development of sports lines for the active ageing (Bieker, 2008). 'Best Ager' describes
the target group of people aged 50 plus but deliberately ignores an age limit
as this group represents a way of life rather than a time span within which
sports and activity play an important role. ISPO predicts that 'this topic is
not short-lived but it is based on a changing society.' Constituting one-third
of the German market, the study found that 'eleven million potential sports
customers are not addressed by the sports industry'. The German market
is said to be comparable to that of the UK. This group is not restricted
financially and feels free with regard to buying decisions. 'Their broad
experience has made them critical about new offers. Style codes and advertising slogans don't impress them much.' Statistics identify that only an
approximate 30% of older people participating in sports find and choose
suitable products in specialised sports retail outlets with only 7% admitting
to understanding the product. Initial characteristics found in the survey
indicate that quality is more important than price and that sales staff should

be friendly and recognise customer competence. Proprietary articles that promise quality are more important than bargain prices. 'Best Agers' feel much younger than they are and, as a result, people in sports marketing tend to believe that the active ageing will be attracted by the same offers in relation to communications and advertising as younger people. This is not the case and many older consumers criticise the fact that advertising is instigated by younger generations, who do not understand their wishes and needs (Bieker, 2008). The term 'The Third Age' 'captures the growing recognition of the opportunities open to people after their main period of working life is concluded' (Metz, 2005, p. 27). This group is motivated by sport to promote a healthy lifestyle with individual or group exercise for fitness as prevention of ill health.

18.4   The culture of the user

18.4.1 Self-actualisation

'By age 60, people generally, there are exceptions as always – become relatively impervious to peer influence on their buying behaviour, especially those who are well along the path of self-actualisation' (Wolfe, 2003, p. 200). The weakening of peer influences and of urges to impress others, free people to make life-changing decisions. 'People begin a search for the real self after years of catering to the needs of the social self' (Wolfe, 2003, p. 45). The focus changes from becoming someone to being someone. The inner self, long submerged by an outer-world-directed agenda, aches for a simpler life. The quest for life satisfaction in full shifts progressively away from a focus on things to a focus on experiences. 'People do not willingly buy products that make them look bad or that are in conflict with their self-image' (Wolfe, 2003, p. 200). To make sure that they look good to others, people in the first-half markets depend on cues issued by their peers. However, as people move into mid-life and beyond, looking good to others, at least in the sense of making favourable impressions, begins to be less important. 'For the first time ever, most adults are in the years when the forces of self-actualisation needs exert decisive influences on lifestyle aspirations, buying decisions, and overall consumer behaviour' (Wolfe, 2003, p. xiii).

18.4.2 The influence of fashion

The Baby Boomer generation has grown up alongside rapid developments in textile and garment technology. In the 1940s, man-made fibres were launched such as nylon, acrylic and polyester; and synthetic dyes. The Second World War prompted women to wear trousers. Post-war, the

American designer Claire McCardell introduced simple casual wear in contrast to Dior's feminine 'New Look'. The 1950s saw the use of the new fibres in lightweight, easy-care fabrics, aided by the introduction of the domestic washing machine. The emphasis of fashion turned from the sophistication of couture to the casual wear of the new youth market and the use of denim. Art school trained designers emerged in the 1960s promoted by fashion photographers using young models such as Twiggy. The 1970s embraced the contrasts of hippies, power dressing, the advent of designer jeans and flamboyant men's styling (Jenkyn Jones, 2005). Bonnington's ascent of Annapurna, shown on colour television, promoted outdoor sport (Lack, 1992) alongside the advent of waterproof breathable membranes and insulating fleeces. The wearing of black, both in punk and Japanese designer fashion, began in the 1980s and the growth of international high street labels. This decade saw the influence of sportswear in menswear and in 'dress down Fridays' in the workplace, while sports fashion emerged in skiwear and subsequently snowboard and skate ranges. The importance of branding and designer labels grew in 1990s in conjunction with deconstructed styles and an awareness of ecologically friendly fibres (Jenkyn Jones, 2005). The recycling of polyester fibres was introduced for knits, weaves and trims throughout the sports layering system, by companies such as Vaude and Patagonia. The year 2000 saw a revival of craft techniques contrasted with, most recently, the emergence of smart textiles and wearable technologies alongside the digital revolution in manufacturing techniques.

Increasingly, global media exposure of designer fashion and quick-response sourcing of mass-market fashion have resulted in greater freedom of choice and less emphasis on the rules of convention (Jenkyn Jones, 2005). This choice is primarily restricted to the youth market in respect of shape, proportion and fit. 'Designer labels are about making social statements, and as people move into and through the second half of life, interest in designer labels falls off because they are not as compelled to make social statements by their brand choices' (Wolfe, 2003, p. 93). Purchasing behaviour of older consumers may be related more to health, independence and self-sufficiency than to other factors normally considered in market segmentation (Metz, 2005, p. 164). In recent times, fitness clothing and sportswear have dominated the leisure-clothing market and become fashionable as indicators of health and youthful stamina (Jenkyn Jones, 2005, p. 24). Sports and fitness clothing styling is often unsympathetic to the changing figure types of older wearers. 'People feel some insecurity about revealing their physical imperfections, especially as they grow older; clothing disguises and conceals our defects, whether real or imagined. Modesty is socially defined and varies among individuals, groups and societies, as well as over time' (Jenkyn Jones, 2005, p. 24).

Cultural considerations and related peer group pressure continue to impinge on the feeling identity, and in particular for men, where occupational clothing has been used to differentiate and recognise profession, religious affiliation, social standing or lifestyle as an expression of authority. There is now a feeling of lack of identity recognised for those who embark on retirement from a role where they were accustomed to wearing formal dress or a corporate uniform (Jenkyn Jones, 2005, p. 27). An overall feeling of comfort inspires confidence with a neutral concept of comfort that may be achieved through addressing an appropriate balance of aesthetic and style, social and cultural and physiological concerns. The 'feel-good' factor may be enhanced through sensorial and aesthetic attributes and through reliability, or perception of product reliability (McCann, 2000).

18.4.3 Recent market research

The 'In Season for Season' project, in 2003, encouraged small manufacturing enterprises in the East Midlands clothing industry to look at the less transient demands of the ageing market as opposed to continuing to chase sales within the relentless youth fashion trade. Market research provided feedback from independent retailers on key product considerations for the 40+ consumer (Oxborrow, 2006). Common drivers and attitudes pointed to this group being more concerned about the appropriateness of their clothes to fit their lifestyle than in their 'fashionability'. They would rather buy one good quality item than several lower quality items with styling that best expresses the wearer's personality and with appropriate 'fit' and shape. The findings revealed a desire for a subtle 'smart casual' version of main-stream fashion rather than extreme styling. This group values originality in design that is simple but effective, with restrained 'tasteful' detail, and product that cannot be found in the main chains. Many women are very sensitive about size; if a size label is not perceived to reflect the 'body shape', this can affect customer confidence in the retailer or brand.

An important consideration was the desire to mix the sizes of tops and bottoms to fit the changing shape of the older wearer, covering sizing from 10 to 20+. Textiles for everyday and casual wear for the older market should be relatively crease resistant and machine washable. Prints should be subtle and not overpowering. Upper garments should not cling, especially in the large sizes. Trousers should contain some stretch for retaining shape and utilise neutral colourways, such as beige, cream, black, with additional colours appropriate to the trend. Jackets should also be in neutral colours. Manufacture and trim should be of good quality and ensure reliable zips and securely sewn-on buttons and fastenings. This research was intended to be an indicative rather than specific guide to new market opportunities and not an absolute specification.

The consumer survey above provides a useful background to subsequent investigations. In addition, findings in the USA further elaborate the profile of the ageing consumer. Older people's behaviour involves making purchases to replace things that they already have, and still need, that are in disrepair or obsolete, to maintain the lifestyles to which they have become accustomed, to make gifts to others, and to gain access to experiences and expectations made possible by the purchase (Wolfe, 2003, p. 211). 'Older people will generally spend more – sometimes much more – for products and services that serve as a gateway to experiential pleasures they covet. While they may be constrained by issues of affordability, they are not as concerned by price as they were earlier in life.' 'Designer labels are about making social statements, and as people move into and through the second half of life, interest in designer labels falls off because they are not as compelled to make social statements by their brand choices' (Wolfe, 2003, p. 93). Purchasing behaviour of older consumers may be related more to health, independence and self-sufficiency than to other factors normally considered in market segmentation (Metz, 2005, p. 164). 'Perhaps mature customers are forced into more rational purchase behaviour because brands that could be aspirational for this segment do not (yet) exist' (Metz, 2005, p. 160).

18.4.4 History and tradition

Successful functional clothing design is the result of designers becoming thoroughly conversant with the culture, history and tradition associated with the particular end-use or range of activities. The development of activity-specific technical clothing, in the culture of performance sport, is a prime example of end-user driven product development. This has often originated in expert practitioners being dissatisfied with ranges available.

Sporting heroes such as Douglas Gill and Musto, sailing, Jean Claude Killy, ski-ing, and Ron Hill, athletics, have initiated the development of their own clothing brands, with heritage that continues today (McCann, 2000). Chris Bonnington has been associated with the UK outdoor company, Berghaus, and performance sports brands such as Nike promote their products through the endorsement of sporting 'heroes'. Many of the outdoor brands have added travel wear ranges to their portfolio. The functional attributes of performance sportswear may be adopted to address the everyday clothing demands of the older market but with reference to the subtleties of particular life-style trends that affect the style and mood of the clothing. Social and cultural issues, participation patterns and levels, status, demographics, and the general health and fitness of the wearer will impinge on design development. An understanding of the lifestyle demands of the active ageing in terms of behaviour, environment and peer group pressure is needed to provide an awareness of clothing and technology requirements that have appropriate functionality and true usability for this diverse group.

18.5   The demands of the activity

18.5.1 Walking

Physical exercise can improve physical performance in later life, as well as aspects of intellectual performance (Metz, 2005). 'Older people have the greatest need to maintain their exercise levels, and those with some disease related impairment may have the most to gain'(Metz, 2005, p. 35). 'Surveys find that walking is by far the predominant activity reported in surveys of older people, half the men in their 60s and 40 per cent of women reporting this' (Metz, 2005, p. 45). This can range from more extreme hill walking to more moderate exercise. There is the suggestion that the older people are taking more active holidays than the younger age groups, who seem more content with the beach. Older people want to experience the world for themselves, spending the money they have accumulated during a lifetime of work, before it is too late to enjoy it (Metz, 2005).

Gardening is reported to be the UK's favourite pastime and for older people it is the most popular form of exercise after walking. Increasing numbers of older people are joining health clubs and gyms with the result that the industry claims that pensioners are the fastest-growing membership group – encouraged by discount rates and specific off-peak programmes tailored to their needs (Metz, 2005).

Of those over state pension age, 9 per cent are still in work, of whom a quarter are self-employed (Metz, 2005). Initial research indicates that the active ageing have a varied routine from day to day or week to week, embracing walking and a mix of the above activities. There is a current focus on walking as a means of increasing personal fitness levels with, for example, local authority walks. The charitable organisation 'Walk the Walk', is staging Moon Walk marathons to raise awareness in breast cancer (www. walkthewalk.org) and a company, 'WalkActive', based in London, (http:// walkactive.co.uk) is conducting studies into walking to make it more enjoyable, more accessible and more effective for all age groups. The practice of sport and healthy exercise demands attractive and multi-purpose clothing that may be enhanced through the functionality of wearable technology.

18.5.2 Setting the scene

Designers should carry out primary research in observing and obtaining feedback from wearers to identify their needs for the chosen activity or task, or for multifunctional end-use. In respect of patterns of use or participation levels, the clothing may be worn regularly for everyday wear, for leisure, at weekends or once a year on holiday. The activity may be performed quickly or be of medium or long duration. For competitive sport, the 'rules of the game' may include a dress code or restrictions to do with safety or fair competition that impinge on apparel design. An appreciation of the impact of the environment in which the activity takes place is required. There may be the demands of an extreme climate or a range of temperatures and degrees of humidity. Designing for very cold, dry conditions or hot humid conditions, is less challenging than catering for both extremes. The sport or leisure activity may be seasonal, carried out indoors in a controlled environment, or outdoors, in unpredictable conditions or in rough terrain. The clothing system may be required for protection from an unknown hostile environment or for a known predictable environment. The wearer may be subject to contact or non-contact activity with regard to aspects such as body protection and abrasion resistance. Transportation is also a consideration in terms of bulk and weight. The wearer may have a vehicle to travel to a destination or may have to transport heavy equipment manually over long distances. An appreciation of the functional needs of the end-user will impact on a breadth of design considerations with regard to comfort, protection, durability, weight, ease of movement, identification and aftercare (McCann, 2000).

18.5.3 Sports layering system

The concept of the tried and tested military type 'layering system', commonly adopted in the performance sports and corporate wear areas, provides a reference point for the identification of design requirements for functional clothing development (McCann, 2005). This system normally comprises a moisture management 'base-layer' or 'second skin', a mid insulation layer and a protective outer layer. Elements of personal protection may be incorporated into the system. The base layer is normally of knitted construction and, most recently, seam-free garments have become prevalent. Varied knit structures, often with elastomeric content, may be placed around the body to aid wicking and offer increased support and protection. Mid layers incorporate textiles such as knit structure fleeces, woven fibre or down filled garments and sliver knit constructions, known as fibre pile or fake fur, to provide insulation by means of trapping still air. The outer layer, or 'shell garment', provides protection for the clothing system microclimate from the ambient conditions by adopting a range of variations on woven or knit structure protective textile assemblies. Personal protective inserts within the system consist of knitted or woven spacer fabrics, wadding and foams or non-woven composite structures in varying degrees of flexibility or rigidity. A balance of aesthetic considerations remains key to the acceptability and wear-ability of the final product. To function effectively, the garments and components within the layering system must coordinate in terms of style, fit, silhouette, movement and closures (see section 12.3).

18.6   The demands of the body

18.6.1 Health and wellness

Physical deterioration is often due to reduced levels of physical activity that may be halted and reversed through exercise. It is recognised that a rise in obesity, diabetes and heart disease is resulting in an ever-increasing financial burden on government and private organisations. The worldwide growth for Biophysical Monitoring Wearable Systems has grown from $192 milion in 2005 to $265 million in 2007, within three distinct market areas. These are Health/Fitness, Medical and Government/Military (Venture Development Corporation, 2007). In order to reduce the level of illness and death associated with obesity and heart disease, governments are recognising the link between sport and fitness and the health of the nation. These trends stress the need for smart wearable textile products that help to make self-monitoring more accessible and positive for those who wish to keep fit, or for those who find themselves gradually or sometimes rather abruptly becoming unwell.

18.6.2 Anthropometry

The size and shape of the changing human physiology of the older body raises practical issues to do with garment cut and textile selection and the placement of wearable technologies that demand appropriate fit and positioning to optimise their potential functionality. Measurement and fit, in relation to different figure types, predominant postures and the ergonomics of movement, directly relate to the comfort and psychological 'feel good factor' of functional clothing for everyday use. Gradual or dramatic changes in older figure types that impact on clothing design will be evident in terms of less erect posture and the onset of conditions such as osteoporosis. Aspects of physical performance that decline with age include mobility, dexterity and the ability to reach and stretch (Metz, 2005). Particular needs, for instance, of older women suffering from arthritis demand functional clothing with enhanced ease of movement for dressing and undressing (Metz, 2005). Arthritic joints may also affect dexterity in use of fastenings. Designers must investigate changes in the size and shape of the ageing body. What is considered as 'small', 'medium' and 'large' sizing for one activity will be inappropriate for different age ranges and end-users. For extreme predominant posture, a complete measurement chart may be needed for the wearer. A visual record of the subject(s) in action will support verbal feedback and observation. Clothing should, ideally, enhance support for the body where needed without restricting movement.

SizeUK, has been the first national survey of the UK population since the 1950s and the first time that the changing shape and size of the population has been captured and analysed by means of 3D scanning technology (Sizemic, 2007). Garment design engineering may now address customised fit for different figure types and predominant postures, providing opportunities for the technical design development of more comfortable prototypes for selected age groups in terms of size, shape and proportion. A particular target market may be identified and a sample user group of men and/or women sought within a representative range of age, approximate height and weight. Data collection and analysis instruments would be implemented that include literature searches, surveys and visual, verbal and written responses from users using questionnaires, semi-structured interviews and workshops supported by observation techniques (Bougourd, 2007). A selected group of willing participants would have their anthropometric data collected using 3D body scanning with results mapped onto existing SizeUK 3D data to automatically generate 2D block patterns of basic garment shapes. Basic blocks such as tops, bodices, trousers and skirts, would be cut and assembled into garment shapes to be evaluated with the user group in terms of fit, balance and comfort using verbal responses from the wearers and from garment experts. The final pattern blocks could be digitally stored for further reference and subsequent adaptation into customised styled patterns. This digital scanning technology will inform designers and, in turn, the general public who are aware of the inadequacies of standard garment sizing charts with inconsistencies across regions, market sectors and brands (Bougourd, 2007). The data will be of considerable value to sectors outside the clothing industry, such as health and medical sectors, particularly with concerns over the increasing level of obesity.

18.6.3 The senses

Physiological changes include a deterioration of the senses. Sensory considerations are key to the effectiveness of the layering system where textile selection and garment features and trims may be designed to address issues such as handle and grip, protection for the eyes and the avoidance of impeding hearing. The skin, the source of most of our tactile sensations, declines in sensitivity with age. Changes in surface texture can be used for both tactile pleasure and for practical end-use and safety. Older people are less tolerant of extremes in temperature. Changes in visual acuity demand special consideration in choosing font and font size, surfaces, colour, degree of glare, back lighting, angle of light, etc. Old age also leads to a gradual loss of hearing, taste and smell. A loss of strength is a consideration for the design of fastenings and packaging. Sensory textiles may now aid the detection and prevention of bedsores. Wearable technologies may provide electronic arrays to enhance visibility. Textile switches and touch-sensitive displays may be used, to aid dexterity, as alternatives to control buttons for wearable devices that may not be further reduced in scale. In respect of olfactory sensing, work has been done on colour for therapeutic scent delivery as a healing platform to address emotional wellbeing. 'Scentsory Design' research explores the mood-enhancing effects of olfactory substances that act on the brain to influence performance, behaviour, learning and mood (Tillotson, 2003).

18.7   Aesthetic concerns

18.7.1 Designing in the round

Functional clothing may be developed 'in the round' with design lines related to fit, proportion, and the positioning of design features and smart attributes around the body, directly responding to the particular demands of the body and end-use. In creating functional garments in relation to the contours of the body, both design lines and ergonomic cutting lines often work in harmony and may merge into clean, minimal styling. An awareness of human movement, the support of muscles, protection, and workload, in combination with the application smart textile innovation, results in garments with design lines that maximise efficiency coupled with meaningful aesthetics. The concept of 'Comfort Mapping Technology', as terminology introduced in the performance sportswear sector (Anon, 2007), directly links the identification of the physiological demands of the body to 2D and 3D garment pattern development with the sourcing of appropriate textile qualities, constructions and properties, selected and positioned to enhance comfort and functionality. Areas or 'zones' of the body are identified and mapped in relation to comfort factors such as ease of movement and articulation, predominant posture, moisture management, thermal regulation, impact and environmental protection. An evaluation of these requirements is captured, and recorded by experts, through direct feedback with users. Design detail may be cross-referenced with the use of a tailor's dummy or transferred directly onto the basic garment blocks developed from the 3D body scans of the user(s). Basic blocks are subsequently manipulated into styled patterns with additional allowances for articulation in relation to the ergonomics of movement. The fabric characteristics and qualities selected must be accurately mapped onto the 2D pattern drafts, with clear specifications provided for garment assembly.

18.7.2 Textile selection

In textile-driven clothing design, aesthetic choices, balanced with technical design considerations, begin with fibre and yarn selection that impacts directly on the properties of the fabrics and trims. Smart textiles may be sourced in a variety of knitted, woven and non-woven constructions, assemblies and finishes, for application within the sports-type garment layering system. Textile structures are discussed in greater detail in other chapters with this publication. A moisture management base layer may incorporate separate panels with attributes such as smart antimicrobial fibres, moulded support, and mesh 'zones' for ventilation. Sports bras and men's base layer garments, in seamless engineered knit structures, incorporate sensor networks with biomedical devices linked to communication systems and display devices to monitor vital signs (Numetrex, 2008). A mid insulation layer may feature engineered fleece construction, with variable insulation, or attributes such as phase-change thermal regulation or electronic heated textiles. Outer layer protective garments may support tracking and positioning devices, soft switches for electronic devices and flexible displays. Protection that changes state from soft to solid on impact may be integrated within the garment system (d3o, Dow Corning, 2007). The durability and performance of smart fabrics are being addressed through the adoption of emerging textile joining techniques such as heat bonding, moulding, and laser welding, initially used in intimate apparel and sportswear, and now adopted for the encapsulation of wearable technologies and flexible displays.

18.7.3 Colour

Concerns such as colour, fabrication, cut, proportion and detail will contribute to the psychological 'feel good' factor of the wearer. The colour and silhouette of clothing gives a first impression that may be influenced by fashion, peer group trends, corporate image, health and safety requirements, codes of culture and tradition, as well as individual preference. The changing complexions and hair colour of older wearers will influence colour preference and selection, more than transient fashion predictions. There are colour consultancies that offer a personal service to individuals to enhance their confidence in colour selection. Colour may be introduced through technologies that include digital printing and the potential for large-scale repeat jacquard weaves and engineered knits that enable designers to position specific colours and textures in relation to the desired garment style. Colour changes may be specified in combination with differentiation in fibre properties and fabric constructions. Further embellishment may be added by means of digital embroidery and laser finishes. Conductive fibres and yarns are used in targeted digital embroidery to create electronic connections. Conductive inks may be used in digital print to create electronic printed circuits. Camouflage effects, colour change and high visibility finishes may be enhanced by smart textile finishes. Colour may attract attention or give warning in complying with safety standards. Certain darker shades can prevent or reduce UV penetration and some colours may repel or attract insects and wild animals.

18.7.4 Stretch

The attributes of stretch fibres have revolutionised the technical and aesthetic design of clothing and may be incorporated, in varying percentages, in warp and weft knitted structures, laces and nets, woven constructions, narrow fabrics and shock cords, and in some 'hook and loop' closures. The properties of elastomeric and mechanical stretch are of particular relevance to garment design for older figure types. A varied percentage and direction of stretch enhances comfort and fit in providing engineered areas of support or enhanced movement that enable the wearer to put on and take off the garment with greater ease. Stretch is a characteristic of Neoprene and of certain waterproof breathable laminates. Recent innovation embraces 'soft shell' garments that incorporate a hybrid mix of the attributes of the protection of outer shell garments with the comfort and insulation of mid layers. Stretch contributes directly to fit and to the embedding of textile sensors and wearable electronics to be held in appropriate locations within garments, as in intimate apparel and base layer garments, for sport and medical applications. Variable stretch textiles are used in compression garments that offer targeted support in maintaining muscle alignment and a reduction in the loss of energy in athletic performance. Additional reinforced taping, outlining and supporting muscle groups, contributes to the aesthetic appearance of the garment. Body scanning is being used to directly inform the design and fit of customised compression hosiery (William Lee Innovation Centre, 2007).

18.7.5 Moisture management and thermal regulation

Key aspects of the maintenance of comfort are moisture management and thermal regulation. For effective thermal regulation in clothing, the design of the system must address the moisture management of sweating through the application of special fibres and fabric constructions and the provision of ventilation within the garment(s). 'Workload', the relationship between physical work rate and body heat production, and the rate of perspiration for a given activity or task, will inform appropriate fabric selection in terms of moisture wicking and, where necessary, the use of spacer fabrics for ventilation. Wearers may be at risk in relation to survival in extreme cold if a high moisture level is retained within the garment base layer and the temperature drops. For heat retention, the clothing system must be designed to prevent wetting and combat the wind chill factor in conserving appropriate levels of still air for insulation. The ambient conditions, and climatic variability, impact on all other aspects of thermo-physiological regulation. There are obvious advantages in the application of high moisture wicking, quick drying textiles for the onset of incontinence. Older people are susceptible to cold and may benefit from further adaptations of technologies initially adopted in performance sportswear. Smart phase-change textiles can help to regulate temperature, such as Outlast technology, which absorbs heat when the body is subject to heavy workload and releases heat when the body cools down. W.L.Gore has developed a channelled construction, 'Airvantage', that may be inflated or deflated to regulate thermal insulation (Gore, 2008). The Berghaus 'Heat Cell' jacket incorporates a polymer panel that may heat the lower back of the garment as required (EXO^2, 2008). Similar technology is available within ski gloves and boots.

18.7.6 Protection

Protection from the ambient environmental conditions may be provided by waterproof, breathable textiles with additional properties and finishes, such as abrasion resistance for more extreme requirements. The design of lightweight outer layer garments has become increasingly fitted and stylish as a result of enhanced comfort, through textile moisture management in hydrophilic or micro porous, waterproof, 'breathable' membranes. In terms of safety, textiles and clothing may be designed to offer varying levels of body protection, dependent on the end-use. Consideration must be given to vulnerable areas of the body in anticipating injurious hazards and commonly occurring accidental incidents particular to the activity. Textile structures such as spacer fabrics provide lightweight personal protection in body armour for performance sport and as impact protection for the elderly. Recent developments in inherently flexible phase-change materials harden on impact (d3o, 2008) and then revert to a flexible state once the impact has passed. Tracking and positioning devices may be used for avalanche detection, and to monitor activity, movement and posture in sports' training and in practice. These devices may also be used, with ethical issues addressed, to track those who may be at risk. The psychological 'feel good factor' is directly related to the reliability, or the perception of the reliability, of the garment system.

18.7.7 Biomimicry

Technical textiles benefit from tried and tested concepts derived from the natural world. In studying biomimicry, innovators of smart textiles are adopting structures and finishes, found in plants, insects, animal skin and furs. Terminology such as 'second skin' is used for base layer garments where textiles are designed to enhance moisture wicking and ventilation in areas or 'zones' related to the demands of the human skin. The mimicking of shark skin informed the development of knit structure textiles for competition swimwear (Speedo, 2008). Fleece pile fabrics are being engineered to incorporate variable patterning, in developing fabric structures replicating animal fur and to provide areas of added protection and ventilation relative to the demands of the body (Patagonia, 2008). 'Body mapping' has also been used as terminology to market outer-layer garments, where features such as breathability, abrasion resistance and thermal regulation have been incorporated in combination with garment cut and construction. Fir cones have been studied in creating a balance of breathability and waterproofness in phase-change textiles that expand and contract, for mid and outer layer garments (Schoeller, 2008). The lotus leaf has been mimicked in nano-scale technology for moisture repellent and stain-proof finishes.

18.7.8 Enhanced aesthetic

Novel manufacturing techniques have become prevalent, predominantly in the performance sportswear sector as an early adopter of novel garment engineering techniques. Textile heat bonding (Sew Systems, 2008) and moulding and laser welding (Jones, 2005) may be applied primarily to technical materials made from a high percentage of synthetic fibres. Functional design details such as zip openings, garment edges and perforations for ventilation, may be laser cut to provide a clean and non-fraying finish (Arcteryx, 2008). Waterproof zips for main closures and pockets, sleeve tabs and hood reinforcements may be bonded without stitching. These joining methods may be combined to produce 'sewfree' garments with clean design lines. Seam-free knitting techniques are prevalent in the structuring and shaping of intimate apparel, base layer garments, in medical hosiery (William Lee Innovation Centre, 2007) and in heavier gauge knitwear (Falke, 2008). These novel textile and garment manufacturing techniques are adopted for both the permanent embedding of miniaturised traditional and textile-based wearable electronics, and for the incorporation of specific enclosures for removable wearable technology, in garments constructed from knitted, woven and non-woven textile assemblies. Textile moulding creates control buttons for electronic switches. Conductive metallic fibres and polymers are used in light-weight textile assemblies for soft key boards and as controls for wearable devices (Eleksen, 2008). Flexible solar panels harvest power on outerwear jackets (Zegna, 2008) and fibre optics provide flexible displays on sports backpacks.

18.8   Technology design and development

18.8.1 A neglected market

The emergence of smart textiles and wearable technologies impacts on the whole textile and garment chain, from fibre production to product launch. This constitutes the focus of this publication, with issues discussed in detail in chapters that correspond to different stages in the design and development critical path. To date, design-led smart clothing has been targeted primarily at athletes and the youth market in areas such as snow sports, mountain biking, motor biking and running. Little has been done to address physical and cognitive limitations when developing these new products and services to ensure that they are appropriate to the culture and real-world needs of the variously described rapidly growing 'Grey', 'Silver', 'Third Age' or 'Rainbow Youth' market. Today's older people have become competent users of high-technology ICT products where they perceive that those products deliver something of value to them (Metz, 2005). They are said to be capable at using technology but are slower. Currently, people over 50 are the fastest growing group of Internet users in UK and, as a group, spend more time on line than any other age group of the population. Continuing advances in microelectronics create new opportunities for assistive technology devices (Metz, 2005).

18.8.2 Technology platform

Continuing advances in microelectronics create new opportunities for assistive technology devices (Metz, 2005). In accordance with user requirements, a technological platform may be developed to realise the 'intelligent' element of the garment layering system development to support a range of functionality such as the recording of vital signs and assessment of behavioural changes. The incorporation of technology into garments to offer a form of improved means of user support has led to alternative approaches and paradigms to support self-monitoring and improve user empowerment (Manning, 2007). A major benefit, from the user's perspective, is that an understanding of their healthcare, lifestyle monitoring and wellbeing becomes possible, as opposed to a required form of treatment (Nugent, 2007). The successful deployment of this technology has the potential to revolutionise the concepts of home-based care; and beyond, by allowing for more clinical information to be recorded in an unobtrusive manner without restricting normal daily activities, movements and location. The adoption of technologies that self-monitor fitness and communicate information, such as heartbeat and temperature, posture, speed, distance, duration and repetition of movements, have important implications for the wearer and others such as peers, family and, subsequently, carers. This technology would provide personal comparison of performance and medication reminders, as required (Nugent, 2007).

18.8.3 Self-monitoring and communication

In relation to healthy exercise and outdoor pursuits, potential needs that may be addressed might include information and communication, security and safety with the ability to summon help. Data would be mapped, as relevant, to do with identification, location and environments in which the activity takes place. The self-monitoring of fitness would provide a personal comparison of performance, such as distance, speed, heartbeat and temperature, and medication reminders, as required. Other services could provide information on the environment, location and route finding, transport and entertainment, etc. In addition, features might include the daily management of the security of the home as well as shopping and banking (Manning, 2007). The devices will also require the support of relevant technology providers and services (Ryan, 2007). The monitoring and feedback of information requires the development of algorithms to process the recorded information, in addition to the long-term monitoring of data to assess changes in behaviour (Nugent, 2007). There are, however, a number of key challenges to be addressed: namely, where are the optimal places to embed the technology within the garment and what is the best means of providing feedback to the user (Armitage, 2007)? The means by which this information can be fed back in a positive manner must also be investigated and optimised in collaboration with users (Ryan, 2007). The user interface must be designed in a format that is easy to understand by the intended audience. There is potential for the development of the product hardware, software and information content to be customised in relation to a user's needs by modifying its characteristics; for example, by increasing the font size of visual displays. At present, customisation is usually done by the user. In future, information held on a smart card inserted could affect the customisation automatically (Metz, 2005).

18.9   Conclusion

18.9.1 A gap in the market

In relation to understanding the needs and wants of the active ageing, the author, as an older user, identifies with a statement that 'Some attitudes remain largely unchanged from youth – hence the common feeling that what is odd about growing old is that you don't feel any different' (Metz, 2005, p. 48). However, it is evident from the market surveys that reported that 'the New Customer Majority' is mostly 'experiencing an ebbing of materialistic influences and rising influence of experiential aspirations on their behaviour' (Wolfe, 2003, p. 101). There is evidence of general dissatisfaction in the dehumanisation of customer experiences in that 'many gains in productivity over the past decade have been at the expense of the quality of the customer experience' (Wolfe, 2003, p. 7). This initial overview 'identifies that older people want less "stuff" as they seek more balance in reprioritizing their lives to become more self-reliant' (Wolfe, 2003, p. 20). Wolfe and Snyder suggest that 'the world views of older people tend to be more cohesive and stable than those of people in the first half of life' (Wolfe, 2003, p. 85). 'The younger mind is more linear, literal, and categorical, making it easier to render what they say into statistical statements with few greys and few in-betweens, because perceptions are more sharply defined – more broadly etched in unambiguous black and white' (Wolfe, 2003, p. 14). Older consumers tend to be more concerned with authenticity, value and the usability of the product, so consequently do not usually figure among the early adopters of a product unless there is considerable advantage to them. 'It follows that new products are not designed initially for older people: they are expected to follow in the early or late majority, when the technology is becoming more mature and the selling points no longer stress the novelty of technology, but rather other aspects such as ease of use (Metz, 2005).

18.9.2 Hybrid design process

The success of assessing market needs and bringing new products to a new customer will depend on reassessing all the links in the chain contributing to the product cycle. The product range to meet those needs, the design and methods of manufacture and supply logistics, pricing, advertising and, point-of-sale marketing, all need to be rethought. A recent product developer for a leading outdoor brand has written about 'the homecoming' in bringing the manufacture of value added products back to Europe as offshore production 'blunts the competitive edge'. Designers must work closer to prototype development. The implementation of a collaborative design approach, more accepted in the disciplines of product and industrial design, is required to promote meaningful engagement with older users in identifying their real needs. With a rapidly developing selection of smart textiles and wearable technologies, and supporting powering and communication devices and their services emerging, designers and the product development team must adopt a new hybrid design methodology to inform and guide the design specification and guide the design development process. An age-diverse team needs to co-design with older users throughout the product development process, from design research to the creation of point-of-sale and retail staff training. Evaluating the outcome of the various stages of the design process will inform the fine tuning, adjusting or rethinking of the outcome. Older people are the experts in making mature choices (Metz, 2005).

18.9.3 Individualisation

Mass customisation in clothing is emerging as a result of body scanning technology (Bougourd, 2007). A three-dimensional body image is processed to derive body measurements to determine sizing off-the-peg garments or specify the measurements for a made-to-measure garment (digital tailoring). A virtual mannequin (three-dimensional computer image) enables the customer to try the chosen virtual clothes before trying on the real clothes. Body measurements can be stored on a database to be retrieved on future shopping from home, enabling a customer to try on clothes using software running on their own home PC. This will also make it easier for people with disabilities or restricted mobility to purchase clothes that fit (Metz, 2005, p. 79). The fundamental concept underlying inclusive design is that all products should be designed to be useable by as broad a range of users as possible. In terms of the outdoor clothing market, the brand 'New Balance' has adopted the signature tagline 'Achieve new balance' that resonates with the growing desire in mid-life to achieve life balance after the frenetic years of early adulthood when unbalanced devotion to career and acquisitiveness dominated lifestyle. As New Balance has not used famous personality endorsements, it has made it easier to shift attention to older markets because it is not seen primarily as a youth-oriented brand, presenting a kinder, gentler set of values than its biggest competitors. It has introduced 5 shoe widths versus the industry standard of 3 – a decisive competitive advantage, especially among older people whose feet may have spread a bit (Wolfe, 2003). In relation to sports clothing, the company Odlo has launched an Active Jacket for Winter 08/09, available in a range of sizes from XXS to XXL and in short sizes and in 13 colour variations (Bieker, 2008). This approach to less transient but more considered styling in a broader range of sizes and colour selection is an obviously valid, but not common, approach within the current market.

18.10 Future trends

This chapter sets the scene for the application of smart textiles, in a clothing system with embedded wearable technologies for the active ageing community. It introduces the focus of preliminary research carried out by a 'Preparatory Network' within the UK joint research council's 'New Dynamics of Ageing' programme. It considers the potential for a new generation of monitoring and responsive devices that can enhance the autonomy and independence of a rapidly growing older market. An active and independent lifestyle may be supported by combining leading-edge textiles, communications and computing technologies to help the ageing population enjoy an extended safe and healthy life, wherever they choose to be. Traditional stakeholders in clothing, technology and care provision intend to engage in cross-disciplinary research, development and evaluation processes, with representation from older user networks, to inform a new 'shared language' and vision that may be easily communicated between these sectors and the end-users. Effective partnerships between researchers and potential users of that research can result in better quality outcomes for all involved. The successful development of innovative functional clothing for the 'new old' demands experts from disparate disciplines that include garment design, technical textiles, wearable electronics, information and communication technologies (ICT), and social and health sectors, willing to engage in collaborative design practice with older users.

This new hybrid area of design demands a merging of methodologies, instruments and analysis techniques that are multi-disciplinary in their own right. We are witnessing a paradigm shift, where stakeholders other than researchers are becoming engaged with all aspects of the research process, from developing the research questions to designing and implementing data collection and analysis strategies and disseminating research findings. In response to these demands, an engaged, pro-active, person-centred research and design process is proposed. Such a shift requires significant changes in thinking and practice to enable new and innovative ways of working in partnership with stakeholders, and necessitates appropriate support and development for all involved, including professional researchers (Ryan, 2007). Meaningful end-user engagement would focus on the social model, with users *wanting* to wear and enjoy wearing assistive technology rather than being told what to wear.

18.11 References

ANON. (2007) 100 Innovations (Comfort Mapping Technology). *Future Materials*, December.

ARC'TERYX (2008) [URI http://www.arcteryx.com/, accessed May 19th 2008]

ARMITAGE, R. (2007) Personal communication, Textronics Inc, September.

BIEKER, C. (2008) Best Ager within the Sportsmarket: Market, potentials, starting points. Munich: *ISPO*, Winter 2008, Presentation [URI www.generation-sport.de, accessed May 19th 2008]

BOUGOURD, J. (2007) 'Sizing systems, fit models and target markets.' In: Ashdown, S. *Sizing in Clothing*. Cambridge: Woodhead Publishing Ltd.

D3O (2008) *Intelligent shock absorption*. [URI http://www.d3o.com/, accessed May 19th 2008]

DOW CORNING (2007) Active protection system. [URI http://www.activeprotection-system.com/, accessed 1st December 2008]

ELEKSEN (2008) *ElekTex textile touchpads*. [URI http://www.eleksen.com/?page=products08/elektexproducts/index.asp, accessed May 19th 2008]

EXO² (2008) *Berghaus Heated Clothing*. [URI http://www.exo2.co.uk/berghaus.html, accessed May 19th 2008]

FALKE (2008) *FALKE Ergonomic Sport System* [URI http://www.falke.no/sports/2008ss/index.php?t=news&g=runnersworldaward, accessed May 19th 2008]

GORE (2008) *Airvantage™ adjustable insulation*. [URI http://www.gore.com/en_xx/products/consumer/airvantage/index.html, accesssd May 19th 2008]

JENKYN JONES, S. (2005) *Fashion Design*, 2nd Rev Ed. London: Lawrence King Publishing.

JONES, I. (2005) Laser welding methods for textiles (Presentation). *Leapfrog Workshop, Cambridge*. [URI http://www.leapfrog-eu.org/leapfrog-ca/resultDetails.asp?Id=58, accessed May 19th 2008]

LACK, T. (1992) Fifty Years of Walking Equipment. *Footprint* 3(3):10–16.

LACKNER, H.B. (1998) Perceptions of the importance of dress to the self as a function of perceived age and gender, *IAA Proceedings*.

MANNING, B. (2007) Personal communication, University of Westminster, Autumn.

MCCANN, J. (2000) *Identification of Requirements for the Design Development of Performance Sportswear*. M.Phil Thesis. Derby: University of Derby.

MCCANN, J. (2005) 'Material requirements for the design of performance sportswear.' In: Shishoo, R. (ed) *Textiles in Sport*. Cambridge: Woodhead Publishing Ltd.

METZ, D. and UNDERWOOD, M. (2005) *Older Richer Fitter: Identifying the Customer Needs of Britain's Ageing Population*. London: Age Concern Books.

NUGENT, C. (2007) Personal communication, University of Ulster, September.

NUMETREX (2008) Strapless heart monitor clothes by Numetrex. [URI http://www.numetrex.com/, accessed May 19th 2008]

OXBORROW, L. (2006) *The Right Trousers: A Fashion Weblog by 'In Season For Season'*. Derby: Derby Chamber of Trade [URI http://inseason4season.wordpress.com/, accessed May 19th 2008].

PATAGONIA (2008) *Outdoor Clothing, Technical Apparel and Gear*. [URI http://www.patagonia.com/web/eu/home/index.jsp?OPTION=HOME_PAGE&assetid=1704, accessed May 19th 2008]

RYAN, J. (2007) Personal communication, Salford University.

SCHOELLER (2008) *The Bionic Climate Membrane*. [URI http://www.schoeller-textiles.com/pdf/eng_c_change_Basis_PR_def_200106.pdf, accessed May 19th 2008]

SEW SYSTEMS (2008) Designers and developers of sewing systems and machinery for the textile industry. [URI http://www.sewsystems.com/, accessed May 19th 2008]

SIZEMIC (2007) *SizeUK Data Set*. [URI http://www.humanics-s.com/uk_natl_anthro_sizing_info.pdf, accessed 8th Nov 2007]

SPEEDO (2008) Speedo UK [URI http://www.speedo.co.uk/, accessed May 19th 2008]

TILLOTSON, J. (2003) *Scentsory Design. Smart Second Skin*. [URI http://www.smartsecondskin.com/main/scentsorydesign_english.htm, accessed May 19th 2008]

VENTURE DEVELOPMENT CORPORATION (2007) 'Mobile and Wireless Practice. A White Paper' In: *Wearable Electronic Systems, Global Market Demand Analysis, Third Edition*. Natick: VDC Research Group, Inc.

WILLIAM LEE INNOVATION CENTRE (2007) *Knowledge for innovation*. [URI http://www.k4i.org.uk/, accessed May 19th 2008]

WILLIAMS, J.D. (2004) *Fashion without limits*. [URI http://www.simplybe.co.uk/, accessed May 19th 2008]

WOLFE, D.B. and SNYDER, R.E. (2003) *Ageless Marketing*. Chicago: Dearborn Trade Publishing.

ZEGNA (2008) *Solar JKT Ermenegildo Zegna* [URI http://www.zegna.com/solar?lang=en, accessed May 19th 2008]

19

Smart clothing and disability: wearable technology for people with arthritis

S. UNDERWOOD, University of Wales Newport, UK

Abstract: If a product is in no way pleasurable, it will be hard to encourage its use with education alone. This chapter examines the design of orthotic devices for people with rheumatoid arthritis. It summarises the disability models, people-centred design methods, presents successful products designed for people with arthritis and concludes with a case study of an inclusive design approach of wrist orthotics for rheumatoid arthritis sufferers.

Key words: user centred, inclusive design, wrist splints.

19.1 Introduction

In the area of health and wellness, the future of design is important. For the critical end-user to increase the usage and effectiveness of a device, pleasure with the product and good design are essential. The history of products designed for the average person with disabilities has been medically structured. Usage of products is suggested as an education issue (Spoorenberg, 1994). However, if a product is in no way pleasurable, it will be hard to encourage its use with education alone.

This chapter examines the design of orthotic devices for people with rheumatoid arthritis. It summarises the disability models, people-centred design methods, presents successful products designed for people with arthritis and concludes with a case study of an inclusive design approach to wrist orthotics for people with rheumatoid arthritis.

19.2 Disability models

There has been a change in the perspective taken in looking at disability, which can be seen in the development of models – from the medical to the social model. The development of the social model is seeing a gradual improvement within society; however, there is still some way to go to experience a fully socially inclusive process in the design of products for people with disabilities.

The *medical model* of disability (1900s–1970) is a model described by Owen (2003, p 60), which defines people through their disability. It implies that people are disabled because of their impairment. The disability is the problem not society. This model depicts the disabled person as dependent, being cared for or seeking a cure. Health professionals identify the physiological needs as a consequence of the condition. These needs are met through product availability and are usually limited in options. The rehabilitation process may involve the use of assistive devices, daily living aids or equipment that is often supplied through disabled living centres. Disability aids are developed through the same medical model mindset. The products address the physiological problem with no consideration of the psychological implications of how these devices may affect the individual. The built-up shoes, walking aids, NHS spectacles and beige splints – the patients had no product choice. They were expected to use the medical devices that were supplied.

The *social model* of disability (UPIAS, 1976; Barnes and Mercer, 1996) was developed by and for differently able people. The Social Model identifies disability as being accentuated by society and proposes that it may be the ill-considered design of products and amenities that primarily disable a person. For example, a person who is left-handed may be able to cut fabric with left-handed scissors, but if they were provided with scissors for a right-handed person they may be disabled through the design of the scissors. The social model of disability has helped the disabled community to challenge society and educate people that disability is more than a medical condition.

In America, the 1990 Americans with Disabilities Act (ADA) and in the UK, the 1995 Disability Discrimination Act (DDA) (http://www.uk-legislation.hmso.gov.uk/acts/acts1995/ukpga_19950050_en_1), which later became the extended DDA in 2005, (http://www.uk-legislation.hmso.gov.uk/acts/acts2005/ukpga_20050013_en_1) has enabled individuals to challenge society. The result of these new laws is slowly having a positive effect on society.

As the laws were enacted and effected, thinking slowly took place towards 'universal design' or 'barrier free design' in the USA and 'inclusive design' in the UK became increasingly recognised. Universal design, described by Coleman (2003a), was a subject promoted by architect and wheelchair user Ron Mace in 1985.

The 'seven principles' guideline was developed at the Center for Universal Design between 1995–1997 (http://design.ncsu.edu/cud/). Many derivatives of these principles have been developed since. The *seven principles* are:

(i)   *Equitable* use – the design is useful and marketable to people with diverse abilities.

(ii)   *Flexibility* in use – the design accommodates a wide range of individual preferences and abilities.

(iii) *Simple and intuitive* to use – use of design is easy to understand, regardless of the user's experience, knowledge, language skill or current concentration level.

(iv) *Perceptible information* – the design communicates necessary information effectively to the user, regardless of ambient conditions or the user's sensory abilities.

(v) *Tolerance for error* – the design minimises hazards and the adverse consequences of accidental or unintended actions.

(vi) *Low physical effort* – the design can be used efficiently and effectively with a minimum of fatigue.

(vii) *Size and space for approach* and use – appropriate size and space are provided for approach, reach, manipulation, and use, regardless of user's body size, posture or mobility (http://design.ncsu.edu/cud/about_ud/udprinciples.htm).

In the UK this theory is known as 'inclusive design'. The inclusive approach to design aims to expand the target group and to create inclusion at a social level. In 2003, Keates and Clarkson discussed 'countering design exclusion', a concept that identifies how end-users cannot readily use a product or service. An understanding of the approach and relationship people have with a service or a product needs to be established. If a product is not intuitive, it may exclude a portion of society.

19.3   People-centred design

A 'people-centred' approach has the same definition working with a 'critical end-user', which is being used at the Helen Hamlyn Centre (HHC). A people-centred approach involves engaging with people who use the product from the earliest stages of a design development. It is relevant to all design development (Black, 2006). People-centred design helps designers to understand people's existing experience and project forward to products and services. It is essential in the development of existing products, as well as in developing new products.

Effective design development is conducive to the people's input in wearable products; the one size fits all attitude is not people centred. The person should be at the centre of any design, which also needs to balance physiological demands and emotional needs to be compliant with disability legislation.

19.4   People-centred needs

To develop an understanding of how to empathise with the target market of a product, a people-centred design approach is appropriate (Fig. 19.1).

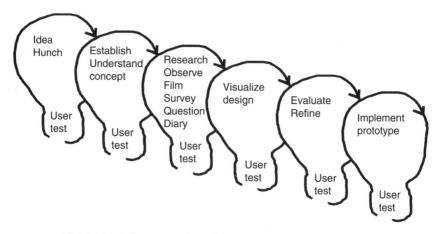

19.1 Author's interpretation of people-centred design.

Pioneering designers will increase the market of a product, service or system through good people-centred evaluation. People-centred research is increasing and, with the population ageing and cultural diversities expanding throughout society, there is an opportunity for innovative inclusive design. Inclusive design had formerly been regarded as restrictive design, described by Coleman (2003b) as 'seen as a curb on creativity'. However, the rich data and design challenges that arise through good, people-centred input are resulting in expanded and cross-market sectors, thus producing exciting new designs. Design consultancies such as IDEO (http://www.ideo.com/) and Smart Design (http://www.smartdesignworldwide.com/) are leading companies within this area of people-centred design.

Smart Design is the strategic design and development partner of OXO International, which was established by Sam Farber in 1989. The OXO Good Grips i-series is a range of comfortable, easy-to-use kitchen utensils. Sam Farber needed to develop a potato peeler for his wife who had arthritic hands and had difficulty using a standard potato peeler. The collaboration of Smart Design with an end-user has led to the success of this product (Fig. 19.2) and the principle has led to eight more award-winning products with the universal design principle that keeps everyone in mind.

For a designer to understand the needs of a person with arthritic hands, observation of a person with arthritis demonstrating how utensils are used in practice is necessary. Filming this process is very useful and insightful for the designer/researcher, because knowledge and understanding are increased through the experience, generating a clearer picture and creating more inclusive designs. The knowledge developed through experiencing this method can trigger many ideas, therefore increasing a design market, business potential and extending areas of design and creativity.

19.2 OXO i-Series Y Peeler.

19.5   Emotional design

Every product evokes an emotion. Author of *Designing Pleasurable Prod-ucts*, Patrick W. Jordan (2002, pp. 4–5) states 'It is emotion that is integral to user-experience'. Donald Norman author of *Emotional Design* (2004) discusses three levels at play in design: visceral, behavioural, and reflective. Norman (p. 6) states, 'it is not possible to have design without all three.' His definitions of the three dimensions of design are:

(i)   *Visceral design* is instinctive. It is about how something looks, feels and sounds.
(ii)   *Behavioural design* is about getting products to function well.
(iii)   *Reflective design* is a message about what the design says about you. It depends on age, background and culture.

A similar alternative approach to Norman's model is a framework devised by Tiger (1992) and developed by Jordan (2002, pp. 13–14); an approach to designing pleasurable products, 'The Four Pleasures' in design, which are:

(i)   *Physical pleasure* (physio-pleasure) includes pleasure connected with touch, taste and smell.
(ii)   *Social pleasure* (socio-pleasure) this is a person's role in society.
(iii)   *Psychological pleasure* (psycho-pleasure) is pleasure derived through both a cognitive and emotional reaction to a product.
(iv)   *Ideological pleasure* (ideo-pleasure) relates to people's values. These values could relate to the aesthetics of a product or the environmental issues it may raise.

The design of assistive devices is still generally applied through a medical mindset. Rehabilitation engineering is defined by Newell (2003) as 'assistive technology that is used for short-term recuperation of illness or injury or for long-term functional support'. The term 'functional' can be disputed here; if the person using the device is unable to perform a function because of the device rather than the impairment, the device is not functional. The medical mindset of the design decision makers in this area has been explored by Zimmerman (2005) who states that the 'people making the design decisions consider emotional and aspirational needs in relation to the products to be perhaps not relevant'. The emotional consequence of this approach to design today has not been researched in the area of orthotic devices.

In 1996, the UK National Health Service (NHS) voucher scheme was introduced (www.wheelchairusers.org.uk/content/voucher.htm). The aim of this scheme was to give wheelchair users more choice as to which wheelchair they would prefer. This scheme changed the status of a patient from 'the supplied' to 'the consumer' and consequently gave him/her power over the choice of the wheelchair manufacturer (Zimmerman, 2005). This assistive technology has been evolving with the application of new materials and manufacturing techniques since the introduction of the voucher scheme and new laws brought about by the DDA, with more independence and therefore influence, creating more financial incentive for manufacturers to improve.

In 1970, psychologist Abraham Maslow described a hierarchy of human needs. 'The model observes the human as a wanting animal who rarely reaches a state of complete satisfaction' (Jordan, 2002 pp. 4–5). This is represented when a basic need is met, such as a physiological and safety need; once this is satisfied, people will still meet with frustration if their higher goals are not met (Fig. 19.3, see also page 97).

The design of products to enable the activities of daily living has been mostly met with the application of a medical model within the design. The physiological and safety needs are (hopefully) met by the designer; however, the end-user frequently feels excluded from areas of society because of the

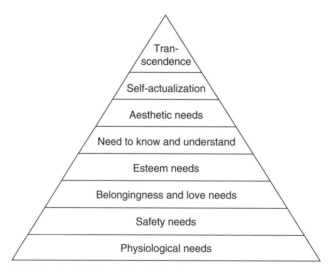

19.3 Maslow's Hierarchy of Needs.

stigma attached to the products. The design of the products may imply the end-user is disabled because of the 'medical' aesthetic. This has implications of poor health and injury that may lead to low self-esteem. This is paramount to people feeling *dis*abled or excluded by design. However, Coleman advocates (2003a) 'people can be enabled and included by thoughtful user led design.'

Inclusive design creates new markets for design. However, it is the design of the device that makes the user feel socially included. Payling (2003) 'Inclusion is not about producing solutions to meet a change in our needs; it is about a change in our thinking' such as OXO good grips i-series, discussed earlier in the chapter; when it was observed that these products were excluding an end-user, an inclusive principle was applied thus including more people.

The history of spectacles can be linked with Maslow's hierarchy of needs. The end-user's primary needs have been met with the 'physiological and safety needs'. Initially the end-user would feel satisfied with the 'gift of sight'. However, the lack of choice and stigma of the NHS-provided spectacles, which are depicted throughout history as 'swotty and bookish' are in contrast with the 'belongingness and love needs' in Maslow's hierarchy of needs or Tiger's Ideological four pleasures if the user is perceived as, and is not, an 'intellectual' or 'bookish' individual. 'Glasses alter not only the wearer's perception of the world, but the world's perception of the wearer' (Busch, 1991).

In the 1930s, NHS spectacles were labelled as medical appliances (Lewis, 2001) and therefore function was the over-riding factor, with the choice of frames being limited and stigmatising. In 1969, the Consumer Association complained about outdated spectacles and a lack of choice, but it was not until 1976 that the UK government acknowledged the importance of styling for glasses (Zimmerman, 2005).

In July 1986, the NHS optical voucher scheme was introduced. It provides eligible people with a voucher which they can use to buy spectacles of their choice. Although there had always been a private market in the design of spectacles, the effect of consumer influence on the design of spectacles finally rewarded the consumer in the 1990s. The NHS spectacle profile was discarded and spectacle wearing was raised by fashion companies as they realised the market potential (Busch, 1991), reaching people on all levels. In 1991, Busch declared that eyeglasses had become stylish with 15–20% of people choosing to wear glasses without prescription lenses.

19.6   Assistive devices

Sufferers from rheumatoid arthritis are often prescribed orthotic devices. These devices are used to protect the weakened joints. In a questionnaire

(Underwood, 2005) to a critical end-user group, callipers, wrist splints, neck braces, walking sticks, knee supports, elbow supports and foot orthotics were investigated. The device that dissatisfied most end-users, and that was most noticeably or commonly prescribed, was the wrist splint. This quantitative survey identified, with statistical significance, that the area of most concern to this group was the inadequacy of the device, which led to inconsistencies with its use.

The usability of the wrist orthotics is limiting the function and produces a medical satisfaction but not a social satisfaction. A medical device or an orthotic device is only functioning if it is being used. Ninety per cent of people prescribed orthotic devices are dissatisfied with them (Underwood, 2005). The questionnaire identified that the aesthetic and social appeal was unsatisfactory. To find out why these devices were unsatisfactory, further information was needed. A more empathetic and holistic approach to the daily experience of the use of the device would outline issues of concern and suggestions may prompt new design solutions.

19.6.1 The design of wrist orthotics

Standard wrist orthotics are constructed (usually) in a beige, Neoprene material. Neoprene is a synthetic rubber, which is very abrasion resistant, waterproof and stretchable. The Neoprene is laminated between two layers of a synthetic knitted fabric retaining the stretch of the material. The device is wrapped around the wrist and fastened on the dorsal (back) side of the hand. The device is fastened with five separate hook and close straps. It has a pocket on the palmar side to hold a metal bar to support the wrist in the correct position (see Fig. 19.4).

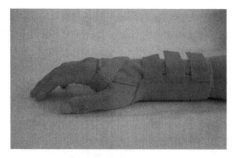

19.4 Standard splint.

19.6.2 Prescribers

The philosophy of occupational therapy (OT) is to enable people to 'do'. The principal concern is to help the patient gain independence. Therefore the visceral design of the enabling device is secondary and not the responsibility of the occupational therapist. How are the concerns, complaints or issues regarding the device going to evolve if the experience remains tacit and unshared? The usage of the device is a concern for the OT. Accepted wisdom to date has suggested that usage can be increased through patient education with no recognition for the improvement using sympathetic or user-centred design.

The primary concern in the design of the wrist orthotic is the management of pain, hand function and grip strength (see Fig. 19.5). The least acknowledged research is cosmetic acceptability and donning and doffing. The last two enquiries will affect the critical end-user emotionally. As this is seen as a secondary concern, research has not been substantial within this area. In relation to Maslow's hierarchy, the medical function of these devices was satisfied in principle; however, without a patient qualitative study on the user-acceptability, these devices hold less functional potential as they are seen as a stigma and a personification of ill health. The third and fourth levels of the hierarchy are applicable here, 'belongingness' and 'esteem'.

If a less medical approach to gain understanding of the end-user is applied, usage may increase. The end-user is utilising the device in a non-clinical environment and to identify these needs requires a non-medical process – new methods of gaining understanding of the end-user experience are necessary.

19.6.3 Silver ring splints

Silver ring splints (SRS) were developed in 1985 by Cynthia Garris (Adams *et al.*, 2008), an American therapist who suffered from arthritis (www. silverringsplint.com/about_srsc.html) (Fig. 19.6). Garris noted that, while the plastic splints were appropriate for short-term trauma, no-one would willingly use the cumbersome and embarrassing solution for the remainder of their lifetime. Rejecting the plastic splints, Garris developed a device in sterling silver; because of its low profile and strong, yet malleable qualities, this material was appropriate for the application. Garris created a double loop with patented elliptically shaped rings that fit to the finger at an angle. Patients willingly wear the splints and feel better about needing them. Garris' people-centred approach resulted in greater utilisation of the therapy. This theory and therapy gives the patient a much-needed sense of control about something that has a great impact on their life. This design

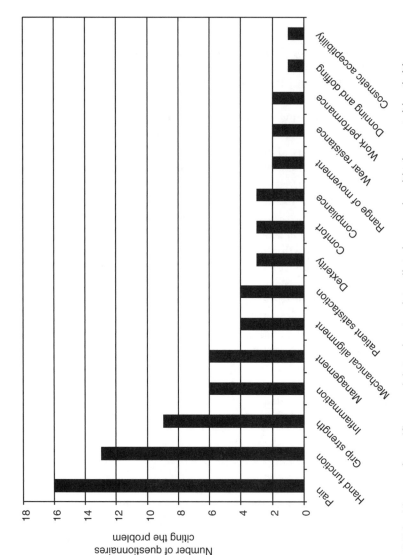

19.5 Incidence of specific complaints about wrist splints by people with rheumatoid arthritis (Underwood 2007).

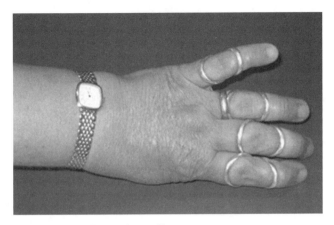

19.6 Patient using a ring splint.

considers function and appearance. Obviously, a splint that functions well will only be successful if the patient wears it.

The plastic splints did not last long and were not enjoyable to wear. The SRS have increased user acceptance and enjoyment. These devices have changed from 'medical looking' objects, which were discarded in resentment, to an elegant, silver bespoke device. This has altered the user's relationship with them. They are no longer negative pieces of plastic representing 'ill-health': they are now a beautiful unique accessory with a functional purpose.

19.7   Case study of wrist splint for people with rheumatoid arthritis

19.7.1 Overview

To understand the people's requirements for a wrist splint for rheumatoid arthritis, a volunteer group needed to be formed. Data protection is essential and ethical approval mandatory in any form of research within the NHS. The author carried out 23 interviews with key volunteers. A triangulation of research was appropriate and affordable for the size of this study.

Triangulation uses different methods of research (often three) to understand the same issue. This can be a combination of quantifiable surveys (statistical methods) and qualitative data. Quantifiable research finds out who, what, and when, and qualitative research finds out why and how. The use of more than one method strengthens the research from varying perspectives. It gives clarity to the 'grey' areas of the design (Fig. 19.7).

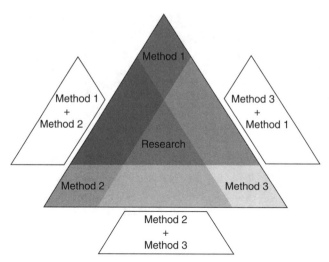

19.7 Triangulation method (illustrated by Sally Underwood), adapted from Gray and Malins (2004).

The critical volunteer group in this initial study were all people with rheumatoid arthritis. As the group had raised a similar pattern of concerns with the wrist splint in the initial survey (section 19.6), the triangulation method suited this study as a more in-depth approach to people-centred design.

Three methods were used in this study:

- Questionnaire 1: a short qualitative and quantitative survey
- Questionnaire 2: an interview for 45 minutes filmed and audio recorded using a qualitative survey
- Diaries supplied for completion over a two-week period

19.7.2 Building a user group

Ethical approval

As it was desirable to work with patients through the NHS, data protection and ethical approval were required for the study. An application was made to the South East Wales Research and Ethics Committee, which covered the local area. To keep costs to a minimum, it was effective to work locally. This study approached volunteers through an outpatient clinic in the region and was therefore classed as a local study. If the study were to extend to two different regions it would be more appropriate to either apply for a national application or to approach the research and ethics committee in each location: www.nres.npsa.nhs.uk/.

Outlines of the enquiry, the questionnaire and the proposal are required for submission to the ethics panel (every hospital has an ethics department for advice and procedure. A complicated study can take up to 18 months for approval). The South East Wales Research and Ethics Committe, after consideration of this research, suggested that the researcher could approach the rheumatology departments and that the rheumatology manager and rheumatology nurse consultant were in agreement.

Where to find volunteers?

Volunteers may be found via the public sector and statutory bodies or via the voluntary sector:

- *Public sector NHS.* With the ethical approval granted, to find a larger number of people with rheumatoid arthritis who use wrist splints, the local hospitals with rheumatology departments were approached. A short questionnaire was devised. This questionnaire could be used whilst patients waited for their appointments. It was compiled for the patients at the out-patient clinic and they were individually briefed by the rheumatology nurse and asked to give their informed consent. This questionnaire provided a non-medical discussion about the devices with the patients (local response physically accessed).

Hand therapists were also sent questionnaires to reach people with arthritis in their hands and wrists. (national response via remote access)

- *Voluntary sector.* Often a quicker and more responsive group can be identified via the voluntary sector as these are formed from bodies who are knowledgeable in the subject and already volunteering their time. Voluntary organisations are usually established by a person or people affected by the lack of resources appropriate and available within the statutory sector. For example, the National Rheumatoid Arthritis Society (NRAS) is a patient–led charity focussing specifically on rheumatoid arthritis, which was set up and is run by people with RA or who care for people with RA. The questionnaire was given to such organisations (national response physically accessed).

19.7.3 Questionnaire 1: a short qualitative and quantitative survey

The short questionnaire applied to this research was initially a structured audit. This involved quantitative and qualitative data.

- What device is used?
- Why is it used?

- What is it used for?
- Is it satisfactory?
- Can you explain why?
- Other comments (provides a free speech area).
- Age?
- Gender?

This questionnaire provided a good basis for initial research because

- It could reach a wide audience.
- It could be postal, electronic, interview or telephone.
- It could provide information quickly.
- It could provide statistical information.

19.7.4 Questionnaire 2: an interview for 45 minutes filmed and audio recorded using a qualitative survey

The interview technique was semi-structured with open questions. It had the following features:

- An interview can provide rich qualitative data.
- Filming the interview can provide a researcher with some greater understanding of tacit knowledge.
- An ability to probe questions will provide knowledge though an unstructured process.
- Observation leads to a better understanding.
- Recorded information can be watched and listened over and over and can be heard or seen on the fourth or fifteenth observation.

19.7.5 The diaries

The volunteers were provided with an A6 size notebook to carry around over a two-week period. They were asked to note when the device was put on and removed and why. The notebook was a tool with no limitations around length or structure. The blank pages enable the volunteer to give as much or as little information as suits them, in marked contrast to the requirements and linear guidelines of health questionnaires.

The diary method reveals more of an emotional insight into the effect that the device has on the end-user. It may indicate more tasks and information considered unimportant or irrelevant than suggested by the previous studies. It provides:

- Memory aid
- Planner
- Time on activity

- Time wearing device
- Explanations why the device is removed or used
- Emotional effect in own language
- A real language
- No constraints

The diary method reveals a daily living reality to the inconsistent or consistent scenario for each end-user. Providing notebook diaries to the end-user as a non-medical unstructured format for data information may provide real language, unpredictable comments and honest behaviour patterns. Understanding when the device is put on, for what task and for how long it is worn is unclear in an observational method questionnaire.

19.7.6 Advantages of the triangulation method

- The short questionnaire establishes a user need.
- Approaching people previously unaware of a small study allows for spontaneous responses.
- The longer qualitative questionnaires can provide richer information.
- Adding 'other suggestions/comments' whilst interviewing, does not allow the end-user a long enough time to really remember all of the situations of use.
- The use of a film process can pick up on a tacit knowledge, emotion, body language, interaction and application of the device. While the volunteer is verbally communicating, they may be unaware of the communication they are giving outwardly through being filmed.
- The diary offers a daily understanding of the use and tasks – the amount of times it is put on and removed and why. It demonstrates the *actual* routine.

If the diary method only had been used, the data received through the open conversation in the film would not have been available. If the questionnaires had been used without the diary, the daily activities would have been omitted and this has been extremely useful knowledge in the development of a people-centred design method. To approach the NHS volunteers in an unexpected opportunity study as well as the NRAS volunteers increased the comparison of end-users.

19.8   Future trends

With groups involved in the new dynamics of ageing being established and networks building towards an inclusive and people-centred approach to good design, the development of people-centred products is increasing. This research continues to discuss product concerns with rheumatoid arthritis sufferers and links to the ageing population. Education of new product designers, and international events such as the Biannual 'Include' Conference at the HHC, is increasing networks and collaborations among designers, researchers and manufacturers.

19.9   Sources of further information and advice

Smart Clothes and Wearable Technology http://artschool.newport.ac.uk/smartclothes/
London Metropolitan University Design Research for Disability MA http://www.londonmet.ac.uk/pgprospectus/courses/design-research-for-disability.cfm
The Helen Hamlyn Centre http://www.hhc.rca.ac.uk/
The Design Council http://www.designcouncil.org.uk/en/About-Design/Design-Techniques/Inclusive-design/
Occupational Therapists http://www.cot.org.uk/
University of the Third Age http://www.u3a.org.uk

19.10   References

ADAMS, J, METCALF, C, MACLEOD, C, SPICKA, C, BURRIDGE, J, COOPER, C, COX, N. 535. 'Three dimensional functional motion analysis of silver ring splints in rheumatoid arthritis', *Oxford J. Med. Rheum.* 25th April 2008 Volume 47, Supplement 2 pp. ii151–ii157

BARNES, C, MERCER, G. *Exploring the Divide: Illness and Disability*, Leeds: The Disability Press, 1996, Chapter 1, p 11–16

BLACK, 2006, http://www.designcouncil.org.uk/en/About-Design/Design-Techniques/User-centred-design-/

BUSCH, A. From stigma to Status: The specifications of spectacles, *Metropolis* April 1991 Vol 10, No 8, 35–37

COLEMAN, R. From Margins to Mainstream, in *Inclusive Design. Design for the Whole Population*; 2003a, Springer p 1

COLEMAN, R. *Inclusive Design. Design for the Whole Population*; 2003b, Springer p 19

GRAY, C, MALINS, J. Visualizing Research, A guide to the research process in art and design, 2004, Ashgate Publishing Ltd, p 137

JORDAN, P. Designing pleasurable products An introduction to the new human factors, Taylor and Francis, 2000, Chapters 1 and 2

KEATES, S, CLARKSON, J. *'Countering design exclusion'* in *Inclusive Design. Design for the Whole Population*; 2003, Springer

LEWIS, J. (2001) *Vision for Britain, the NHS, the optical industry and spectacle design 1946–1986* Dissertation, Royal College of Art, London (2001)

NEWELL, A. Inclusive Design or Assistive Technology in *Inclusive Design. Design for the Whole Population*; 2003, Springer, ch. 10, p 173

NORMAN, DONALD. A. Emotional Design Why We Love (or Hate) Everyday Things, Basic books 2004, ch 1, p 21–24

OWEN, K. Lifestyle, design and disability, in *Inclusive Design. Design for the Whole Population*; 2003, Springer

PAYLING, J. 'Disability unplugged', in *Inclusive Design. Design for the Whole Population*; 2003, Springer, p 70

SPOORENBERG, A, BOERS, M, VAN DER LINDEN, S. 'Wrist splints in rheumatoid arthritis: A question of belief?' *Clin Rheumatol*. 1994 Dec;13(4):559–63. Department of Medicine, University Hospital Maastricht, The Netherlands

TIGER, L. *The pursuit of pleasure*. Boston: Little, Brown. 1992. 330 pp. ISBN 0-316-84543-4

UNDERWOOD, S. Orthotics to suit the wearers needs both socially desirable and medically functional p 9 *Include 2005* Proceedings April 2005, www.hhc.rca.ac.uk/resources/publications/include/2005/Include2005.pdf

UNDERWOOD, S. *An investigation of how to apply a more inclusive design method to orthotic wrist devices for people with rheumatoid arthritis*, University of Wales Newport (UWN) and University of Wales Cardiff (UWIC), April 2008

UPIAS. 1976: *Fundamental Principles of Disability*. London: Union of the Physically Impaired Against Segregation pp 3–4 (Also available on: www.leeds.ac.uk/disability-tudies/archiveuk/index.)

ZIMMERMAN, L, HILLMAN, M R, CLARKSON, J P. 'Wheelchairs: from engineering to inclusive design'. *Include 2005* Conference Proceedings. p 1

20
Wearable technology for the performing arts

J. BIRRINGER, Brunel University, UK, and
M. DANJOUX, Nottingham Trent University, UK

Abstract: In artistic contexts one expects integrated wearable devices to have the two-way function of interface instruments (e.g. sensor data acquisition and exchange) worn for particular purposes, either for communication with the environment or various aesthetic and compositional expressions. Wearables in performance surveys the history of wearables in the performance arts and distinguishes stand-alone and performative interfacial garments. It then focuses on current experiments in 'design in motion' and digital performance, carefully examining the prototype development at the DAP Lab in London and Nottingham involving transdisciplinary intersections between fashion and dance, interactive system architecture, electronic textiles, wearable technologies, and digital animation. The concept of an 'evolving' garment design that is materialized (mobilized) in live performance between partners originates from DAP Lab's experimentation with telematics and distributed media addressing the 'connective tissues' and 'wearabilities' of film through a study of shared embodiment and perception/proprioception in the wearer (tactile sensory processing) and the wearer/designer/viewer relationship. Such notions of wearability are applied both to the immediate sensory processing on the performer's body and to the processing of the responsive, animate environment. The chapter concludes with an outlook to the future of embodied wearable performance, new materials and new interface designs that enable the social experience of such performative garments in diverse aesthetic and cultural contexts.

Key words: performance, wearables, interactivity, design-in-motion, emergence, screendress, sensorial garment, digital environments.

20.1 Introduction: wearables in performance

Wearable computers are devices worn on the body giving the potential for digital interaction in the world. A new stage of computing technology at the beginning of the 21st century links the personal and the pervasive through mobile wearables. The convergence between the miniaturization of microchips (nanotechnology), intelligent textile or interfacial materials production, advances in biotechnology and the growth of wireless, ubiquitous computing emphasizes not only mobility but integration into clothing or the human body. In artistic contexts one expects such integrated wearable devices to have the two-way function of interface instruments (e.g. sensor data acquisition and exchange) worn for particular purposes, either for communication with the environment or various aesthetic and compositional expressions.

If mobility is a key element for performance with soft technologies, a second, even more important, aspect is the communicative potential – the particular forms of address, action, and interaction enabled by textiles and clothing which are designed to allow data processing and mediation.

On a functional level, wearables are distinctive from other mobile devices by enabling hands-free interaction, minimizing the use of a keyboard or manual input, for example, by devices worn on the body such as a headset allowing voice interaction and a head-mounted display, which replaces a computer screen; or shifting data control of sensor interfaces directly to bodily motion, muscle activity or breath. In the performing, visual and media arts, 'wireless' interaction with networked, digital environments never really meant 'without wires': sensors, electrodes and data capturing devices had to be applied to the body in order to produce data transmissions that could effect reactions from the 'system'. It is necessary, therefore, to look carefully at smart fabrics to explore whether they can replace the wires on the flesh-body, and how they are used in 'stand-alone' reactive design or as interfacial, affective design allowing performers to engage with a real-time system.

Current interdisciplinary developments in the meeting of soft technology and wearables show a wide range of design approaches for the fashion, sportswear, lifestyle or rehabilitation/medical sectors. Research initiatives such as the Smart Clothes and Wearable Technology Research Group (Newport, Wales) or the Interactive Textiles and Wearable Computer Group at Hexagram (Montreal, Canada) focus on smart clothing that addresses so-called end-user needs from technical, aesthetic and cultural viewpoints, pursuing a vision of sensing/communicating clothing that can express people's personalities, needs and desires, or augment social dynamics through the use of wearables as 'theatre' and as emotional 'tools'. While the latter purpose is of great interest for performers, the artistic experimentation with smart technologies is, of course, not market-oriented and has no end-user. Rather, the mobilization of smart technology concerns sensorial experience and expression (involving transformations of the performer–audience relations), a more experimental and playful adaptation of the digital medium as a *wearable medium.*

The historical precursors to intelligent garments in the performing arts are musical instruments and electronic devices used by musicians and, in some cases, dancers, actors or singers, to connect gesture, movement or voice to media output. Since the 1950s and 1960s, musical interface design added new dimensions of musical expression through, at first, MIDI technology (transmitting data to receivers and synthesizer mappers through a standardized interface) and, more recently, digital signal processing and real-time gestural control, deploying sensor and actuator technologies and haptic or force feedback devices for musical control. Important research on digital musical instrument design, mapping algorithms, and intelligent controllers was conducted at IRCAM, STEIM, and other labs. Engineers at Bell Laboratories and MIT Media Lab designed innovative wearable tools, e.g. Joe Paradiso's Instrumented Dance Shoes (1997); Steve Mann's Wireless Wearable Webcams (1996) or his early wearable computing apparatus (1981). Charmed Technology, an MIT Media Lab spin-off, developed wireless devices that mimicked fashion accessories, and the Brussels-based Starlab began to develop a series of conceptual prototypes for intelligent wearables ('i-Wear') intended to communicate with the Internet. Other artists experimented with wearables as extensions/prosthetic devices for the body (Stelarc, Marcel.li Antúnez Roca) or as implants (Eduardo Kac). Trisha Brown's *Homemade* (1966) stands out as an early example of a dancer 'wearing' a film: Brown literally performed with a film projector mounted on her back. The projection of the film touched the wall, floor, ceiling and audience in synchronization with the 'live' dance. Die Audio Gruppe wore its 'Sounding Bodies' (electro-acoustic interacting garments with integrated sound systems) in public spaces, creating highly memorable characters such as the Audio Ballerinas (1990), Audio Geishas (1997), or the shrieking Audio Peacock (2003).

Brazilian artist Hélio Oiticica, producing his *parangolés* and *penetrables* (1960s) to be worn/inhabited by dancers to effect surprasensory states of absorption, anticipated today's interactive art and smart textile design, emphasizing biofeedback and haptic experience through tactile, auditory, sensorial and kinetic attention. The New York City-based Troika Ranch pioneered the use of wearables in the 1990s with their custom-built Midi-Dancer, a set of accelerometers worn on arms and legs. Thecla Schiphorst, Susan Kozel, Hellen Sky, Ruth Gibson, Tomie Hahn, Carol Brown, Pamela Z, Katsura Isobe, Yacov Sharir, Ermira Goro and other dancers/choreographers have worn sensors to explore soft processes of improvization with fluid, intangible exchanges, composing sound and imagescapes through their motion. Recent dance technology experiments point to an increasing integration of wearable sensor technology and intelligent fabrics with an aesthetic of multi-sensorial, transformative reverie, extending beyond the wearer's immediate self-expression to the entire virtual environment. Immersive performance, in which movement with body sensors creates a constant flow and feedback loop of sensory and perceptual data, which can 'edit' (control) image/sound projections of Virtual Environments and permutations of the time-images, heightens proprioceptive awareness of the physical body moving in space, and stimulates a process of re-experiencing what constitutes self and identity. The acceleration and deceleration of the image, of the human perception of image and sensorimotor logic, has made performance with wearables (e.g. sensors measuring neurophysiological functions) not only significant for artistic experiments with the physiological, the machinic, and the virtual; increasingly, wearables in performance also suggest intimate and varied interfaces for the body, which allow micro-sensory surveying of consciousness and the imaging activity of the human organism.

In general, one can distinguish two categories of wearables used performatively:

(i)   Stand-alone, integrated intelligent garments such as Joanna Berzows-ka's 'Memory Rich Clothing' prototypes, which record intimate actions felt on the clothes, or Barbara Layne's 'Jacket Antics', which can be considered 'reactive' as they respond to surface contact or built-in programming. Layne's garments are woven with light emitting diodes (LEDs), microcontrollers and sensors that can generate a scrolling message when the wearers hold hands; the fabric itself is the circuit board and is completely flexible as it is woven into the structure. Thermochromic and other transformational garments, or Chalayan's shape-shifting, mechanically morphing 'Flip Through' dresses (Spring, 2007) and 'Video Dress' (Fall, 2007) prototypes also belong to this category, as the media or mechanical effects occur *within* the worn materials themselves. Such surface effects have a display character that lends itself to theatrical and intimate dialog, but they do not seem to demand particular performance techniques.

(ii) Interfacial garments transmitting data and 'controlling' the virtual environment through sensorial embodiment, which offer more potential for a relational aesthetics, as performers can effect changes in the audio-visual context. Such 'wearing' of the space, so to speak, produces transformations in the behavior of information environments, and the careful modulation of such interrelations with a reciprocating environment requires a performer technique. It requires, at the least, a familiarity with the dimensions of real-time composition implied by such transductions. Music, theatre, dance, and locative media performances have experimented with presenting transductive embodiment. Installation art (e.g. Thecla Schiphorst's *soft(n)* and *exhale: breath between bodies*) in particular probes a wide range of affective interfaces for users but only very rarely 'gives away' intelligent garments to be worn by participating audiences. Artistic contexts in many cases privilege the trained performer to articulate the wearable as a medium of representation.

20.2   Design-in-motion: the emergent dress

20.2.1 Performance wearables/wearable performance

The notion of *design*, as it was understood in fashion and fine art, film, the graphic arts, and product design, has expanded in many directions and is now infused with new developments in information and communication technology, ubiquitous computing, biotechechnology and nanotechnology. These mediating technologies have profound effects on perceptual systems and the habituated knowledges to which we are adapting today amongst highly technological living patterns. Designing our environments, and the tools through which we communicate with them and experience ourselves in the world, therefore must be considered to have vital cultural, social and political stakes. The role of fashion and clothing, although marginalized in contemporary critical theory and performance studies, cannot be over-looked as it directly relates to complex social as well as theatrical concepts of 'performance'. The latter are influenced today by the impact of com-puter and video games on our understanding of participatory player culture and the various scripts that are followed by gamers to configure settings, characters, plots and the modifications of avatar appearance. The fantasy worlds of games are a primary example of shared design, and in some instances (*Second Life*), all the 3-D content and appearances are entirely user-created. The sharing of fantasies is an activity particularly encapsu-lated in fashion.

To give attention to the 'fashionable' or to performing fashion does not necessarily mean talking about fashion styles or trends. Rather, in the context of highly technological living patterns under late capitalism, perfor-mative fashion can be linked to disciplinary power in Foucault's sense of social organization, insofar as fashion coerces the body to shape and rear-range itself in accordance with ever-shifting social expectations.[1] The skills required to adapt to such internalized expectations – including the ability to diet, apply facial cosmetics, arrange clothes, and wear ornamentation – are in the service of aesthetic innovations that continually reinvent subjec-tivity and body-image. Foucault's notion of the self-regulating 'docile body' indicates how elements of a fashionable lifestyle are techniques for 'wearing the body' as a commodity. One could even connect this performativity to the training standards (doping) amongst competitive athletes and the pub-licity standards of pop stars who embody the fetishized icons our societies like to worship and emulate. Celebrity hero worship is a phenomenon in the fashion industry in recent years (e.g. the Beckhams who have their own line in perfumes and clothing). These celebrities are our avatars: we pay for the entertainment in which our role models perform, and we emulate them. The red carpets are laid out. The transformable body becomes a training site of aesthetic innovation, a projection site or *tableau vivant* for fetishistic desires and, like our other technological accessories, it is subject to periodic upgrading. Redesigning the look of the model is to give it a new lease of life, specifically by submerging its use-value into its appearance-value.

The notion of iterative design, common in prototyping of interactive products as well as interactive artworks, here echoes with ironies as body transformations or body prosthetics, so often displayed in the arts, are now critiqued within the ethical contexts of biotechnology.[2] Redesigning life or human enhancement no longer looks as innocent as the other spring collections on the scientific catwalk. In the artistic context of wearable performance, mobile control of transformability subverts the commodity aspect. The wearable rather points to fashion in the sense of re-fashioning, not just 'controlling' surface functionality in the interface but challenging digital transformation of the materiality of the body to provoke a new language through which discrete representations of the body can be generated and re-invented. We will discuss an example of the second category, interfacial design, which holds this transformative potential for the compositional modes of the performing arts.

The collaborative 'Emergent Dress' project also demonstrates how the performance, media, and fashion design context intersects with computer science, engineering and new developments in human–computer interaction design.[3] The project is fuelled by new material technology – new fibres, fabrics, and innovative processing techniques that allow the integration of sensors or smart functionality into clothing. While the category of the 'wearable,' drawing attention to the sensorial effect as interface, is slowly introduced into the field of performance, this area is still very much experimental. There are few artistic works deploying wearables and reaching a wider audience or influencing other work.[4] This chapter therefore proposes design principles and maps the ground for a speculative description of how performance transforms design strategies for wearables, and how the wearable experience affects highly mediated performances.

The type of mediated performance which stimulated our project – interactive, telematic performance – implies the experience of being present at a location remote from one's own physical location, generally involving a camera-based internet convergence between two or more sites. Someone experiencing telepresence would therefore be able to behave, and receive stimuli, as though at the remote site. The work requires networked audio-video convergence, with the scenes at two distant sites becoming one. The architecture of such convergences in a studio or gallery involves multiple screens and sound diffusion for the live web streams and real-time 3-D Virtual Environments. Such an immersive environment exponentially expands what we normally comprehend as our immediate sensory environment or 'kinesfield'.[5] The kinesfield is extensive of the tactile experience of the garment as well. We suggest that such tactility can interfuse multiple telepresent bodies, making fashion an intersubjective experience. At the same time, the wearer of the wearable acts to enframe digital information, giving body to digital processes and thus to her or his own intimately and affectively experienced sensation of 'wearing the digital,' of becoming digital(ized).

Digital performance marks a significant shift toward a tactile, haptic aesthetic, away from an ocularcentrist mode of perception to embodied affectivity. The wearing, in other words, emphasizes synaesthesia and heightened awareness of the intersensory world. Digital performance is also theorized in terms of the disjunctions it opens up in the experience of reality, time, and digital space which no longer have an analogical basis. In the following, we examine how emergent design works with this and facilitates disjunction *and* continuity. First, we address the notion of wearable in motion, focusing on gesture and body movement. Secondly, we describe the prototypes within telematic environments for performance augmented by particular fabrics, materials and motion analysis as well as by gesturally nuanced computational media.

20.2.2 Digitized movement

The prototypes of the 'Emergent Dress' can neither be industrially mass-produced nor understood as *haute couture*. Rather, they are technically more *bespoke* garments intended for particular performance characters in choreographic and narrative works for which they are cast to have an integral role. They are mobile in a very personal and creative sense, practical and informal, ready-to-wear but at the same time also elusive and precious, evolving and changeable. The garments are meant to have a 'digital' quality.

The research connecting digital performance with new fabrication and interactive textiles therefore requires not only new fashion content for such wearables, but perhaps places the emphasis of design somewhere else entirely, namely to different qualities of 'performance' addressing not functionality but character, emotions, memory, fantasy, and experiential or psychological dimensions along with a heightened kinetic awareness of bodies as intimate communicators. These qualities, once considered a domain of theatre anthropology and social science research, now intrigue product designers looking at how artifacts elicit emotions. Sometimes these are very intimate emotions, but since they need to be articulated and shared, there is a microscopic trend among performance artists to devise very 'private' transactions and exclusive one-on-one encounters with their audience.[6] The collective potential of intimate transformative ecstasy experienced in wearable fabrics (*Parangolés*) enmeshing the participants, enveloping them in carnivalesque play of spatial and social relations, was advocated by Hélio Oiticica in the 1960s. Lygia Clark's sensorial masks were created in Brazil around the same time; her 'relational objects' were primary artistic vehicles for her sustained exploration of the client or receptor's experience of a micro-sensorial intimacy.

More recently, Lucy Orta reinvigorated these principles of intimate architectures of cloth, devising her 'collective wear' into interactive shelters and modular habitations (*Nexus Architecture*) drawing in the collaboration of local community participants. Extending the lineage of sound art, kinetography (Laban), and contact improvisation, recent experiments in dance and music technology point to a stronger interrelationship between gesture analysis (biophysical data) and the design of responsive architectures for emergent behaviors. If psychoacoustics research speaks of the subjective visceral nature of experiencing sound in the body, so is 'Emergent Dress' directed at the intimacy of the wearable experience. It is directed at the desire and erotic sensuality attached to the clothes we wear on our skin, the frivolous, extroverted but also secretive (even anti-aesthetic) dimensions of fashioning appearance, and the physiological processes and patterns through which the proprioceptive systems attend to the body's wearing of itself.[7] This emphasis on intimacy features a design conception opposed to Hussein Chalayan's 'Remote Control Dress' collection, which like his earlier Aeroplane and Kite dresses seems preoccupied with futuristic design shapes inspired by aerodynamics, architecture, furniture and mobility (speed and the conquest of space). The 'Remote Control Dress' was meant to interact with the built environment and was created with hi-tech materials, including glass fibre and resin, with glossy panel-like shapes (the front and back panels were held together with metal clips) and translucent plastics. These dresses look more like armors, façade-like exoskeletons that appeared to be protective shells for the wearer who is envisioned maneuvering through hostile habitats with a smart garment able to detect changing weather conditions or security alerts. These dresses were only shown as prototypes intended to relay information from the outside to the wearer – the wireless system was to be developed by Starlab but was never completed.

Crucial for our concern with close-to-the-skin technology are the affective and perceptional processes working both ways in the interaction, as the wearable here is not only a garment but also the interface. While there is indeed a noticeable tendency in the West towards an 'experience economy' and a cultural privileging of intensities and (emotional) participation, the question of what is meant by 'experience design' needs addressing, as the increasing use of sensor technology in our environment reveals little about *how* people do make use of 'feedback from an information technology' (Baurley), how they integrate the machine intelligence emotionally and cognitively, or how such intelligence influences clothing experienced from an expressive/psychological point of view (as protection, modesty, ornamentation, articulation of desire, etc.). Linking fashion to the pleasure principle, one wonders how the interface becomes charged with elements from the catalogue of eroticism and seduction, and how a particular style of wearing it can be decoded if the 'image-clothing' implies an endless number of ambiguous possibilities not determined through its semiological structure (as myth or message). Roland Barthes's analysis of the 'fashion system' points to the complexities of fashion's mobility, for example, if one looks at the vestimentary system ('dress') and its variant replications of attributes of the body (sexuality, desire), as well as at the circumstantial modes of behavior ('dressing') adopted by wearers.[8]

Dressing-performance in the digital context is all about articulating such mutability and exploring subtle or frightening exaggeration. Some avant-garde fashion designers, such as Rei Kawakubo of Comme des Garçons, have also contradicted the idea of decorative intensities and proposed unexpected, deconstructive designs which complicate the conventional premises of congruence (measurements that follow fit, like the glove to the hand), proportion, vertical axis, and the contouring of the body's outlines. For example, in 1998 Kawakubo produced a collection of disconnecting parts; the clothes were broken up, fragmented and incomplete, sides and backs missing, left and right not matching, etc. This strategy points to a diffusion and metonymic fragmentation of units which capture the sense of the digital we experience in the interface. Similarly, the recent photographic work of Zoren Gold with fashion model/designer Minori (*Object that Dreams*) is a *tour de force* of extraordinarily surreal and fetishistic images, digitally manipulated to the extent that Minori's doll-like appearance either seems textured into the surrounding environment or mysteriously alienated from it. Critical investigations of what might be called 'feedback design' and its effect on 'dressing' in movement are needed, and although Barthes's fashion theory tends to privilege the 'written garment' over 'real clothing,' his inventory ('Variants of Existence') provides many insights.[9] His 'Variants of Movement' are particularly helpful in understanding how the body animates the garment (e.g. he considers movement values such as rising, upsweeping, hanging, plunging, falling, swaying, etc.).

For wearables to become meaningful in artistic or theatricalized social settings, the affective experience in human–computer interaction needs to go beyond simple 'actuators' and those expressive interactions often referred to in terms of sensorial qualities (touch, sound, taste, smell) which are outside of the visual but whose significance is not brought home. How do we make sense in a haptic relationship to a certain material (glass, soft fabrics, leather, metal)? What is sensed? And what cognitive and aesthetic processes are engaged when, for example, a garment can measure the beat of your heart or transmit your emotional disposition towards your partner? When it hangs loose or plunges forward to 'act' like a magnetic or chemical attractor, playfully dominating the partner or the voyeur, toying with excess, then withdrawing, swaying and folding? When it remembers your touch, as in Joanna Berzowska's 'Memory Rich Clothing' prototypes that sense and display visible markers of events such as whispering and groping? How do you perform with a dress that exposes your inner thoughts, your anger or your longing? In opera, theatre, and dance the traditional costume design tends to support or illustrate character, and character motivation is expressed through voice, dialogue and movement. The intelligent garment, therefore, would have to be considered as having a 'voice' of its own or participate equally in immanent expressive role play.

In our group's examination of the kinesthetic wearable, we started out with a particular interest in remote communication (telematics) and the proprioceptive relations to the virtual. We refer to this sensory relationality as 'exo-processing,' and 'ScreenDress' (Prototype 1) was tested in performance improvizations by Helenna Ren and Nam Eun Song with a particular garment material and the motion graphics, which are digitally printed/projected onto it. The graphics represent the continuous, infinitely changeable feedback design, in line with the fluidity of the moving shape of the body. We then focused increasingly on the dancer's immediate experience of sensors on the body, and the wearer's proprioceptive relationship to sensorimotor and internal biophysical data (heart rate, pulse, breath), analyzed and transmitted in collaborative telematic 'play' with interacting partners. This sensory relationality is called 'endo-processing,' but the performance involves both an internal and an externalized dimension, as the dancers are pushing the data into the visual screen displays and image flows. The different outfits of 'SensorDress' (Prototype 3) drive our investigations of the subjective experience of the garment when touched and 'exposed' in dance. The experience of the performer is modeled for the camera but also activates an avatar for a game environment which playfully invites participation in the game-fantasy. The *Klüver* installation of Prototype 3 (exhibited at various festivals including the Prague Design Quadrennial 2007), the completed exhibition of all prototypes in *Suna no Onna,* and the live game *See you in Walhalla* provide the context for reflections on emergent design and the audience's empathetic relationship to the emotional character of the body–garment interface.

20.2.3 Emergent Dress

Our ideas on intimacy evolved from Helenna Ren's telepresence performances with partners in Arizona, Italy and Japan over the past years (2004–2007). She tested all of the versions of the transformable body–garment and the digitally manipulated garment–body. In the development of the new prototypes, the 'corset' became our first vehicle – conceptual metaphor and material object – along with other parts and fragments of pink, yellow and black cloth that Ren put on, assembled and disassembled in performance, and thus composed into her improvized choreography of wearing, gesturing, folding, stretching, unbinding fabrics and needles, moving and re-moving gloves, shifting elements of the garments into and out of focus, while inter-acting with the 'physical camera' (moved in close proximity by the camera operator). In these performances, created in real-time telematic contact primarily with performers Natalie King, Keira Hart and Joe Willie Smith in Tempe, Arizona, 'characters' also emerged gradually as the dancer began to shape the information composed with the *corsetbody*. We called them 'Zorro,' 'Houdini,' and 'Klüver.' While Ren moved with the garment, her video image was sent to Arizona's screen environment (constructed as a garden of large hanging leaves) and projected onto the leaves as well as a luscious white Victorian dress suspended from the ceiling and functioning as the centre-piece of our partners' experimentation with haptic memory (see Figs 20.1–20.3).

Helen Raleigh, AJ Niehaus and Galina Mihaleva devised the conceptual strategies with which our partners in Arizona used this memory-rich garment as a 'stand-alone' sensory sculpture, which could be projected upon, and then stepped inside and worn, which the Arizonan dancers did during the rehearsals, shedding their other clothes and embedding themselves inside an older (historical) garment concept. They inserted themselves into a kind of 'habitat' which enveloped them.

20.1 Keira Hart behind Victorian dress (video still © 2005 J. Birringer).

20.2 Helenna Ren in front of Victorian dress (video still © 2005 J. Birringer).

20.3 Helenna Ren merging her movement with Victorian dress (video still © 2005 J. Birringer).

The telematic performances challenge the double bind of the literal and the virtual garment. We experience the suspended dress as a receptacle, a sensory surface which functions as a mnemonic landscape and an instrument responding to touch, its built-in sensors producing a sonic text (words recorded in Joe Willie Smith's voice). The 'Victorian' dress flares into temporary focus, historical images or fragmented film stills from a natural landscape appearing on it, then all discrete traces disappear again, and now, as Ren's telepresent body-image remobilizes it, distance and proximity become interwoven. John Mitchell, who directed our partner team in Arizona, suggested that such a dress is viewable from *the inside out*, its porous quality 'evocative of ancient shadow plays and early cinematic devices that created viewer intimacy through subtle perceptual and sensual shifts. The resulting experience is expansive, contractive, enveloping and yet non-enclosing' (email, December 1, 2005).

At our end of the telepresence, it became apparent that Ren's movement itself contributed to this evocation of a living dress-sculpture or vessel. Wearing pink corset and rose-colored stretches of fabric, she used her breath to work with contractive and expansive rhythms, continuously changing the shape of her body as the tight-laced corset shifted flesh and muscles, inflating the shoulders, exaggerating the wide–narrow–wide silhouette of the female form. At the same time, the compressions and deformations of fleshy tissue were distracted by the layerings of fabric Ren whirled around herself (see Fig. 20.4). As the camera captured her movements and stretched the fabric into the virtual space of Arizona's suspended dress, we began to see Ren animating the distant garment: her body clothed in undergarment streamed into the latter's faint blue apparition, the outer wear in the distance, silhouette intermingling with silhouette, real and virtual bodies compounded into an illusion of a (digitally) composited whole. These rehearsals prepared the ground for the development of the ScreenDress prototype which elaborates the real-time compositing effect of the digital media.

20.2.4 ScreenDress: the two forms of the garment

ScreenDress is constructed from Chromatte, a technical light-reflecting cloth for chroma key production in TV and film. This material, designed to work dynamically with a LiteRing (a camera-mounted device featuring LEDs) utilizes the retro-reflective properties of its fabrication for live effects/image replacement, fusing motion graphics with onscreen performance. The *technological garment* is a real garment – the physical form of the garment, existing in the real world and in its isolated state (uncoupled from the LiteRing), is a gunmetal grey. Here, the focus is on the material facts of fashion, the applied aspects and technical solutions to produce an artistic result. We are concerned with the cut and the fabrication, the detailing and finishes, and the overall silhouette statement, the structure of the garment and how the body engages with the piece, i.e. how it is worn/performed and choreographed into movement. Pleats, expanding and contracting, layers, seams and a modular approach to garment construction are used. Garment and movement are inseparable, in that one extends the other,

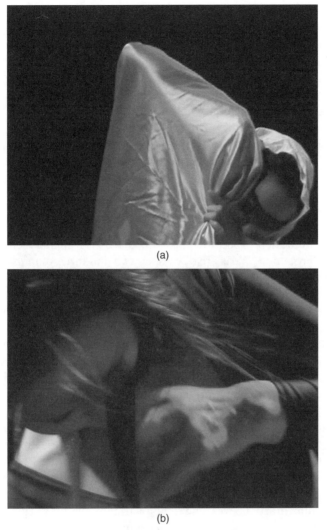

(a)

(b)

20.4 Helenna Ren manipulating her partial garment in performance (video still © 2005 J. Birringer).

becomes the other. The moving body has an impact on the form of the design and the form of the design has an impact on the moving body.

The garment, according to Barthes' analysis of the body–garment relation, is extended as a body.[10] He explains this through his study of Erté's fashion drawings and specifically the Ertéan silhouette, where the woman becomes the garment and is somehow biologically fused into the woman–garment. The body can no longer be separated from its adornment and decoration. The woman and the garment become one. ScreenDress, with the anamorphic shards of motion graphics projected onto it, is extended as a body, poetically shifting its surface of moving patterns and textures. It invents and substitutes, simultaneously masks and reveals, is animated, becomes alive and organic, with multiple juxtapositions of image and color. The live 'camera eye' removes us from one reality into another, one where dancers, organic tissues and animation become fused into expressive visual statements, an indissociable mixture of body, garment and graphics, bleeding forms, one into the other. This is a dialog between the natural and the artificial, brought together in an intimate relationship to create a new object or artifact, the *iconic garment*. The iconic garment becomes a spectacle in its own right, a mechanism for display and experience.

20.2.5 Interaction with cloth and partial design states

Generally, the design process begins with the 2-D design sketch and the static form. From here, ideas are developed two and three-dimensionally in preparation for final 3-D garment realization. However, working with the 'Emergent Dress,' we do not start with the design sketch or static state. Instead, we choose movement narratives, introducing partial garment structures and cloth to the initial frame; inviting the performer to move with the cloth, exploring and experiencing the qualities of the cloth (in front of our eyes/the camera eye), its potential and design possibilities. We observe the movement reactions initiated by the tactile stimulus of the cloth and consider how garment form and structure might begin to emerge.

In the case of the Chromatte cloth, Song was invited to discuss with the designer how she felt about the cloth; what type of movement behaviors it began to generate; structures and scale. Song found the touch of this particular cloth somewhat harsh and aggressive; it felt hostile to her movements and unforgiving. We explored slashing the fabric to ease the sensation of restriction and tried coiling and wrapping cut lengths of the cloth to produce rudimentary sleeves and other garment features. The designer responded to the cloth's structural dimension with pleats, creased stitched folds in the fabric which created an even more structural surface, one that could now expand and contract, open and close with each affected move. There is a fusion of designer and dancer in this process-based methodology of iterative design (see Fig. 20.5).

Song experimented with different movement qualities and energies to explore how the 'digital sketches' – incorporating the digital graphics into her movement consciousness and proprioception – affected her ability to frame the constant flow as time–image and image–movement. A particular screen poetics evolved, and although Song's unconscious experience cannot be verbalized here, it is apparent that she investigated the ScreenDress as an interaction instrument, a membrane between the real and the projective,

20.5 Retro-reflection (green screen), Chromatte and LiteRing work dynamically together. Nam Eun Song with Emergent ScreenDress (photograph © 2006 M. Danjoux).

between herself and her other. However, the relative stiffness of the Chromatte material and its austere appearance in real space gained mysterious textures and luminosities in the animated screen graphics. The fabric material thus created a contradictory pleasure. Or rather, the motion graphics behaved in a paradoxically counterintuitive manner, concealing the true nature and identity of the fabric and revealing a more organic, biological, emotional response. A confusion of sensation, control shifting from fabric to dancer and back, the experience became visibly more visceral and sensually involved as the relationships shifted in motion.

In one particular dance improvization, Song decided to use her voice (to spit words out and suck them in) in a complex breathing pattern to modulate the dressing performance, to emphasize the contradictory interplay between inside–out and outside–in motion. Her movement vocabulary is a mixture of ballet-inflected soft-flowing phrases and harsher martial arts punctuations, and throughout her dance with the wearable, she carefully allowed her sensory perception to guide her. Rather than looking at the screen and the transformed digital images of herself, she performed through being touched (by the motion of the stiff Chromatte fabric) and through kinesthetic sensing of the pulsating digital graphics – almost as if she could viscerally feel the digital animation. The kinesthetic sensation of this dance is primarily proprioceptive (inner), which becomes quite visible in the screen images of her movement. Her 'dressing performance' looks entirely different from Ren's much more extroverted, sardonic and outer-directed movement, which tends to toy with the voyeur-camera and also exaggerate both the stereotypical associations one might have with sensual/erotic gestures and the more grotesque aspects of her volatile characters (see Fig. 20.6).

These character studies or narratives formed the basis of constructing the digital movement composition for *Suna no Onna*, the exhibition we created

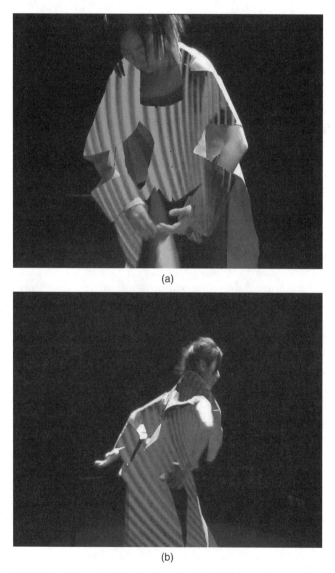

(a)

(b)

20.6 The Animated Dress: A union of design and technology, transforming patterns expand and contract (interstitial forms). Nam Eun Song with Emergent ScreenDress (video still © 2006 M. Danjoux).

from our story-board adaptation of Hiroshi Teshigahara's 1964 film. In test rehearsals involving Ren, we explored further the impact of garment form constructed with Chromatte. Ren shifted her focus from the characteristics of the fabric to the constricting corset structure she wore (tightly laced and reinforced with spiral steels), then to the fragments of cloth that began to adorn her body and the influences these began to have on her movement behaviors. She rotated her hips, displaced her weight seductively and used her hands to further cinch in her already diminished (effects of corset) waist. The clothed body became the eroticized body. Clothes are the form in which the fashioned body is made visible. Ren's response was equally important to the designer, as it revealed the intense kinesthetic stimulation of the garment or accessory, in this case the corset, for her. It demonstrated the 'touch' of the garment, the relationship between garment and body, and how this permeated the surface beyond skin deep. Ren's movement also acknowledged some of the conditioned movement behaviors carried within our bodies and social expectations of dress. She communicated to us her own erotic fantasies, the private-lived experience of wearing the corset, whilst also displaying learned movement behaviors; the way we move in a tightly laced corset, or a loosely fitting pair of combats, an identity-concealing hooded top, a fluidly cut silk dress – all those countless internalized movements and knowledges.

The process is taken one stage further with the integration of the motion graphics and the dancer's intimate engagement with her own animated self, the likely interaction of control and submission. The garment becomes capable of mediating interaction and encouraging new social relationships (between online partners). ScreenDress is ornamental and expressive, a union of design and technology. Its transforming patterns grow and contract, interstitial forms and pulsating rhythms move across the surface, digitally emulating the biological body, the nerves and membranes and natural flows of energy. It is a shape-shifting kinetic form, constantly morphing, moving from concave to convex; one minute the dancer is wearing herself, her own emotions, the next she is displaying a (remote) partner's imagery, becoming interwoven and intermeshed. Positioned as progressive fashion, this is not conventional clothing, but clothing for exploratory experience. ScreenDress is directed at the notion of watching and sensing, the intrigue of knowing more, of seeing beneath the surface. As a 'relational' wearable, it enables us to engage with ourselves and to touch the other (and be touched) intimately at a distance.

20.2.6 SensorDress and Digital Dunes

Barthes' historical theory of the different structures involved in the fashion system – the real garment, the iconic structure, and the written garment –

can now be extended to include the digitally animated garment and, inversely, the intelligently worn film. The shifting between forms of the garment in *Suna no Onna* (The Woman in the Dunes) is effected by digital technology and real-time transformation. The various garments prepared for the production are transposed from cloth to moving graphics and a fluid, sculptural, kinetic and cinematic form that resembles the visual kinetics of earlier experimental film (Moholy-Nagy, Léger) but also breathes the spirit of contemporary futurisms in the world of hip hop, games, power advertising, kung-fu and sci-fi movies. We deliberately chose a classic Japanese film noir to explore such a total digital real-staging of our performers' interpretations of the characters and their behavior in the dunes environment. Our stage adaptation uses color and a range of textures, intermingling real objects with digitally animated environments. The data for the animated landscapes were motion-captured from our performers.

The notion of 'wearing the film' was first used by Jane M. Gaines in her analysis of specific film dresses that draw attention to the elaborate design in excess of the narrative or diegetic function.[11] In our case, SensorDress is a garment constructed with intelligent materials (sensor fabrics), allowing the motion with the garment and the body to transmit wireless signals to the computer. Rather than re-appearing as an iconic garment in the screen environment or virtual space (as with ScreenDress), SensorDress allows the dancer to animate directly all the events that happen in the dunes environment projected as a responsive world now activated, ruffled and moved in real-time. The woman protagonist in *Suna no Onna* spends her entire life in the dunes, her body almost literally fused with the sand habitat. In our treatment, the garments worn by her become the interface with the digital sand. She is embedded, she wears the space in the sense of continually affecting the flow and motion of the generated live filmscape.

What does the dancer control? How is the dancer's experience controlled and affected? In the expanded kinesfield of a three-, four- or five-dimensional 'unstablelandscape',[12] the dancer is wearing and carrying the fabric of the film, so to speak. Her body, clothed in sensor-rich garments (using orient, flex, tilt, rotation, photocell sensors), directly mobilizes the image-movement, and thus also our eyes. As the first-person eye, she controls the 'camera eye' in the programmed environment of the virtual world, through her own body motion and manipulation of fabric on skin, the underfabric of the digital data. The lifestyle and perpetual habits of the female protagonist (an entrapped woman, continually shoveling shifting sands that threaten to bury her home) inform our design and performance processes for sand habitation. Katsura Isobe, who interprets the role, and the worn garments become 'dissolved' into a representational state of the psychological existence of the woman, her labor and expectations. The colors and textures relate to the space as a whole, the surrounding scenic space and the digitally projected space.

For additional design inspiration, the designer studied the organic photographic images of Edward Weston and the multi-layered, multi-textured knowledge based dress systems of the Japanese Samurai Warriors and traditional kimono. The layering and tying of these systems afford design potential which offers the dancer the scope to alter the nature of the garment statements during performance. Garment surfaces were treated with a sandblasting finishing process to give the impression of time and effort and of a wearing-down in the dunes. Color palettes for the woman range from tones of sand, ochre and sienna to rich earth colors. The garment states are deliberately not fixed, but in a fluid state of transition. For instance, a utility coat with hood can become a sculptural, organic and abstract form when wrapped differently around the body. Each shifting form of garment brings about a new form of interaction and movement behavior.

Both underwear or outerwear become the touch of sensors, but in the interactive world of this performance, the cloth also is the camera or the 'controller' of the movement-images, and thus the tilts, the swoons, the falling and rising, the drift of the 3-D world that projects itself. When Isobe turns in a 360°-rotation, the filmic motion enacts the same rotation. The organic wearing of the shifting, moveable, moldable kimono reveals and dispels a world of images that floats, turns, collapses and rises. Motions of floating forms, sometimes congealing to a recognizable Gestalt, sometimes flowing as graphic and geometric abstractions, anamorphoses, distortions of space. Digital insects appear to crawl out of the sand and move about (motion captured from our dancers who 'become' the insect avatars). Each garment part employs a different digital–body interface, allowing different techniques and gestures to be combined to explore the interface between the real and the projected. The layers of cloth become prosthetic impulses, carnal and entomological projections hovering between the phenomenal and the biocultural, grains of sand, grains of sounds, insect antennae, colors and folded cloth and bodily animations feeding back from the screen world, the digital modulations of her figure in a suspended landscape.

The dancer is the sensorimotor actor of the movement–image, but the question of control, in the sense of cause-and-effect, can never be fully answered, as the programming of the interface allows the sensor-rich dress to move the frames (forward, backward, fast forward, slow down, freeze-frame, jitter, etc.) within algorithmic parameters that also allow the computer to interpolate or generate movement/stillness and various 'masks' that act as overlays or filters of the images. The audience has the illusion as if they are floating into and through the landscape, perceiving the dancer's kinetic movement in fluid contact with the audiovisual landscape, but of course the actual sensors are invisible (incorporated into the garment) so

that no particular triggering motion can be discerned. Movement is continuous, subtle behavior, as if becoming one with the film, a diver submerging
in water and becoming indistinguishable from the liquid environment.

Ermira Goro, during separate rehearsals for the live computer game *See
You in Walhalla*, enacts her role of a real/physical avatar in front of a 3-D
projective urban environment. The avatar and the remote players (streamed
into the triptych video projection architecture) are participants in the game,
which consists of hundreds of scenes filmed in several European cities
merged into one (the fictive Walhalla). Goro is wearing 14 different sensors
distributed all over her body-suit, from head to toe. In performance, she is
re-choreographing her movement continuously, incorporating the sensorial
processes on her body which allow her an immediate, direct relationship to
the virtual worlds (digital film and animation) in front of which, into which,
she literally moves and navigates. As one would navigate an avatar in a
game, the performer in this live production is navigating her own character
into the digital world, and she is also responding to the programmed interactions she does not control, such as the appearances of the other players
streamed in from the remote webcams (see Fig. 20.7).

Her vocabulary and the subtlety of her gestures direct her first-person
perspective into the animated world manipulated through her body and
garment. This requires a deep practice and immense dexterity (rehearsing
with the sensorial fabric). At the same time, our production rehearsals with
the SensorDress have shown that the embodied garment-character itself is
partly responsible for the affective and responsive proprioceptional processing of the dancer in action, her physical and affective relationship
to the garment and the image of 'character' created in the wearer. The

20.7 Ermira Goro as Avatar in Walhalla (video still © 2006 J.Birringer/
Christopher Brellis).

emotional relationship and expressional exchange between dancer and garment, in turn, effect the projected world and its behaviors, which can be felt by the audiences. This exchange directly affects alterity, the other person or persons in the same space with the performer, and we do not distinguish here between 'real' temporal space and disjunct 'real-time' space. The consensual process takes place in telepresence. Other persons, creatures or digital landscapes partner the effect, respond to gesture or action, modulate perceptual experience, as in any call and response situation in a social context. The wearer is the caller, responding. The behaviors of the responsive environment are keyed to the motion in wearing (see Fig. 20.8).

New research into the impact of human behavior on design practices suggests that typically we do not realize how and to what extent we are participating in and therefore shaping culture. The products of design engage humans both through their utility and their cultural location – their situatedness. While such context-dependency and the high social variability in the connotations of dressing performance are obvious, the question of how sensorial design is experienced and interpreted has barely been answered. But our performances point to sensorial and tactile affect as primary experience of movement within the body, which inhabits intensive (internal) and also extensive space. The artistic use of fashion and game scenarios can create its intimate complexities precisely through the way in

20.8 Katsura Isobe in *Sun no Onna* (video still © Interaktionslabor).

which it modifies or perverts the codes it seems to be using. At a time when we seem to move casually between real habitats and digital, virtual domains, busily reconfiguring cultural differences and boundaries, it is useful to pay attention to small details, especially when microelectronics become fused with the tissues of the human body as exotic add-ons or when iPods become cool personalized jewelry enabling private listening expression. In the realm of clothing and sensor performance, the small, minimal cues can have large consequences. Blatant erotic messages can turn into their opposites, sand can turn into water or bury you. With the Emergent Dress project, explorations focus on extending the wearables into an understanding of constantly moving environments. Embeddedness means living in a permanent state of emergency.

20.3   Outlook: an overview of cutting edge experiments in intelligent fashion/ wearables and performance

The contemporary fashion/performance hybrid began to emerge in the late 1990s with designers such as Alexander McQueen and John Galliano hosting elaborately staged fashion shows and the more avant-garde designers such as Martin Margiela holding smaller, 'esoteric performances, reminiscent of happenings and performances by artists such as Rebecca Horn and Ann Hamilton.'[13] The development of wearable technologies and intelligent fashion now offers new scope for this genre of exploration, extending the hybrid to include experiments in intelligent fashion/wearable and performance. As discussed earlier, the presence of the wearable in the performing arts offers new creative dimensions and modes for spontaneous expression, blurring the boundaries between internal and external, real and virtual realities. The wearable can, for instance, act as conduit into 'immersive space' (e.g. Char Davis's *Osmose* with head-mounted display and motion-tracking vest) and between networked bodies (e.g. Schiphorst's *exhale*). It can express emotional and physiological states and social interaction patterns in a surface show of relational and visually revealed intimacy of wearing (e.g. Despina Papadopoulos's 'loveJackets).[14]

Like a prosthetic limb, the 'wearable' is given vitality by the wearer and wearer interactions, requiring bodies and language for its expression. Vivian Sobchack describes the prosthesis as being 'organically related in practice'. She states that a prosthetic leg has many components and involves dynamic mechanical and physical processes, as well as a descriptive vocabulary of its own.[15] The wearable also involves similar processes and methods of description and technological interfaces to speak its language (through data processing and mediation). But, just as the prosthetic leg cannot dance meaningfully without the rest of the body, the 'wearable' cannot perform meaningfully without the wearer–performer.

20.4   Sources of further information and advice

Recent developments in wearables/intelligent fashion explore a range of embedded technologies and bring forward an array of new and interactional prototypes, some still at concept stage, others already commercially viable and ready for the market.

Chalayan's 'Video Dress' prototype falls into the category of Video and Photonic Textiles, which have been described as light-emitting clothes built from photonic textiles, currently under exploration by companies such as Philips, where a garment is transformed into intelligent and revealing display through the integration into the fabric of flexible multicolored LEDs. This type of garment works with a reactive and communicative surface, similar to Joanna Berzowska's memory-rich dresses (see Fig. 20.9).

Another category of wearable technology is that which focuses on wearable body monitoring products and technologies/cognitive body suits which use physiological sensing, monitoring and positioning. Such 'smart clothes' house sensing systems to remotely monitor vital health signs, or

20.9 'Video Dress' (Hussein Chalayan, 2007 © courtesy of the artist).

new tracking devices to enable parents to establish the exact location of their offspring. Sensors (orientation and pressure) are integrated and use communication devices such as Bluetooth and GSM for interactivity. Textronics' (leaders in electro-functional fabrics and fibres) *NuMetrex Clothes*, such as the *NuMetrex Heart Sensing Sports Bra* (2005), also fall into this category. Electronic sensing technology for detection of the heart's electrical impulses is integrated into the knit fabric of the bra. These impulses are then transmitted to a wristwatch for monitoring of heart rate whilst exercising (http://www.textronicsinc.com/). Wearable body monitoring is more applicable to the sporting sector and also for healthcare (e.g. other sporting sector examples are Adidas's *Fusion* for sport/training integration, Nike's shoe with integrated sensor for the iPod jogger, and the *Know Where Jacket* with integrated GPS for satellite-assisted positioning). But some artists have made use of such wearables in installations offering the visitor a particular auto-sensing experience which translates physiological information into sound and image visualizations. At the third InBetween Time Festival at Arnolfini, Bristol (2006), Australian artist George Poonkhin Khut's *Cardiomorphologies* invited individual audience members to wear an accessory which enabled them to observe and then interact with their breathing and heart-rate patterns (www.arnolfini.org.uk/ibt) (see Fig. 20.10). It will be interesting to see what audiences or participants think of Berzowska's new *Skorpions*, a series of kinetic electronic garments that shift and change in unexpected ways because of (deliberately) malfunctioning micro-controllers.

The *Hug Shirt* by CuteCircuit is a shirt that enables people to send and receive hugs over distance. Embedded in the shirt there are sensors that feel the strength of the touch, the skin warmth and the heartbeat rate of the sender and actuators that recreate the sensation of touch, warmth and emotion of the hug to the shirt of the distant loved one (http://www.cutecircuit.com/now/projects/wearables/fr-hugs/) (see Figs 20.11 and

20.10 'Cardiomorphologies' (George Poonkhin Khut, 2006 © courtesy of the artist).

20.11 'Hug Shirt' (CuteCircuit, 2007 © courtesy of the artists).

20.12 'Jacket Antics' (Barbara Layne, Studio subTela, 2007 © courtesy of Barbara Layne).

20.12). CuteCircuit claim that the *Hug Shirt* is not meant to replace human contact, but to make you happy if you are away for business or other reasons and you miss your friends and loved ones. While this prototype enters the interrelation category of emotional design and emotional wearing/

bio-responsive wearing with the body-garment, it has not found an artistic context yet. However, Barbara Layne's 'Jacket Antics,' presented at SIG-GRAPH (2007), were first tested in performance by choreographer Yacov Sharir who wore the jacket with embedded LEDs on stage, directing his dancers to respond to words that appeared on his back, while someone off stage was wirelessly changing the scrolling text via a bluetooth device (http://subtela.hexagram.ca/blog/?author=2subTela).

20.5   The future of embodied wearable performance

New developments in biotechnology and genetic engineering take perfor-mance (and human-centered design) into the medical laboratory and oper-ating theatre, with artists and designers using human tissue as a medium for their art. Barthes once referred to the hair as 'promised garment' when discussing the fashion illustrations of Erté, positioning this excreted fila-ment of protein as a potential garment or covering for the body. Today 'fragments of our bodies are potentially becoming part of the extended body and fusing with other semi-living beings,' Oron Catts and Ionat Zurr suggest, as garments and accessories are cultured from the microscopic raw materials provided by ours and other species' flesh and bones.[16]

Catts's and Zurr's Tissue Culture & Art Project (TC&A), a research and development project into 'Victimless Leather' conducted by SymbioticA (Perth, Australia) steals the show and points the direction in terms of where the future might lie for performance and fashion/wearables. 'Victimless Leather' – a 'prototype of stitch-less jacket grown in a Technoscientific Body' – is a miniature garment grown from mouse and human bone cells. These cells have been applied to a biodegradable polymer base and the jacket is grown under laboratory conditions outside the body/inside a glass flask. The jacket measures just 2 × 1.5 inches and is described by TC&A as 'artistic grown' and 'semi-living.' It is one of their 'object-beings' that use a disturbing and ironic approach to genetic (re)production and thus question ethical issues of 'wearing' and these new dimensions of bio-performance (http://www.tca.uwa.edu.au/vl/vl.html) (see Fig. 20.13).[17]

Other examples in the field of bio-engineered wearables are 'Epi Skin' by Marta Lwin, biological jewelry grown from skin cells, and 'Biojewelry' by designers Tobie Kerridge and Nikki Scott, working in collaboration with the bio-engineer Ian Thompson. In the field of performance/body art, Stelarc has experimented with TC&A and surgeons to grow a tissue-engineered 'Extra Ear' (2006), implanted on his left forearm. The ear aug-ments Stelarc's body; it is not a replacement prosthesis but an addition. When equipped with sensor and microchip, it will become a 'speaking' ear. For the artist, performing with the body's evolutionary architecture means rendering it into a 'host' for more biocompatible components.

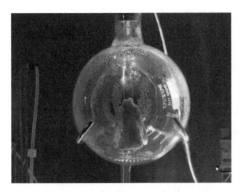

20.13 'Victimless Leather' (SymbioticA, 2007. Photograph courtesy of Oron Catts and Ionat Zurr).

Here the performance wearable becomes the 'object-being': technologically enhanced, it becomes something part organic, part machine. It is designed to extend human capacity, creative expression and 'wearable' function beyond a mere substrate for monitoring and signal. It is not inconceivable that developments in medical science and stem cell research will eventually enable us to relocate the cultivating of the 'grown' garment or the bio-engineered fashion accessory or textile to the surface of the human body and away from the test tube environment. If this does happen and ethics allow, then we could ask how this process of creation might be displayed artistically and over time. Bio-art performances raise provocative questions, namely whether the human body could actually one day grow a garment. An organic extension of the natural body and its various physiological and sensing systems might produce the ultimate in bespoke of the personally tailored body-garment. Such technological and genetically engineered body-garments would, one imagines, clothe the body in an entirely new way, possibly offering protection against predisposed diseases or the ultimate in body enhancement and augmentation. The function would be more complex than the standard 'wearable', with flesh–machine fusion models such as the 'bionic arm' creating new potential in body–object relationships and interaction design.[18]

Blaine Brownell argues that 'we live in a time of unprecedented material innovations that are affecting our lives' and that 'the accelerated pace of these innovations and the breadth of their applications have enhanced our awareness about new products and the ways in which they are transforming our physical environment.'[19] Wearable and bio-technologies and affective computing have paved the way for a new type of interaction in both the physical and the virtual world. Transfers of knowledge are happening more and more between disparate disciplines, with boundaries becoming blurred and possibilities endless. The artists' interest in the body and technology lies not just in utilizing these new and influencing human–technology interfaces, but also in interrogating them for greater human understanding. Performing arts, bioscience, electronic engineering and fashion, together shape the future of embodied wearable performance.

20.6   Textual notes

1. See Gilles Deleuze's reading of Foucault in, 'Society of Control', *L'autre Journal* 1 (1990): http://www.nadir.org/nadir/archiv/netzkritik/societyofcontrol.html.
2. See Joanne Finkelstein, 'Chic Theory,' *Australian Humanities Review* 5 (1997): http://www.lib.latrobe.edu.au/AHR/archive/Issue-March-1997/finkelstein.html; Alphonso Lingis, *Body Transformations. Evolutions and Atavisms in Culture*, London: Routledge, 2005; and Marquard Smith and Joanne Morra, eds., *The Prosthetic Impulse: From a Posthuman Present to a Biocultural Future*, Cambridge, MA: MIT Press, 2005.
3. 'Emergent Dress' is a project produced by the DAP-Lab, a research partnership between Brunel University and Nottingham Trent University (http://www. brunel.ac.uk/dap). The prototype development of the 'Emergent Dress' collection (http://www.brunel.ac.uk/dap) includes 'ScreenDress,' featuring design concept for garment by Michèle Danjoux (fashion) and Jon Hamilton (motion graphics); 'Explay,' 'CaptureDress,' and five versions of the 'SensorDress' conceived by Michèle Danjoux (fashion), Johannes Birringer (choreography, sensordesign), Paul Verity Smith (sensordesign), with Helenna Ren, Nam Eun Song, Katsura Isobe and Olu Taiwo (dance). Additional corset design fabrication is by Susanna Henson of Eternal Spirits. The 'Emergent Dress' collection was featured in the installation *Suna no Onna*, produced for the Laban Center, London, in December 2007. 'Walhalla' is an avatar design for a live game performance developed through 'i-Map,' a European coproduction, with garment concept by Despina Makaruni. *See you in Walhalla*, featuring dancer Ermira Goro, was tested at Interaktionslabor in the summer of 2006 (http://interaktionslabor.de); the premiere was in Athens in September 2006, and the production is now touring internationally.
4. For a provocative concurrent experiment in wearables, especially focussing on the somatic aspect of sensor technology integrated into fabrics, see Thecla Schiphorst's description of her *exhale: breath between bodies* exhibition in 'Breath, skin and clothing: Using wearable technologies as an interface into ourselves,' *International Journal of Performance Arts and Digital Media* 2, 2 (2006): 171–86. Her most recent project, *soft(n)*, shown at *DEAF07: interact or die* (2007), featured intimate tactile and kinesthetic interaction with hand-sewn objects that elicit behaviors such as humming, shaking, sighing, singing, and luminous patterns. Jane Harris' work, on the other hand, explores the presence and portrayal of characters through dress and textiles in the realm of 3-D computer graphic visualization. The digital animations (*Potential Beauty*) she exhibited in the UK in 2002–2003 focused on the poetic and dreamlike movement of the dresses alone, insofar as the actual wearer of the garments is 'deleted' in the final screen version. See http://www.janeharris.org/. The 'Remote Control Dress' collection (spring/summer 2000) created by Hussein Chalayan is described in Bradley Quinn, *The Fashion of Architecture*, Oxford: Berg, 2003, 127–30. See also, Bradley Quinn, *Techno Fashion*, Oxford, Berg, 2002, Chapter 5. For Zoren

Gold and Minori's work, see *Object that dreams*, Berlin: Die Gestalten Verlag, 2006.

5. Gretchen Schiller, 'Interactivity as Choreographic Phenomenon,' in *Dance and Technology/Tanz und Technologie: Moving towards Media Productions – Auf dem Weg zu medialen Inszenierungen*, edited by Söke Dinkla and Martina Leeker, Berlin: Alexander Verlag, 2002, 164–195. In describing the installation architecture of her collaborative work *trajets* (with Susan Kozel), she speaks of the particular 'kinaesthetic responsivity' in highly mediated, sensitive and interactive environments which integrate movement and digital media. Referring to Laban's definition of bodyspace or kinesphere, she suggests that, in interactive installations, *bodyspace* extends from Laban's *kinesphere* or personal reach to a 'kinaesthetic dynamic across material forms, forces, space and time,' and she is not even including the telepresent dimension of co-present remote spaces and actors. But she evocatively argues that the dynamic and interaction between 2-D and 3-D spaces collectively fall into a new conceptual bodyspace (*kinesfield*), concluding that 'digital technologies can bring our awareness to qualitative variations and inhabited dynamics between spaces in this *kinesfield*.'

6. See 'This Secret Location,' an exhibition of performance-installations at the 2006 *In Between Time Festival*, Arnolfini, Bristol, or Kira O'Reilly's performance *Untitled Bomb Shelter* at the 2005 New Territories/National Live Art Review in Glasgow. Similar one-on-one encounters have been devised by choreographers Willi Dorner, Felix Ruckert, John Jasperse, Leung Po Shan, amongst others. Microscopic intimacy was revealed in DS-X.org's *Image-controlled sound nanospheres*, an installation at the 2005 Digital Cultures Lab featuring emergent behavior of cells.

7. Sharon Baurley's essay, 'Interaction design in smart textiles clothing and applications,' in *Wearable electronics and photonics*, edited by Xiaoming Tao, Cambridge: Woodhead Publishing, 2005, 223–43, provides a broader context by introducing some of the advances in technical textiles production but especially in pervasive computing and the shift towards wearables, mobile devices, and the embedding of computer intelligence within everyday objects and environments. Referring to scientific research in affective computing and interaction design, Baurley points out that pervasive computing indicates the dissolution of electronics into the material environment where the interface is constant, while 'affective computing,' grown out of wearable computing, aims at educating intelligent systems to recognize physical and physiological patterns and translate these into emotions. The growing interest in the sensorial and emotional affect in design was widely reflected in the 2006 Design and Emotion Conference at Göteborg's Chalmers University of Technology: http://www.de2006.chalmers.se/.

8. Roland Barthes's *Système de la mode* first appeared in 1967 and has been much overlooked in performance studies. Within the emerging context of digital performance and wearable technologies, re-viewing Barthes's semiological study seems long over-due, while it must be kept in mind that Barthes himself later revised his 'brutally inelegant' structuralist approach, aware of the complexity of ambiguous undercoding through which fashion continuously modifies what it seems (not) to be saying. In the 1970s, Barthes's interest also increasingly turned to the body and away from the 'written' garment. Cf. Roland Barthes, *The Language of Fashion*, edited by Andy Stafford and Michael Carter, Sydney: Power Publications, 2006, 9.

9. Roland Barthes [1967], *The Fashion System*, trans. Matthew Ward and Richard Howard, Berkeley: University of California Press, 1990, 111–43.

10. Whereas Hegel seems to have preferred a formless surface as 'ideal' in clothing the body for the expression of the 'spirit,' Barthes voices his critique of Hegel by way of the silhouette in Erté's alphabet-drawings of women. 'Hegel has noted that the garment is responsible for the transition from the sensuous (the body) to the signifier; the Ertéan silhouette (infinitely more thought out than the fashion mannequin) performs the contrary movement (which is more rare): it makes the garment sensuous and the body into the signifier; the body is there (signed by the silhouette) in order for the garment to exist; it is not possible to conceive a garment without the body' (*The Language of Fashion*, 153). The body, in other words, is the support for the garment.

11. Jane M. Gaines, 'On Wearing the Film,' in *Fashion Cultures*, edited by Stella Bruzzi and Pamela Church Gibson, London: Routledge, 2000, 159–177.

12. The term 'unstablelandscape' is used by Marlon Barrios Solano to describe generative hybrid performance systems (for humans and computers) for digital real-time interaction in which performers or participants complete the feedback loop improvisationally, and all the elements in the environment are inherently changeable and unpredictable within computational parameters. See Marlon Barrios Solano, 'Designing Unstable Landscapes,' in *Tanz im Kopf/Dance and Cognition*, edited by Johannes Birringer and Josephine Fenger, Münster: Lit Verlag, 2005, 279–91.

13. Ginger Gregg Duggan, 'The Greatest Show on Earth. A look at contemporary fashion shows and their relationship to performance art,' in *The Power of Fashion. About Design and Meaning*, edited by Jan Brand and José Teunissen, Arnhem: ArtEz Press, 2006, 222–243.

14. For Char Davies's writings on 'Virtual Space,' see http://www.immersence.com/. Despina Papadopoulos's 'loveJackets' were shown at *Wearable Futures: Hybrid Culture in the Design and Development of Soft Technology*, 14–16 September 2005, University of Wales, Newport, Wales, UK. For more information on Papadopoulos, see http://www.5050ltd.com/peppy.html.

15. Vivian Sobchack, 'A Leg To Stand On: Prosthetics, Metaphor and Materiality,' in *The Prosthetic Impulse. From a Posthuman Present to a Biocultural Future*, edited by Marquard Smith and Joanne Morra, Cambridge: MIT Press, 2006, 17–41.

16. For a thought-provoking essay on biotechnology and living fragments of bodies, see Oron Catts & Ionat Zurr, 'Towards a New Class of Being: The Extended Body': http://www.tca.uwa.edu.au/atGlance/pubMainFrames.html.

17. With SymbioticA's 'Victimless Leather,' there is also moral wearing attached to garments and technology, not quite wearable technology but more to do with new materials (for the fashion industry). They claim that the 'artistic grown' garment will confront people with the moral implication of wearing parts of dead animals for protective and aesthetic reasons. They are not providing another product but raising questions about the exploitation of other beings (http://www.tca.uwa.edu.au/vl/vl.html).

18. The development of the 'bionic arm,' a replacement arm and flesh–machine fusion, is a feat of neuro-engineering, linking the functioning of the prosthesis directly to the body's own neurological and information processing systems. The wearer's brain controls the movements of the newly integrated limb.

19. Blaine Brownell, 'Introduction: A Material Revolution,' in *Transmaterial: A catalog of materials that redefine our physical environment*, New York: Princeton Architectural Press, 2006, 6.

20.7   Bibliography

http://www.nime.org/
http://www.hexagram.org
http://www.xslabs.net/theory.html
http://artschool.newport.ac.uk/smartclothes/
http://www.wearablecomputing.com/
http://www.media.mit.edu/wearables/
http://www.eee.bham.ac.uk/wear-it/intro.htm
http://www.interactive-wear.de
http://www.research.philips.com/newscenter/archive/2005/050902-phottext.html
http://www.eyesweb.org/
http://www.anat.org.au/
http://www.brunel.ac.uk/dap

BIRRINGER, JOHANNES, *Performance, Technology and Science*. New York: PAJ Publications, 2008.

BRAND, JAN and JOSÉ TEUNISSEN, eds., *The Power of Fashion. About Design and Meaning*. Arnhem: ArtEz Press, 2006.

BROWNELL, BLAINE, ed., *Transmaterial: A catalog of materials that redefine our physical environment*. New York: Princeton Architectural Press, 2006.

CATTS, ORON and IONAT ZURR, 'Towards a New Class of Being: The Extended Body,' ISEA 2006. http://www.tca.uwa.edu.au/atGlance/pubMainFrames.html.

CORIN, FLORENCE (ed.), *Interagir avec les Technologies Numériques*, *Nouvelles de Danse* 52 (2004).

MASSUMI, BRIAN, *Parables for the Virtual: Movement, Affect, Sensation*. Durham: Duke Univ. Press, 2002.

MCQUAID, MATILDA, *Extreme Textiles: Designing for High Performance*. London: Thames & Hudson, 2005

MITCHELL, ROBERT and PHILLIP THURTLE, *Data made Flesh: Embodying Information*. New York: Routledge, 2004.

OSTHOFF, SIMONE, 'Lygia Clark and Hélio Oiticica: A Legacy of Interactivity and Participation for a Telematic Future,' *Leonardo* 30:4 (1997).

PREECE, JENNIFER, YVONNE ROGERS, Helen Sharp, *Interaction Design: Beyond Human–Computer Interaction*. Hoboken, NJ: John Wiley & Sons, Inc., 2002.

QUINN, BRADLEY, *Techno Fashion*. Oxford: Berg (2002).

QUINZ, EMANUELE (ed.), *Interfaces*, *Anomalie digital_arts* No. 3 (2004).

SMITH, MARQUARD and JOANNE MORRA (eds), *The Prosthetic Impulse. From a Posthuman Present to a Biocultural Future*. Cambridge: MIT Press (2006).

STELARC, 'The Involuntary, the Alien, and the Automated: Choreographic Bodies, Robots & Phantoms,' Digital Performance, *Anomalie Digital_arts*, No. 2 (2002).

VINKEN, BARBARA, *Fashion Zeitgeist. Trends and Cycles in The Fashion System*. Oxford: Berg (2005).

XIAOMING, TAO, *Wearable Electronics and Photonics*. Cambridge: Woodhead Publishing (2005).

21

Branding and presentation of smart clothing products to consumers

W. STAHL, Peninsula Medical School, Exeter, UK

Abstract: Sophisticated branding and graphic techniques are used to give products their own identity and to provide the necessary information to a potential customer. Even if a person knows of the existence of a new item of clothing, they need to know how it benefits them and how to get the most out of it. This chapter attempts to provide a starting point for the representation and presentation of smart clothing and wearable technology based on the current situation, which it is hoped may also prove interesting to other collaborative design disciplines. It discusses briefly the differences in the way that electronics and fashion brands present themselves through their visual identities. This provides a basis for looking at the possible difficulties faced by graphic designers attempting to create identities and provide information on new smart clothing and wearable technology products. The techniques employed here are to identify some of the most challenging aspects of future clothing for the graphic designers, and to look both inside and outside the fashion and electronics areas for inspiration. The discussion concludes with a concept for bringing together established design techniques as a possible approach to the situation.

Key words: smart clothing, wearable technology, graphic design, presentation, branding, identity, display.

21.1 Introduction

The area of smart clothing and wearable technology can be both confusing and exciting. There is a fantastic range of possibilities for new clothing – heart monitoring garments, temperature sensors, changing colours and embedded communication technologies. Clothes can begin to receive data from our bodies and the world around us, and react accordingly. This could provide benefits for such diverse end uses as sports wear, business clothing, the military, consumer goods, the services, and many other areas. However, this wide range of potential functions and uses that make this such an exciting new area of design, can also be somewhat confusing (see Fig. 21.1).

From the very start of the design process, collaboration is necessary between the fashion industry, electronics engineers, software programmers and textile manufacturers. Many of the specialists in these areas will not have had contact with each other before, and will not be used to the

21.1 Smart Clothes and Wearable Technology, sketch (Will Stahl, 2006).

methods employed in other design disciplines. But it is not only designers by any means who need to be informed of the possibilities for smart clothing and wearable technology. The end-users themselves will need a certain amount of information if they are to identify how smart clothing and wearable technology can be of benefit to them. Between the designers and end-users of smart clothing and wearable technology come the intermediaries such as retailers, health care professionals, and other advisors, who will be the ones who have direct contact with these end users. All of these people will need to be exceptionally well informed to be able to design, provide, use and support the use of the garments of the future.

Visual communication can be useful in providing information in such complex and diverse areas. This chapter suggests that graphic, visual communication can be a useful, effective method of approaching the communication that is needed, without supposing that it is the only method that might be used. If these designers, suppliers and end-users need increased knowledge, a process of learning is required. There is a wealth of material to be found on how people learn. It has been suggested by writers such as Howard Gardner (1983) that there are many different learning styles, to which some are better suited than others. Visual communication is often taken to be one of these learning tools, which can be used to impart information quickly and effectively.

In an ideal situation, we would see increased awareness of the possible uses and functions of smart clothing and wearable technology, as well as confidence in their reliability and useability. The manufacturers and retailers would no doubt like such products to be seen as desirable. Graphic communication can be used to assist in all of these things, as described in this chapter. Information graphics, such as signage, transport information, and many others can effectively communicate even complex information and ideas. Branding and identity work is frequently used to reassure customers, to give an idea of values or qualities of a product or organisation. This chapter is aimed at giving a very brief overview of how graphic communication is used in areas relevant to smart clothing and wearable technology. It will discuss how visual elements and identities might be approached in this area, attempt to show examples of these ideas, and end by suggesting a possible way forward.

This should prove informative and interesting to anyone working in the field of smart clothing and wearable technology, but is particularly aimed at designers. Care has been taken to ensure that, while no particular knowledge of or training in graphic design is necessary, the chapter should hold relevance to anyone who is interested in visual, graphic communication in the area of smart clothes and wearable technology.

As a note of warning, however, the lists of possible technologies and applications are growing at a fast rate. At the time of writing, the number of electronic devices within clothing that have reached the markets are few, but is growing quickly. This rate of change in terms of available technologies makes it impossible to predict with any surety how the field will develop. Writing for publication in this area is therefore difficult. It should be acknowledged that all a chapter such as this can contribute is a snapshot in time, a historical record of the particular state of the industry at the time of writing (2007).

21.2 Shopping for future clothing

Within the model of our current consumerist society, once a point is reached at which one's coat talks to one and to the outside world, it is likely that a graphic or visual communication system will have been found. A person's experience of wearable technology might be expected to involve engagement with graphic communication techniques at two main stages.

Visual communication will probably appear on and around the garments themselves; in the display of information using embedded flexible screens, printed onto the garments themselves, or designs picked out in lights across a piece of clothing. This could provide environmental data to the wearer, or to people around them. It could provide information on the care or use of a garment, or help wearers to locate their position in an unfamiliar city. The list is a long one, and will, of course, extend as new technologies advance to the stage where they can easily be incorporated into clothing.

However, before a person even gets to wear the item in question, it is almost certain that they will encounter graphics in the form of advertising, branding, visual identity, and other informational/instructional materials that relate to the item in question. New smart clothing might be expected to come from the same outlets as traditional clothing. We will consider here the graphics involved in the experience of buying wearable technology from 'bricks and mortar' shops, printed catalogues and online shops.

21.2.1 Shops

Perhaps the most obvious encounter with graphics in a clothes shop, after the initial branding and identity of the store window, is the point-of-sale materials. These may help in navigating a densely packed shop, provide information about the clothing on offer, or be purely decorative, offering 'lifestyle' options or reinforcing the identity of the shop itself and its products. At the time of writing, these displays tend to be static, but they are becoming increasingly media-rich, incorporating video, animation or sounds.

When electronic systems are incorporated widely in clothing, the possibility appears that the shop itself will be talking to our clothes, especially if they are electronically enhanced products that have come from the same shop some time before. This might make it possible for a previous garment purchased to show the owner where the latest upgrade is in the shop, or give information about choices. The graphics, in this case, move from the physical environment of the shop rails and walls, into the changing and virtual space of the screen or display on a worn garment.

One potentially exciting benefit of electronic communication and clothing is the opportunity for two-way communication. Instead of the shopper being bombarded with information, but without their own voice, what happens when their clothing is fully able to reach out and share information with the shop? Shops and shop assistants are in a unique position. They are the point of contact for those who are acquiring new clothing. They, of course, need to be able to inform the shopper, but can also receive valuable feedback directly from the end user. The graphics in shops could well assist in both.

Displays, both static and media-rich, can certainly inform about the functions of clothing. However, the inclusion of interactive methods could well provide opportunities to record patterns of use, and to gauge interest in smart clothing possibilities.

21.2.2 Packaging

Should the physical high street shop continue to be prevalent, the packaging of new devices and clothing will be important. 'Packaging' for clothing tends to come in the form of swing tags, at least where it is important for clothing to be tried on. One of the classic questions asked in the area of electronics and clothing is: 'Does wearable technology come on a hanger or in a box?' In the early days of electronic clothing, however, it is quite likely that both methods will be useful. The use of modular electronic components that fit together with items of clothing are a likely 'low-tech' solutions to issues such as washing garments and the need to upgrade devices, or transfer them to different garments.

The Nike/Apple collaboration is an early example of this. The electronic components are here sold completely separately from the shoes themselves. In this situation, graphics can tie the two parts of a purchase together – something of which Nike and Apple are keenly aware. But this idea extends readily to other possible clothing technologies. When you get the same symbol on your swing ticket as the box that the electronic devices come in, you are reassured. The two share a common look, or brand, and therefore the consumer can be confident that they will work well together.

21.2.3 Catalogues

Printed mail-order catalogues have offered clothing, delivered on a 'direct to door' basis, for over 100 years. Despite the obvious problems of not being able to try clothing on until it arrives, this method of shopping has survived to this day, due perhaps in part to the rising overhead costs associated with traditional shops. These printed catalogues use graphic techniques, along with descriptive and informative text, to impart a great deal of information; however, this is ultimately static, one-way communication.

Catalogue sales may look more attractive in the world of electronic clothing, where function becomes more and more important. The fit of a garment will, of course, continue to be fundamental to the enjoyment of wearing future clothing. However, it needs to be said that the technical specifications of clothing may one day become more important to some shoppers than being able to try on the garments before buying them. As printed catalogues are an entirely visual medium, graphic design might be expected to play a fundamental part in imparting the necessary information on price, fit, size, capability and compatibility of future clothing.

21.2.4 Shopping online

While online shopping for clothes currently suffers from some of the same difficulties that are found in printed catalogue shopping, the future may be different. A good quantity of the consumers of new, technologically advanced clothing might be expected to be fairly comfortable with the idea of virtual shopping. As technology advances, it is not so hard to imagine an online shop in which you might be able to construct a representation of yourself to try on the clothes that you are about to buy. You would literally be shopping through graphics. Rudimentary forms of this are already appearing; for example, My Virtual Model – www.mvm.com (see Fig. 21.2).

These initial systems might offer only a couple of choices from a set of limited features, ending up with an avatar of 'someone that you think looks a bit like you'. However, from this initial stage, think what might happen once three-dimensional body scanning becomes easily achievable. In the

21.2 My Virtual Model – http://www.mvm.com.

end, we may get clothing that is ordered online, viewed on a virtual model/avatar with photo-quality features, then laser cut/welded to fit your exact body size, and delivered to your door.

While we may still be some way from such 'seamless, 1-click clothing', the individual stages are beginning now. We may expect more graphics-based avatar shopping in the near future. The advantages to both customers and retailers make such a system seem very appealing.

21.3   Graphics in the area now

The field of smart clothes and wearable technology can be seen as a blend of clothing, electronics and other technologies, combined to be used in a wide variety of end-uses. This chapter will look at how graphics are used in the current areas of electronics and fashion. This should help to indicate possible difficulties and opportunities that may arise when they are combined.

21.3.1 Electronics

Three very branding-aware companies in the electronics arena are Apple, Sony and IBM. They all have very clearly recognisable company identities, which can give us some information about what they are trying to achieve. They all share similar difficulties in marketing their products. Electronic devices, particularly computer systems, can be perceived as complicated and difficult to use. Having a strong brand identity can help to reassure customers.

Apple are a case in point here. They are renowned for having a large design and branding budget. While this can lead to criticisms along the lines of 'style over substance', there is no denying that they have very successfully marketed their computers and electronic devices as the logical choice for those for whom design and ease of use is important. The company also inspires fervent loyalty among its small, but significant, customer base. To these people, the Apple branding represents a high-quality product. Also, importantly, the consumer gets to know that an Apple monitor, iPod, or other peripheral device, is likely to work seamlessly with their computer, if they are all relatively up to date. The Apple brand serves to reassure the customer, in terms of how their computer will work, and also how it will link to other purchases. Their approach is to suggest their products as 'the friendly face of technology', allaying common fears in the area.

Sony have a similar approach; their range of products is far wider than Apple's, encompassing all manner of electronic devices, from TVs, portable audio devices, computers, stills and video cameras, games consoles, mobile phones, and many more. The Sony brand, therefore, is perhaps a little more diluted. However, it still stands for the higher price bracket, more reliable products.

IBM is a slightly different case. While they have made approaches into the consumer sphere with their personal computers during the 1990s, they now again operate, by and large, solely in the business world. IBM, the name under which they have traded since 1924, in fact, stands for International Business Machines. Their branding is also very well recognised, however. Their striped logo particularly, used from 1972, speaks strongly of reliability and a sensible, business-minded approach. That this logo represents their company aims so well is perhaps not surprising, given the fact that it is the work of the well-known designer, Paul Rand.

All these companies have their own aims and target customers, but they all serve as examples of how electronics companies use graphics, particularly corporate branding, to reassure their customers. There are many others, such as Sharp, Philips, Canon, Samsung, to name but a few. All seem to share this idea that the brand must give a solid, dependable authenticity to an electronic product, to reassure the consumer that it is well made and reliable.

21.3.2 Fashion/clothing

For fashion houses, almost exactly the reverse of the approach used by the electronics companies is prevalent. In the fashion world, particularly high fashion, style becomes the most important element, rather than function. Instead of the reliable corporate approach, fashion brands are often linked to individual designers (think of Armani, Givenchy). These brands are more likely to challenge than reassure their customers, representing change and up-to-the-minute style (see Fig. 21.3). However, in this lies what might be the only real similarity between the values promoted by fashion house brands and electronics manufacturers. They both must strive to be absolutely up to date. Computer manufacturers must be mindful of competing companies' progress. In the world of fashion, to be out of date is commercial suicide and the seasonality and turnover is far quicker than that possible in the electronics industry.

However, this is not the whole of the clothing market by any means. There are the more traditional clothing retailers and work wear suppliers, such as Moss Bros, and department stores, which seek to promote a more everyday image. Also, one potentially interesting area is that of sportswear. Brands such as Nike and Adidas are far more function or activity focused than the fashion market. They do use personalities to advertise their products, but this tends to be along the lines of using 'heroes' to reinforce their brand, rather than actually using a particular person as the brand itself (see Fig. 21.4). Aside from this, such companies use similar techniques to the

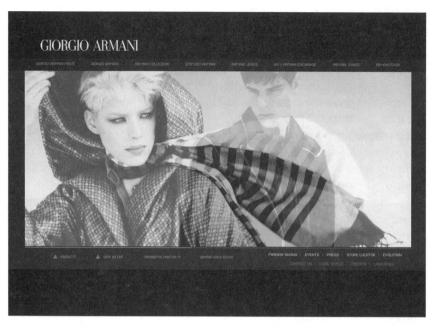

21.3 Giorgio Armani – http://www.giorgioarmani.com/ga_menu/EN/home.html.

21.4 Nike desktop wallpaper – http://inside.nikebasketball.com/news/.

corporate electronics brands, using a company brand to indicate a certain level of quality and value to their customers, intending to reassure them. In the case of sportswear, this often gives added value to a product. It is almost as though wearing a branded product has a status value in itself – to be seen wearing a 'genuine' product.

21.4 Graphics for smart clothing

The question still remains: how can we bring electronics and fashion together? What happens when the style-conscious fashion world would like to sell electronic devices, and vice versa?

One difficulty faced here is the differing lifecycles of clothing and electronic products. Fashion items quite often are brought in for 3- to 6-month cycles, at the end of which styles/colours are changed for a new season. Sportswear lifetime is a little longer than this, and electronics are often seen as too expensive to replace more than every couple of years. However, portable devices such as mobile phones are increasingly being offered as 'upgrades' as often as every year in the case of a typical mobile contract. It may be that more frequent minor upgrades to technological components could be possible, even given that electronic devices have necessarily longer lead times than clothing. It is, however, beyond the scope of this chapter to delve too deeply into this, for fear of replicating work of earlier contributors.

There are other problems, however, more particularly concerned with the graphical representation of smart clothing and wearable technology: When clothing and electronics converge, can there still be a clear identity for the new clothing? For such a collaboration between so many different areas, what could the brand represent? If it is possible, what might it look like? There are many issues still to be discussed.

Of course, we are already assuming that there is such a thing as 'a smart clothing and wearable technology company.' Can there be a smart clothing and wearable technology shop? It may be more likely, at least initially, that smart clothing and wearable technology will appear in small amounts in, for example, sportswear shops. It is also probable that different items will emerge simultaneously, aimed at different end uses.

Items for different end uses are going to need different styles. A heart monitor for snowboarding is necessarily different from one that is going to be depended upon for healthcare reasons. The question here might be: 'Can a single brand work across such differing end uses?'

21.5 Branding

It might take some time before new brands can emerge that can offer consumers the reliability of a technology company along with the desirability of a fashion brand. For the present, it seems that component technologies may acquire their own brands, which exist independently of the fashion labels. But this is not the only branding model that is emerging in the industry at the time of writing. Also, there have been several brand collaborations between companies that are keen not to miss out on this exciting new area. This section discusses three main branding models that appear to be used at present. We will refer to them here as: '*Technology as branded component*', '*Dual branding*' and '*Going it alone*'.

21.5.1 Technology as branded component

Some wearable electronics manufacturers are appearing, giving the technology itself a brand that works independently of the clothing's brand identity. An example of this is Eleksen's 'ElekTex' brand (see Fig. 21.5). This is a simple textile keyboard technology which can be stitched into the sleeves of garments, etc. Brands currently using the ElekTex branded controls include O'Neill, Spyder, Bagirm, Craghoppers, Marks & Spencer, and several others. This may be one way of supplying a trusted 'technology brand', while still retaining the desirable fashion brands, perhaps similar to

21.5 ElekTex website – http://www.elektex.com/.

21.6 Eider website – http://www.eider.com/.

what was attempted by the "Intel inside" mark for personal computers, but with some degree of style.

Electronics is not the first technology to be brought into a clothing range complete with a corporate identity. As far back as the early 1990s, Berghaus were proudly displaying the Gore logos alongside their own, and this practice continues to this day with many other companies (see Fig. 21.6). It does seem that long-standing clothing brands are not quite sure how to incorporate smart attributes. This leads to a reliance on other established technology brands to reassure the clothes buyer. New electronic technology may use similar branding methods to past technological advances in clothing, but this is not the only option available.

21.5.2 Dual branding

Adidas and Polar were amongst the first big names to collaborate on an electronic technology embedded within a garment (see Fig. 21.7). Their heart monitoring tops are a fairly simple application of an existing technology, basically stitching in a Polar heart monitor into a tight-fitting Adidas vest, with a couple of soft sensors in the lining. However, this collaboration does allow the two companies to retain their distinct identities.

Apple and Nike have recently collaborated to create a running shoe that records footsteps in an attached iPod; this then links to a comprehensive online service. The website, and the branding and identity, is particularly interesting. Instead of keeping the two identities separate, Apple and Nike have worked together to produce a visual style that is easily recognisable as a combination of both corporate identities (see Fig. 21.8). The Nike+

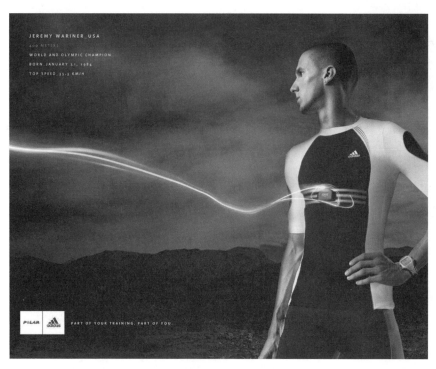

21.7 Adidas-Polar wallpaper – http://www.adidas-polar.com/.

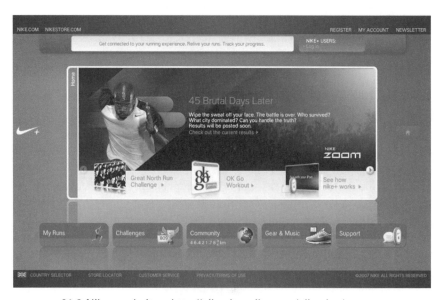

21.8 Nike+ website – http://nikeplus.nike.com/nikeplus/.

website in particular uses Nike's black and red colour scheme, but with the rounded edges and friendly type that so often appear in Apple's product information. Only Nike's corporate symbol appears on the site, but Apple's wouldn't look out of place on the page.

More such brand collaborations seem likely to keep appearing that maintain the companies' separate identities. Using established brands in this way does have the advantage of not requiring so much brand proliferation work, which means the products can be brought to market much more quickly. However, no such alliance is without its complications, and a certain amount of freedom must be compromised for two brands to be brought together.

21.5.3 Going it alone

There may well be a space for new brands to appear that can far more elegantly represent reliable technology and have enough style to be worn. Until then, we will continue to see brands trying to find common ground.

21.6   Creation of brand/identity – a starting point

To look for a brand that works across areas with different visual styles, we might look at graphics, which have been used to communicate where spoken language barriers are a problem, or where complex information needs to be communicated. This chapter will briefly look at the Munich Olympic Games in 1972, the London Underground map, and emergency exit signs.

21.6.1 International graphics

In 1972, the Olympic games were held in Munich, Germany. This was an international event, which drew an audience from around the world. The decision was made to appoint the graphic designer Otl Aicher to help direct people around the games (アイデア (Idea) magazine, 2005). His symbols that represented the different events not only proved useful in directing and informing an international audience possessing differing language abilities, but have passed into public consciousness as an example of good design (see Fig. 21.9).

His symbols were perhaps so successful because he took a familiar visual language, the body language of the sports, and used it to communicate independently of spoken language.

21.9 Three of Otl Aicher's pictograms from the 1972 Munich Olympic Games – http://www.pictogram.de/erco_piktogramme/geschichte/en/ en_geschichte_1.html?piktos.

21.6.2 Information design

The London Transport underground map is another highly popular example of good design. It took a very complex system and broke it down to its essential elements. The branding here is also highly recognisable. Once a passenger on the London Underground passes that famous red divided circle, they enter the virtual space of coloured lines, taking them to their destination. The London Underground brand is linked to the clear communication of hugely complex transport networks (see Fig. 21.10).

21.6.3 Signage

Even the UK's health and safety authorities have picked up on the idea of using visual communication. In the UK, it is no longer legal to provide just the word 'EXIT' to mark an emergency exit. Such signs must include a pictogram, as it cannot be assumed everyone can read the text (see Fig. 21.11).

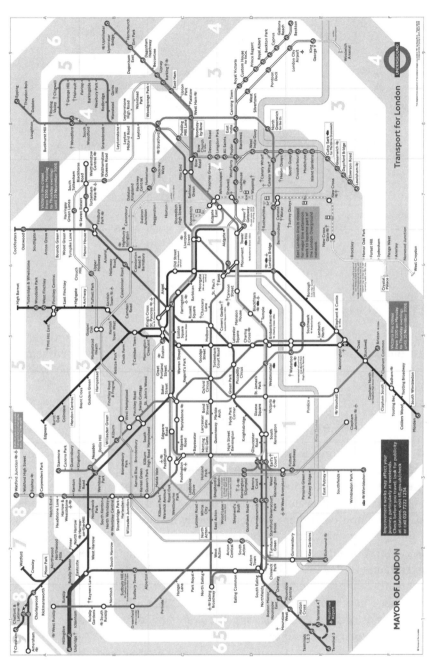

21.10 London Underground map – http://www.tfl.gov.uk/.

21.11 Exit sign.

21.7   Rules for presentation?

Graphic designers like rules, mainly because once they know what the rules are, they can start to work out ways of breaking them. In the case of smart clothes and wearable technology, there may be a few things that we can recommend as a starting point.

21.7.1 Target end-user

Much of the existing design models that this chapter has looked at do share one thing in common. From Otl Aicher's Olympic Games symbols, to the branding of Apple's computers, a clear idea of the target audience is essential. Knowing what visual language your potential customers will respond to, you can approach them with clear, easily understood communications.

Of course, most successful brands have a clear awareness of their end-user. Marketing departments continually gauge public response to brands, and attempt to steer a company in the right direction. Re-branding exercises frequently happen when a brand wants to appeal to a new audience. Guinness, once perceived as an 'old man's drink', had a huge marketing campaign, transforming and renewing the company, thus appealing to a far wider and younger audience.

Carhartt, once seen as a work-wear company, was successfully re-branded to apply to a completely different audience on its transition to the European market. The American Carhartt brand is still mostly hard-wearing work and protective clothing, whereas the European brand is much more street fashion based. Both of their brands often feature the same symbol as

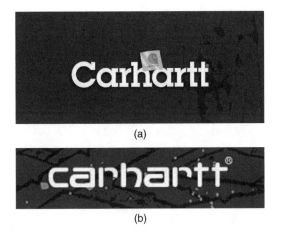

(a)

(b)

21.12 Carhartt logo. (a) US logo, (b) European logo.

part of their visual identity, but the type in the two logos is completely different (see Figs 21.12a and b).

21.7.2 Adaptability

For smart clothes and wearable technology, adaptability is particularly important. It seems unlikely that hundreds of different brands will be created for individual products, with one target end-use, and therefore one target audience. If individual components of clothes continue to be branded, as in the case of ElekTex and GoreTex, these brands will have to be applied within a wide range of end uses. If clothing brands appear that can bridge the gap between a desirable fashion brand and a reliable technology brand in one neat package, the company will more than likely have to supply a wide range of products. From a visual point of view, this will still require an adaptable identity.

One particularly interesting example of an adaptable symbol is the universal recycling symbol (see Fig. 21.13). This icon enjoys worldwide recognition, and yet it is difficult to find two appearances of it that are exactly the same. The simplicity of Gary Anderson's design, along with the fact that the symbol is in the public domain, and not a trademark, has ensured the symbol's enduring status as one of the most widely recognised symbols today. The downside of this for the designer is perhaps obvious: They can make no money out of it. However, this quality of adaptability and constant change in appearance is still potentially interesting in terms of developing symbols that can appear in differing situations.

21.13 A selection of recycling symbols.

21.7.3 Use of familiar visual language

We have already mentioned, in section 21.6.1, the use of existing visual language to approach communication where common spoken languages cannot be assumed. It may well be the case that visual languages will prove useful in many ways in the area of smart clothing and wearable technology. In a global market, this way of communicating without relying on shared spoken languages has advantages. It may be employed in describing the use of garments, creating awareness of possibilities in the area, reassuring customers, and creating a desirable brand to be bought into.

We have looked at Otl Aicher's use of the familiar visual language of body position in sport, but there are many other examples of similar techniques. Just one of these might be the use of colour in identity design. Almost all fast food chains have red and yellow logos. This makes viewers think of hot food, and increases their appetite. The red also attracts attention. In contrast, pale colours or desaturated colours can project an image of quality and/or relaxation. Designers are more fond of creating corporate logos with strong blues which emphasise reliability. These 'design rules' are things that we hardly ever think about consciously as consumers, but can all be used to speak to us.

21.8   Bringing it together

Smart clothing and wearable technology are certainly very large fields. Indeed, there is so much variety that it must be stressed again how essential it is that the end-user be very clearly understood. It seems from researching the area at the moment that certain questions need to be answered for each item of new clothing technology: 'Why is the technology needed?', 'Who needs it?', and 'How can this be clearly communicated?'. For this last question, a further addition of: 'What languages does the audience speak?' may prove to have helpful results. By language, this should not be restricted to spoken, but also visual languages. It should be possible to communicate clearly and effectively with end-users, and those that have to inform them, while still maintaining clear brand identities. This points towards an adaptable information system for technologies that work within clothing.

21.9   A concept

Of course, to truly understand the end-user, a great deal of research is necessary. The concept suggested for approaching clear visual communication while still leaving space for established brands, is to use a research-based design strategy – a system of triangulation, or hybrid strategy, combining qualitative and quantitative research data to access and utilise existing visual languages. This information can then lead to the creation of a communication system based on a particular visual element (such as a set of symbols) to describe the use and functions of new clothing technologies. The idea is to move on from the point at which we seem to have remained since McLuhan suggested that clothing can be looked at as a second skin (McLuhan, 1966). The possibilities of future clothing suggest that garments can become even closer to a second skin with their new sensory and reactive roles. More information on the categorised display of smart clothing information (*the web of meaning*), while leaving *space for design play* (adaptability) can be found in the paper: '*Design Play in the Web of Meaning: Towards a Methodology for Visual Communication in Emerging Collaborative Design Disciplines*' (Stahl, 2007) (see Fig. 21.14).

21.10   Conclusion

Smart clothing and wearable technology need a lot of work in terms of awareness building and information provision. Visual presentation is key to

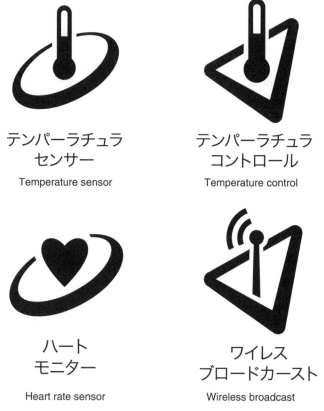

テンパーラチュラ
センサー

Temperature sensor

テンパーラチュラ
コントロール

Temperature control

ハート
モニター

Heart rate sensor

ワイレス
ブロードカースト

Wireless broadcast

21.14 Functionality symbols, Smart Clothes and Wearable Technology Research Group, University of Wales Newport (Will Stahl).

this. While corporate brands are a part of living in a consumerist society, and are likely to remain relatively unchanged, they seem to struggle to adapt to such complex, collaborative ventures as smart clothing and wearable technology. There may well be advantages in developing a more 'seal of approval' style of branding for wearable technology. Think of the Soil Association organic standard, but more adaptable. Perhaps a cross between this and the recycling symbol, plus the textile care symbols that already appear on garments are the way to go (see Fig. 21.15). This would be able to provide identity, information, and possibly even quality assurance in one symbol.

Whatever happens, it is likely that we will see new brands, new visual identities and innovative graphical communication helping to provide awareness and information on new smart clothes and wearable technologies.

	Cotton	Linen	Wool	Silk	Viscose Modal	Acrylic	Poly-ester	Nylon
Washing	95° Whites	95° Whites	30°	30°	40° Viscose	40°	50°	40°
	60° Colours	60° Colours	(hand wash)	(hand wash)	60° Modal			
	40° Dark Colours	40° Dark Colours	(hand wash tub)	(hand wash tub)				
Bleach	△ a	△ a	⨂	⨂	⨂	⨂	⨂	⨂
Ironing	⊞ •••	⊞ •••	••	•	••	•	••	•
								⊠ Without Steam
Dry cleaning	Ⓐ	Ⓐ	Ⓟ	Ⓟ	Ⓟ	Ⓟ	Ⓟ	Ⓟ
Drying	[••]	[••]	⊠	⊠	⊠ Viscose	⊠	[•]	⊠
					[•] Modal			[•]

21.15 Universal textile care symbols. The care of textile products depends on the fibre content and fabric finishes used.

21.11 References

GARDNER, H. 1983. *Frames of Mind: The Theory of Multiple Intelligences.* New York, USA: Basic Books.

アイデア (Idea) magazine, 2005. *Otto Aicher: West Germany.* No. 313: Nov., pp 104–105, Japan.

MCLUHAN, M. 1966. *Understanding Media.* London, UK: Routledge.

STAHL, W. 2007. *Design Play in the Web of Meaning: Towards a Methodology for Visual Communication in Emerging Collaborative Design Disciplines.* Newport, UK: University of Wales, Newport.

21.12 Bibliography

BOURRIAUD, NICHOLAS. 2004. *Relational Aesthetics.* France: Imprimerie Darantière.

CHANG, A. 2006. *Engineers are from Mars, fashion designers are from Venus: Bridging the gap between two opposing industries.* Wearable Futures Conference, University of Wales, Newport, 14–16th September 2005.

COBLEY, P. 1999. *Introducing Semiotics.* London, UK: Icon.

DE KERCKHOVE, D. 1997. *The Skin of Culture.* London, UK: Kogan Page.

EMBREE, L. 2002. *Reflective Analysis: A First Introduction into Phenomenological Investigation,* Florida Atlantic University, USA.

FLETCHER, A. 2002. *The Art of Looking Sideways.* London, UK: Phaidon.

GLASER, B. G. 1992. *Basics of Grounded Theory Analysis: Emergence Vs. Forcing.* Mill Valley, California, USA: Sociology Press.

GLASER, B. G. and STRAUSS, A. L. 1967. *The Discovery of Grounded Theory: Strategies for Qualitative Research.* New York, USA: Aldine de Gruyter.

GRAY, C. and MALINS, J. 2004. *Visualizing Research.* Burlington, USA: Ashgate.

JENSEN, K. B. (Ed.) 2002. *A Handbook of Media and Communication.* London, UK: Routledge.

JOLLEY, N. 2005. *Leibniz.* London, UK: Routledge.

KING, D. B. and WERTHEIMER, M. 2005. *Max Wertheimer & Gestalt theory.* London, UK: Transaction.

KOFFKA, K. 1925. *The Growth of the Mind.* London, UK: Kegan Paul, Trench, Trubner & Co.

LEE, S. 2005. *Fashioning the Future.* London, UK: Thames & Hudson.

LIVINGSTON, A. and LIVINGSTON, I. 2003. *The Dictionary of Graphic Design and Designers.* London, UK: Thames & Hudson.

LYNCH, P. J. and HORTON, S. 2001. *Web style guide: Basic design principles for creating web sites.* Yale University Press.

MCLUHAN, M. and CARSON, D. 2003. *The Book of Probes.* California, USA: Gingko.

MERLEAU-PONTY, M. 1958. *The Phenomenology of Perception.* London, UK: Routledge & Keegan Paul.

NISBETT, R. E. 2005. *The Geography of Thought.* London, UK: Nicholas Brealy.

PARRINDER, M. 2006. Part of the Process. *Eye Magazine,* Vol. 15, Spring 2006.

RAMACHANDRAN, V.S. and HUBBARD E. 2003. Hearing Colors, Tasting Shapes. *Scientific American,* Vol. 288, Issue 5 (May 2003), 42–49.

ROBSON, C. 1993. *Real World Research.* London, UK: Blackwell.

ROSCH, E. and LLOYD, B. 1978. *Cognition and Categorization.* New Jersey, USA: Lawrence Erlbaum Associates.

RUSSELL, B. 1978. *The Problems of Philosophy.* Oxford University Press.

THORLACIUS, L. 2002. *A model of visual, aesthetic communication focusing on web sites.* London, UK: Digital Creativity, Routledge.

http://www.ship.edu/~cgboeree/gestalt.html (28 Apr 2006)

http://gestalttheory.net/ (28 Apr 2006)

http://www.groundedtheory.com/ (03 May 2006)

http://www-cdr.stanford.edu/ICM/icm.html (03 May 2006)

http://en.wikipedia.org/wiki/Characteristica_universalis (03 May 2006)

http://en.wikipedia.org/wiki/Jakobson (04 Nov 2005)

http://www.kommunikation.aau.dk (13 Dec 2005)

http://www.cenorm.be/BOSS/supporting/guidance+documents/gd030+-
+graphical+symbols/index.asp (19 May 2006)

VINES, G. 1995. Get Under Your Skin. *New Scientist* [Internet] Vol. 145, Issue 1960,
14 January, 01. (18 Mar 2006) http://www.newscientist.com/article/mg14519607.400.
html

http://www.elektex.com/ (17 Sep 2007)

http://nikeplus.nike.com/nikeplus/ (18 Sep 2007)

http://www.berghaus.com/ourworld/heritage.aspx (18 Sep 2007)

http://www.carhartt.com/webapp/wcs/stores/servlet/HistoryView (18 Sep 2007)

http://www.tfl.gov.uk/ (18 Sep 2007)

http://www.adidas-polar.com/ (18 Sep 2007)

Index